METHODS IN
MICROBIOLOGY

METHODS IN
MICROBIOLOGY

Volume 16

Edited by

T. BERGAN

Department of Microbiology, Institute of Pharmacy and
Department of Microbiology,
Aker Hospital, University of Oslo,
Oslo, Norway

1984

ACADEMIC PRESS

(Harcourt Brace Javanovich, Publishers)

London   Orlando   San Diego   San Francisco   New York
Toronto   Montreal   Sydney   Tokyo   São Paulo

ACADEMIC PRESS INC. (LONDON) LTD.
24–28 Oval Road
London NW1 7DX

U.S. Edition published by
ACADEMIC PRESS INC.
(Harcourt Brace Jovanovich, Inc.)
Orlando, Florida 32887

British Library Cataloguing in Publication Data

ISBN   0–12–521516–9
LCCN   68–57745
ISSN   0580–9517

Filmset in Monophoto Times New Roman by Latimer Trend & Company Ltd, Plymouth
Printed in Great Britain by St Edmundsbury Press,
Bury St Edmunds, Suffolk

CONTRIBUTORS

A. Bauernfeind Max v. Pettenkofer-Institut der Universität München, Pettenkoferst-
rasse 9a, 8000 Munich, West Germany

T. Bergan Department of Microbiology, Institute of Pharmacy, P.O. Box 1108,
University of Oslo, Blindern 0317 Oslo 3, Norway

U. Berger Department of Bacteriology, Hygiene-Institut der Universität, Im
Neuenheimer Feld 324, 6900 Heidelberg 1, West Germany

R. C. W. Berkeley Department of Microbiology, University of Bristol, The Medical
School, University Walk, Bristol BS8 1TD, UK

A. G. Capey Department of Microbiology, University of Bristol, The Medical
School, University Walk, Bristol BS8 1TD, UK

G. R. Carter Division of Pathobiology and Public Practice, Virginia–Maryland
Regional College of Veterinary Medicine, Virginia Polytech Institute, Blacksburg,
Virginia 24061, USA

M. J. Corbel Ministry of Agriculture, Fisheries and Food, Central Veterinary
Laboratory, Weybridge, Surrey, KT15 3NB, UK

T. J. Donovan Cholera Reference Laboratory, Public Health Laboratory, Preston
Hall Hospital, Maidstone, Kent ME20 7NH, UK

K. B. Døving The Institute of Zoophysiology, University of Oslo, Blindern, Oslo 3,
Norway

Y. Fukazawa Department of Microbiology, Medical College of Yamanashi,
Tamaho-mura 409–38, Yamanashi, Japan

E. I. Garvie National Institute for Research in Dairying, Shinfield, Reading RG2
9AT, Berkshire, UK

H. Gyllenberg Department of Microbiology, School of Agriculture, University of
Helsinki, Finland (Present address: Academy of Finland, Drumsövägen 1, 00200
Helsingfors 20, Finland)

K. Holmberg Section for Medical Mycology, National Bacteriological Laboratory,
Stockholm, Sweden

R. Lallier Faculté of Médecine Vétérinaire, Université de Montréal, Case Postale
5000, Saint-Hyacinthe, Québec J25 7C6, Canada

N. A. Logan Department of Biological Sciences, Glasgow College of Technology,
Cowcaddens Road, Glasgow G4 0BA, UK

C. E. Nord Department of Oral Microbiology, Huddinge University Hospital,
Karolinska Institute, 10401 Stockholm, Sweden

M. Popoff Unité des Entérobactéries, Institut Pasteur, 28 rue du Docteur Roux,
75724 Paris Cedex 15, France

R. Sakazaki Enterobacteriology Laboratories, National Institute of Health, Tokyo,
Japan

T. Shinoda Department of Microbiology, Meji College of Pharmacy, Tanashi 188,
Tokyo, Japan

L. A. Shute Department of Microbiology, University of Bristol, The Medical School,
University Walk, Bristol BS8 1TD, UK

O. Solberg Department of Bacteriology, National Institute of Public Health,
Gietmyrsveien 75, 0462 Oslo 2, Norway

R. Solberg   Department of Microbiology, Institute of Pharmacy, P.O. Box 1108, Blindern 0317 Oslo 3, Norway

J. Stringer   Central Public Health Laboratory, Division of Hospital Infection, 175 Colindale Avenue, London NW9 5HT, UK

M. Taguchi   Department of Microbiology, Kanagawa Prefectural College of Medicinal Technology, Yokohama 241, Japan

T. Tsuchiya   Department of Microbiology, Kanagawa Prefectural College of Medicinal Technology, Yokohama 241, Japan

PREFACE

Volume 16 of "Methods in Microbiology" continues the Volumes containing methods of epidemiological typing methods of microbes. In this volume the characterization by serological and diagnostic and more refined chemical procedures are presented for bacteria outside the enterobacteriae and for yeast and other fungi. The significant advances in the characterization of yeasts is presented in a thorough review prepared by the Japanese collaborating group, to which we owe much of our systematic knowledge of these fungi.

Oslo
January 1984

T. Bergan

CONTENTS

Contributors   v

Preface   viii

1 **Phage Typing of** *Streptococcus agalactiae*
  J. Stringer   1

2 **Phage Typing of** *Brucella*
  M. J. Corbel   23

3 **Serological Characterization of Yeasts as an Aid in Identification
  and Classification**
  T. Tsuchiya, M. Taguchi, Y. Fukazawa and T. Shinoda   75

4 **Biochemical and Serological Characteristics of** *Aeromonas*
  M. Popoff and R. Lallier   127

5 **The Separation of Species of the Genus** *Leuconostoc* **and the
  Differentiation of the Leuconostocs from other Lactic Acid
  Bacteria**
  E. I. Garvie   147

6 **Fatty Acid and Carbohydrate Cell Composition in Pediococci and
  Aerobocci and Identification of Related Species**
  T. Bergan, R. Solberg and O. Solberg   179

7 **Epidemiological Typing of** *Klebsiella* **by Bacteriocins**
  A. Bauernfeind   213

8 **Serology of Non-Gonococcal, Non-Meningococcal** *Neisseria* **and
  Branhamella species**
  U. Berger   225

9 **Serotyping of** *Pasteurella multocida*
  G. R. Carter   247

10 **Serology and Epidemiology of** *Plesiomonas shigelloides*
   R. Sakazaki   259

11 **Serology and Epidemiology of** *Vibrio cholerae* **and** *Vibrio
   mimicus*
   R. Sakazaki and T. J. Donovan   271

12 **Identification of** *Bacillus* **Species**
   R. C. W. Berkeley, N. A. Logan, L. A. Shute and A. G. Capey   291

13 **Automated Identification of Bacteria: An Overview and Examples**
   H. Gyllenberg   329

14 **Application of Numerical Taxonomy to the Classification and
    Identification of Microaerophilic Actinomycetes**
    K. Holmberg and C. E. Nord   341
15 **Stimulus Space in Olfaction**
    K. B. Døving
Index   369
Contents of published volumes   383

1

Phage typing of *Streptococcus agalactiae*

J. STRINGER

Central Public Health Laboratory, Division of Hospital Infection, London, UK

I.	Introduction	2
	A. Qualities required of a good typing system	2
	B. Typing systems of pathogenic streptococci	3
II.	Bacteriophages of streptococci	3
III.	Development of a phage-typing system for Group B streptococci	6
	A. Current situation	6
	B. Need for an additional typing method for Group B streptococci	6
IV.	Technical methods	8
	A. Media and conditions for growth	8
	B. Isolation, propagation and testing of phages	8
	C. Typing	12
V.	Evaluation of the combined sero-phage typing system	14
	A. Reproducibility	15
	B. Discrimination	16
	C. Storage and stability	17
	D. Interpretation and reporting of results	17
	E. Sero-phage type associations	18
	F. Virulent sero-phage types	18
VI.	Application	18
	A. Longitudinal epidemiological studies	18
	B. Neonatal cross-infection investigations	19
VII.	The future	20
	A. International significance of phage typing and modification of the existing set of typing phages	20
	References	20

I. Introduction

The application of Lancefield's serological method to the haemolytic strepto-cocci has led to the recognition of some major pathogenic groups for man including Groups A, B, C, D and G. Groups A, B and C can be equated with species described on the basis of biochemical characteristics:*Streptococcus pyogenes* (Group A), *S. agalactiae* (Group B) and *S. equisimilis* (Group C). Group D streptococci can be subdivided into three main species: *S. bovis, S. faecalis* and *S. faecium*, which all possess the same group antigen.

To trace the spread of these organisms or relate strains to particular diseases, further identification of the organisms beyond the species is required. Subdivision on the basis of differences in a single set of characteristics, for example susceptibility to bacteriophages or bacteriocines or the presence of additional antigens, has permitted the recognition of "types" within members of one species. For example, the Group A streptococci have been subdivided into more than 70 distinct serotypes on the presence or absence of type-specific M-antigens. When two organisms belonging to the same type are isolated at the same time and from the same place, it is often assumed that they are epidemiologically related; this assumption, however-may not necessarily be true and it is dependent on the characteristics of the typing system used.

Qualities required of a good typing system

1. *Reproducibility*

An ideal typing system should reproducibly identify related strains as being the "same", and unrelated strains as being "different". However, factors such as genetic variation and experimental conditions influence the reproducibility of any typing method. Systems based on the recognition of one characteristic, for example an M-antigen, are more stable than those based on the recognition of pattern reactions, such as most phage- and bacterio-cine-typing methods. In the latter methods, the differences in patterns are used to establish the probability that two strains are different. The wider the differences in pattern allowed between two cultures thought to belong to one strain, the greater the errors in calling organisms the same or different. Blair and Williams (1961) have described in detail the use of this difference rule in the phage typing of *Staphylococcus aureus*.

2. *Discrimination*

The second quality required of a good phage-typing system is that it is discriminatory, that is it should subdivide members into a large number of types, no one of which predominates in an unselected population.

B. Typing systems of pathogenic streptococci

The Lancefield serological typing system for Group A streptococci (Lancefield, 1928) has allowed these organisms to be widely studied. The typability of clinical isolates is high (87% in the Streptococcus Reference Unit, Colindale). The method is both reproducible and, by virtue of recognizing many types, highly discriminatory. The serotyping system for Group B streptococci, however, permits the recognition of only five major types (Lancefield and Freimer, 1966), and it is consequently of limited epidemiological value. Methods for further subdividing Groups C, D and G are rarely applied.

Phage typing has been used in the construction of highly discriminatory systems for other common pathogens, notably *S. aureus* (Wilson and Atkinson, 1945; Williams and Rippon, 1952; Blair and Williams, 1961), but streptococcal bacteriophages have been studied much less intensively than staphylococcal bacteriophages and Gram-negative bacilli. Phage sensitivity has been used as an additional means of subdividing certain types of Group A streptococci (Skjold and Wannamaker, 1976), and lysotyping schemes have been proposed for Groups A, C and D (Mihalcu and Vereanu, 1978; Plecas, 1979).

This chapter, however, is concerned primarily with the development of a phage-typing system for Group B streptococci.

II. Bacteriophages of streptococci

Until the 1950s the only extensive studies on phages active against haemolytic streptococci were those of Evans (1933, 1934, 1940; Evans and Sockrider, 1942). She worked mainly with four phages and in general they appeared to show some degree of specificity for streptococcal groups. In 1955 Maxted showed that it was possible to alter some of the cultural characteristics of Group A streptococci by exposing them to phages, for example he found that the amount of extractable M-antigen could be greatly increased or that non-mucoid strains could become mucoid. Kjems (1955) and Maxted (1955) isolated temperate phages from Group A streptococci either by the cross-culture technique of Fisk (1942) or by ultraviolet light induction and found that lysogeny occurred frequently, although there was no evidence of any spontaneous lysis. Kjems (1958a,b, 1960a,b) further characterized some of these temperate phages and also a virulent phage for Group A streptococci isolated from sewage. He was able to distinguish them by their host range, growth kinetics and phage neutralization tests.

Attempts have been made to subdivide Group A streptococci on the basis of their reaction to virulent or temperate phages or to phage-associated lysis (Evans, 1940; Evans and Sockrider, 1942). In 1952 McKenna attempted to group and type 74 strains of streptococci using three of his own and four of Evans' virulent phages. He found that 43 of the 60 Group A strains could be identified as types 3, 17 or 19, but 13 of the strains were not classifiable. Kjems (1955) tested 188 strains of Group A streptococci representing 18 different T types with four temperate phages and found that although lysis by either two or three phages occurred with strains from nine of the 18 T types, four T types (34% of the strains) were not lysed by any of the phages and no individual types could be distinguished by any of the phage patterns. Leonova *et al.* (1968) induced 14 temperate Group A phages with ultraviolet light and applied them to 325 strains of Group A streptococci. They found 52% of the strains were sensitive to phage attack and could distinguish five distinct groups of phages. They attempted to correlate the patterns of lysis with the disease-producing ability of the strains; one phage was specific to some of the nephritogenic strains, all but one of which were type 12. Strains from cases of tonsillitis and scarlet fever were lysed by four of the five phage groups, and strains isolated from "rheumatism" cases were lysed by phages from all five groups.

These attempts to subdivide and classify Group A streptococci by phage susceptibility have proved largely unsuccessful, partly because an inadequate number of different phages was available, but also because a good alternative serological typing system is available. More recently work with Group A phages has been directed towards the division of individual M types of particular interest. Pitt (1980) reviewed a similar hierarchical typing system for *Pseudomonas aeruginosa* in which phages were used to subdivide serotypes. Wannamaker *et al.* (1970) investigated the phage susceptibility of Group A streptococci of two M types, namely 12 and 49, sometimes associated with nephritis. They succeeded in distinguishing between the two types and recognized several patterns of lysis of type 49 strains. However, as this depended on obtaining and propagating phages from all the strains tested, the method was laborious. In 1976 Skjold and Wannamaker described two additional approaches. The first was prophage typing by mitomycin C induction of lysogenic test strains followed by inoculation directly onto a set of natural indicator strains. The second was to use a set of induced lysates from selected strains to produce specific patterns of lysis on lawns of test strains. The latter method proved useful for subdividing members of both M type 12 and M type 49, and geographical associations among some of the phage types of M type 49 were shown.

Spanier and Timoney (1977) have described the subdivision of Group C streptococci by bacteriophages. They isolated phages from 12 lysogenic strains of *S. equi* (the animal pathogen of Group C) and found that these phages subdivided the strains into two distinct but related groups. All 12 phages changed the colony appearance of exposed cultures of *S. equi* from mucoid to matt.

Vereanu *et al.* (1977) examined the morphological characteristics of nine Lancefield Group A and Group C phages. The former conformed to Bradley's Group B classification (Bradley, 1967). They had isometric heads with long non-contractile tails, whereas the Group C phages corresponded to Bradley's Group C classification, that is they had polyhedral heads and short tails. The plaques produced by the Group A phages were smaller (0.2–0.5 mm) to those produced by the Group C phages (1–1.5 mm). Mihalcu and Vereanu (1978) later proposed a provisional lysotyping scheme for Group A and C streptococci. This overlapped with Lancefield's serological classification but was reported to subdivide more than 90% of the Group A strains tested. In 1979 Vereanu and Mihalcu proposed a lysotyping scheme for Group C streptococci using Group A streptococcal phages that were active on Group C strains, and temperate and virulent phages isolated from Group C streptococci. The 16 phages they selected lysed 77% of the Group C strains from man, and they examined and subdivided them into seven groups.

Phage activity in Group D streptococci has been described by Brock (1964) and Brailsford and Hartman (1968). They characterized the phages they had isolated, both virulent and temperate, by host range, morphology and serum neutralization tests. Subdivision of Group D streptococci by phage typing was described in 1959 by Cinca *et al.* but it was not improved until 1974 when Plecas and Brandis used mixtures of phages to establish an enterococcal group and species identification in a single test. Plecas (1979) has recently proposed a phage-typing system for *S. faecalis* and *S. faecium*, with the aim of distinguishing strains involved in hospital cross-infection. She found that using 20 phages, about 90% of the 439 Group D strains examined could be subdivided into 20 types with 15 subtypes. The patterns produced also permitted the recognition of *S. faecalis* and *S. faecium*.

The application of the proposed phage-typing schemes has been limited, possibly because of their restricted discriminatory power and problems of reproducibility. However, the major reason for their apparent failure may be that good alternative serological typing systems exist for those groups of greatest epidemiological interest, namely Group A (Rotta, 1979) and Group D (Sharpe and Shattock, 1952). In recent years Group B streptococci have emerged as important human pathogens in neonates and adults (Bayer *et al.*, 1976; Baker, 1977; Parker, 1979; Wilkinson, 1978). The shortcomings of the existing serological typing system have been emphasized as the need for detailed epidemiological studies of this organism in the hospital environment have become more acute. Serotyping in the UK and USA has shown that three of the serotypes each account for approximately 20% of all strains in unselected populations (Parker, 1979; Wilkinson, 1978), and that serotype III comprises 60–90% of isolates from neonatal meningitis (Baker and Barrett, 1973; Parker and Stringer, 1979). The need, therefore, for an additional means of identifying these strains has led to the development of a phage-typing system which subdivides the serotypes (Stringer, 1980).

III. Development of a phage-typing system for Group B streptococci

A. Current situation

Streptococcus agalactiae (Lancefield Group B) has until recently been regarded as almost exclusively a pathogen of animals. It is an important causative agent of bovine mastitis. Until the late 1950s there were few accounts of Group B streptococci causing serious human infections, but since then the number of reports of neonatal and adult infections has increased dramatically. The question arises whether this emergence is real or can be attributed to improved diagnostic methods. According to Baker (1977) the evidence from extensive reviews of the literature suggests an increase in the frequency of disease, at least in neonates.

   In neonates, Group B streptococci can cause either an acute septicaemia or meningitis. By the early 1970s it was realized that there were, in fact, two patterns of disease related to the age at onset (Quirante and Cassady, 1972; Franciosi *et al.*, 1973; Baker and Barrett, 1973): early-onset disease, which is septicaemia either without localization or combined with meningitis, occurs within the first week of life; late-onset disease, which is almost always meningitic, occurs after the first week of life with a peak around the second and fourth weeks. Epidemiological studies of these diseases have revealed that in early-onset disease, the causative streptococcus is invariably found in the vagina of the mother, but that in late-onset disease the causative streptococcus is rarely isolated from the mother (Paredes *et al.*, 1977). Acquisition rates of 58–62% have been recorded for babies born to colonized mothers, compared with 5–12% of babies born to apparently non-colonized mothers (Anthony *et al.*, 1979; Pass *et al.*, 1980). Nosocomial infection rates of up to 40% have also been reported in infants during their stay in hospital (Baker, 1977). Early-onset disease of either clinical type is usually associated with obstetric complications such as premature rupture of membranes, prolonged labour and low birth weight. No such predisposing factors can be implicated in the late-onset neonatal disease that appears to affect normal babies. However, mortality is lower in late-onset neonatal disease (14–18%) if prompt treatment is given, whereas early-onset infections are more often fatal (55%) even if treated promptly (Anthony and Okada, 1977).

This leads to the question of treatment versus prophylaxis. There have been many attempts to eradicate Group B streptococci from the female genital tract, most of which have failed. The reasons for this may be that either insufficient antibiotic reaches the mucosal surface of the vagina or that Group B streptococci are part of the normal flora of the bowel and consequently reinfect the genital tract even after treatment with antibiotics (Parades *et al.*, 1976). Another disadvantage of prophylactic treatment of Group B streptococci is that, because carriage rates in pregnant women can be as high as 25%, the number of women requiring treatment is far in excess of the number at risk of delivering an infected, in distinction from a colonized, infant.

If it is accepted that in early-onset disease the mother is the source of the infecting streptococcus, it is important to determine the precise location of carriage sites, both genital and extragenital, in adult females. Similarly in late-onset meningitis, it is necessary to search for non-maternal sources of the infecting streptococcus, among, for example other babies in hospital, other persons in hospital and the immediate family in contact with the baby after leaving hospital. However, the restricted serotyping system has never enabled the precise identification of strains required for these studies.

There is some evidence that Group B streptococci isolated from animals are a distinct population from human origin Group B streptococci (Pattison *et al.*, 1955; Butter and de Moor, 1967) but this has never been formally established and warrants further investigation before the possibility of zoonotic infections can be dismissed.

B. Need for an additional typing method for Group B streptococci

A serotyping scheme for Group B streptococci uses five type antigens, Ia, Ib, Ic, II and III, to which can be added the two protein antigens R and X. The techniques for serotyping this organism have been described by Rotta (1979). Although combinations of polysaccharide and protein antigens may be observed, the proteins R and X are often associated with animal strains. A system which recognizes only five major types is insufficiently indiscriminating for most epidemiological studies because the number of types that can be recognized is small, and the percentage frequency of certain types in unselected populations is unacceptably high.

Stringer (1980) using a set of 24 phages found that approximately 80% of

Group B streptococci of human origin could be typed, that is lysis by one or more of these phages was easily observed. She also found that numerous distinct patterns of lysis could be observed but that these were not specific for any one serotype. Thus by first serotyping and then phage typing, the discriminatory power of this combined system was high with few types exceeding 3% of any unselected population. The combined typing method was also found to be highly reproducible and has been used to advantage in investigating the spread of Group B streptococci in hospitals (Anthony *et al.*, 1979; Boyer *et al.*, 1980).

IV. Technical methods

A. Media and conditions for growth

The medium for the growth of propagating and test bacterial cultures is composed of:

Todd Hewitt Broth Base (Oxoid)	30 g l^{-1}
Yeast Extract (Oxoid)	2 g l^{-1}
Calcium chloride (dried)	0.012 g l^{-1}
L-Tryptophan	0.01 g l^{-1}

The pH is adjusted to 7.6 and the medium autoclaved at 115°C for 10 min. Solid medium for phage typing is prepared by adding 0.7% agar (Oxoid No. 1) to this broth. Plates poured from this medium are dried at 37°C before use.

All phage cultures are incubated in air at 30°C overnight.

B. Isolation, propagation and testing of phages

1. *Isolation*

Phages can be isolated from mitomycin C induced cultures of Group B streptococci. Cultures to be tested in this way are incubated in broth at 30°C for 2 h and mitomycin C is added to each to a final concentration of 0.1 μg ml^{-1}. Incubation is continued for a further 2 h. These cultures are then centrifuged at $1500 \times g$ for 15 min and cross-spotted onto seeded dried lawns using a multiloop applicator (Lidwell, 1959). A uniformly dense broth culture for flooding the lawns is obtained either by diluting an overnight broth culture 1 in 10, or from a 2-h broth culture inoculated from an 18-h culture on a blood agar plate. Both methods result in a broth culture with an optical density of 0.2–0.4 when read at 600 nm (Unicam 600). All plates are incubated in air for 18 h at 30°C.

2. *Propagation*

Zones of lysis, either individual plaques or areas of confluent lysis, indicate phage activity. For the initial propagation of any phage, single plaques are picked and inoculated in 5 ml of the modified Todd Hewitt Broth with a few colonies of the propagating strain. It is important that the specified propagating strain is used because if another strain is substituted the host range of the phage may be altered. The broth is incubated at 37°C for about 4 h and the supernatant, after centrifugation, titrated on a lawn of the relevant propagating strain. This procedure is repeated twice more to ensure that the phage is pure.

Bulk phage preparations can be propagated by several methods.

1. A cube of agar encompassing a zone of lysis and a portion of lawn growth is inoculated into 50 ml of the modified Todd Hewitt Broth and incubated at 30°C for 4 h, or until lysis can be seen to have taken place. The broth is centrifuged for 15 min, the supernatant titrated in ten-fold dilutions and spotted onto a lawn of the propagating strain by means of 0.02-ml droppers. (The propagating strain for any phage is the strain on which the phage was detected and is subsequently used for the propagation of that phage.) The plates are incubated at 30°C for 18 h and the number of phage particles is expressed in plaque-forming units per millilitre (p.f.u. ml^{-1}) (calculated by multiplying the number of plaques in one 0.02-ml drop by the amount it has been diluted).
2. A "sweep" through the zone of lysis and across the lawn of propagating strain is inoculated into 5 ml of broth and incubated and harvested under the same conditions as in method 1.
3. Phage and cells in a ratio of ten plaque-forming units (p.f.u.) to one colony-forming unit (c.f.u.) of bacteria are inoculated into broth and incubated and harvested as in method 1.

We have found that phage suspensions reaching the required titre can be filtered through membrane filters of 0.45-μm porosity and kept at 4°C for short-term use or below -70°C or in the lyophilized state for long-term storage.

3. *Testing of phages*

An original set of 31 phages has been reduced to 24 by the elimination of phages that were either duplicated by another in the set, or of such restricted host range that they were of little value. These phages and their respective propagating strains have been deposited with the National Collection of Type Cultures (NCTC), Colindale, London and the catalogue numbers are given in Table I.

TABLE I

Characteristics of strains used to isolate and propagate the 24 typing phages

Phage No.	Carrier strain			Propagating strain			Plaque		Routine test dilution (p.f.u. ml⁻¹)
	No.	NCTC No.	Serotype	No.	NCTC No.	Serotype	Diameter (mm)	Appearance[a]	
1	75/601	11255	IIR	75/785	11234	IIR	1.0	Clear	2×10^6
2	76/200	11256	IIIR	75/609	11235	II/Ic	0.5	Clear	3×10^6
3	75/1133	11257	IbX	75/609	11235	II/Ic	0.5	Clear	5×10^5
4	75/201	11258	Ic	75/609	11235	II/Ic	0.5	Clear	5×10^5
5	75/621	11259	IIIR	75/609	11235	II/Ic	0.5	Clear	3×10^6
7	76/196	11261	Ic	76/200	11237	IIIR	0.5	Clear	2×10^7
8	76/193	11262	IIIR	76/1685	11238	IIIR	0.25/0.5	Clear	1×10^7
11	"Phage 1"	11265	—	76/1875	11239	IIR	0.5	Clear	1×10^7
12	76/200	11266	IIIR	75/2483	11240	III	0.25	Clear	4×10^6
13	75/621	11267	IIIR	76/195	11241	IIIR	0.25	Clear	2×10^7
14	76/32	11268	IIIR	76/2924	11242	Ic	0.25	Clear	2×10^6
15	76/2971	11269	Ic	76/2978	11243	Ic	1.0	Clear	5×10^6
16	76/200	11270	IIIR	76/742	11244	Ib	0.5	Clear	1×10^7
17	76/3135	11271	Ic	76/2969	11245	Ic	0.25	Clear	8×10^5
18	76/3084	11272	Ic	76/2969	11245	Ic	0.25	Clear	5×10^6
19	76/2971	11273	Ic	76/2969	11245	Ic	0.25	Clear	2×10^6
20	77/445	11274	II	77/537	11246	IIR	0.25	Clear	3×10^6
24	77/1932	11278	IIIR	77/1766	11248	Ia	0.25	Clear	8×10^5
26	77/2081	11280	Ic	77/2684	11250	Ic	0.5	Clear	5×10^6
27	77/2082	11281	IIIR	77/2336	11251	III	0.25	Clear	3×10^6
28	77/2081	11282	Ic	77/1879	11252	Ic	0.25	Clear	8×10^6
29	77/2078	11283	Ic	77/1879	11252	Ic	0.25	Clear	7×10^6
30	77/2684	11284	Ic	77/2078	11253	Ic	0.25	Clear	6×10^6
31	77/2082	11285	IIIR	77/1878	11254	III	0.25	Clear	3×10^6

[a] By transmitted light.

(a) *Routine test dilution.* The routine test dilution (RTD) for phage typing Group B streptococci is the highest dilution for each phage that produces the widest spectrum of lysis of bacteria without producing inhibitory reactions on a wide selection of these strains. This dilution is not the same for all the phages and is expressed as a titre in p.f.u. ml^{-1}. In most other phage-typing systems, for example those used for pseudomonads or staphylococci, the RTD is the dilution of phage producing just less than confluent lysis of its propagating strain. Phages for Group B streptococci used at this dilution are restricted in their spectrum of lysis. The predetermined RTDs of Group B phages are approximately 100 times the dilution producing semiconfluent lysis. The RTD for each phage is given in Table I. Phage filtrates after propagation should have a titre 10 to 100 times greater than RTD.

(b) *Lytic spectrum.* After a plaque count has been carried out on the filtered lysate, the lytic spectrum of the phage must be determined to ensure that no mutations or other changes have occurred during propagation. The lytic spectrum of any phage is the characteristic pattern of lysis produced by this phage on a selected set of test strains. For the 24 Group B typing phages, 21 propagating strains are used as the test cultures. Eighteen of these are listed in Table I and the additional three are the strains 3A (NCTC 11236), 77/331 (NCTC 11247) and 77/2090 (NCTC 11249). The method used was devised by Rippon (1956) for staphylococcal phages.

The phage filtrate under examination is first tested undiluted, but at known titre, for its ability to lyse any of the 21 test strains. If however the phage filtrate has a titre greater than 1:10 000 it is diluted to a level at which semiconfluent lysis is produced by a 10^{-4}-fold dilution. This procedure is necessary to avoid the effects of inhibition, that is the prevention of bacterial growth without lysis caused by high-titre phage suspensions.

The phage filtrate is then titrated in ten-fold dilution steps on the test strains that were lysed by the neat suspension. The titre on each strain lysed, relative to that on the homologous propagating strain is then calculated and scored as shown in Table II. The lytic spectrum of each newly propagated phage is compared with that of the standard stock phage suspension. Provided that the two phage suspensions show similar lytic activity the newly propagated phage is considered acceptable. A variation in score from 4 to 5, 3 to 4 and so on and the loss or gain of reactions scored as 1 or 2 are considered permissible. The lytic spectra for the 24 typing phages are shown in Table III.

TABLE II

Example of determination of the lytic spectrum of a typing phage: phage No. 30, homologous propagating strain No. 20

Test strains	N^b	10^{-1}	10^{-2}	10^{-3}	10^{-4}	10^{-5}	10^{-6}	Score^c
				Lysis^a of the indicated test strains by the phage at a dilution of				
1	—	—	—	—	—	—	—	0
2	—	—	—	—	—	—	—	0
3	0	—	—	—	—	—	—	0
4	+ + +	+ + +	+ + +	+ +	+	+	±	5
5	—	—	—	—	—	—	—	0
6	+ + +	+ +	+ +	+	±	—	—	4
7	±	—	—	—	—	—	—	0
8	—	—	—	—	—	—	—	0
9	0	—	—	—	—	—	—	0
10	+ + +	+ + +	+ +	+	±	—	—	4
11	—	—	—	—	—	—	—	0
12	+ + +	+ + +	+ +	+	±	—	—	4
13	—	—	—	—	—	—	—	0
14	—	—	—	—	—	—	—	0
15	—	—	—	—	—	—	—	0
16	—	—	—	—	—	—	—	0
17	±	±	—	—	—	—	—	1
18	—	—	—	—	—	—	—	0
19	+ +	+	±	—	—	—	—	3
20 (Prop. strain)	+ + +	+ + +	+ + +	+ +	+	+	±	5
21	—	—	—	—	—	—	—	0

^{a} + + +, confluent lysis;
  + +, 51 plaques—semiconfluent lysis;
  +, 21–50 plaques;
  ±, 1–20 plaques;
  O, inhibition.
^{b} undiluted.
^{c} 5, titre on homologous propagating strain;
  4, 10^{-1}–10^{-2} of titre on propagating strain;
  3, 10^{-3}–10^{-4} of titre on propagating strain;
  2, 10^{-5}–10^{-6} of titre on propagating strain;
  1, very weak lysis.

C. Typing

1. *Preparation of reagents*

Phages may be reconstituted from freeze-dried ampoules by adding distilled water and then incubating discrete drops of phage suspension on a confluent

TABLE III
Lytic spectra of the 24 typing phages

	Score of lysis on test strain No.																				
Phage No.	785	609	3A	200	1685	1875	2483	195	2924	2978	742	2969	537	331	1766	2090	2684	2336	1879	2078	1878
1	5	0	0	0	0	3	0	0	0	0	0	0	0	0	0	0	0	0	0	0	0
2	0	5	0	0	0	0	0	1	1	0	1	0	0	0	0	0	0	0	0	0	0
3	0	5	0	0	0	0	0	0	0	0	0	0	3	0	0	0	0	0	0	0	0
4	0	5	0	0	0	3	0	4	3	0	2	0	2	0	0	0	0	0	0	0	0
5	0	5	0	0	0	3	0	4	0	0	0	0	0	0	0	0	0	0	0	0	0
7	0	0	0	5	0	2	2	0	0	0	0	0	0	0	0	0	0	0	0	0	0
8	0	0	0	0	5	0	0	0	0	0	1	0	0	0	1	0	0	0	0	0	1
11	5	0	0	0	0	5	0	0	0	0	0	0	0	0	0	0	0	0	0	0	0
12	0	1	1	0	0	4	5	4	2	0	3	0	3	0	0	0	0	0	0	0	0
13	0	3	0	1	0	1	0	5	0	0	0	0	0	0	0	0	0	0	0	0	0
14	0	0	0	3	0	3	0	0	5	0	3	0	5	0	0	0	0	0	0	0	0
15	0	0	0	1	0	0	0	0	0	5	0	3	0	0	0	0	0	0	0	0	0
16	0	5	1	1	1	0	0	3	4	1	5	0	3	0	0	3	1	0	0	0	0
17	0	5	0	3	2	0	0	0	0	3	0	5	0	0	0	0	0	0	0	0	0
18	1	2	0	3	0	0	1	0	0	0	4	0	5	0	0	0	0	0	0	0	0
19	0	5	0	3	0	0	0	0	0	2	0	5	0	0	0	0	0	0	0	0	0
20	0	4	3	0	0	0	0	4	4	0	4	0	5	0	0	0	0	0	0	1	0
24	0	5	0	3	0	0	2	1	0	1	0	0	0	0	5	0	0	0	0	0	0
26	2	0	0	4	0	3	0	1	0	4	0	3	3	0	0	0	5	0	3	0	0
27	1	0	0	5	1	4	0	0	0	0	0	0	0	0	0	0	0	5	0	0	3
28	1	0	0	2	0	0	0	0	0	5	0	4	2	0	0	0	3	0	5	0	0
29	0	0	0	4	0	0	0	0	0	3	0	5	3	0	0	0	3	0	5	0	0
30	0	0	0	5	0	4	0	0	0	4	0	4	0	0	0	0	1	0	3	5	0
31	0	0	0	5	1	4	0	0	0	3	0	0	0	0	0	0	0	4	0	0	5

[a]5, degree of lysis on homologous propagating strain (Table II).

lawn of the propagating strain. Each phage is then propagated until a titre greater than RTD × 10 is achieved (Section IV.B.2). These propagations should be kept to a minimum to reduce the possibility of genetic variation. Blair and Williams (1961) have suggested guidelines for the propagation of staphylococcal phages, and so the reader is advised to follow the procedures recommended in their paper. The phage lysates are stored at 4 C and the titres should be checked routinely every six weeks. Before phage typing aliquots of each phage are diluted to RTD with broth (approximately 5 ml total volume). These diluted suspensions will only maintain their titre for a short time and should be discarded if not used within four weeks.

Freeze-dried propagating strains should be reconstituted in broth containing blood (5%) or Todd Hewitt Broth and subcultured on blood agar. They should be stored either on blood agar slopes at 4°C or in glycerol blood broths at − 18°C.

Test strains for typing should be inoculated into 5 ml of broth from 18-h cultures of blood agar plates. The broths are incubated at 37°C for 2–3 h or until light growth is visible (approximately a D of 0.2 at 600 nm). Petri dishes (90 mm) containing modified Todd Hewitt agar are flooded with the young cultures and any excess fluid is pipetted off. The plates are dried open and inverted at 37°C for 30 min.

2. *Application of phage*

The 24 typing phages at RTD are applied to dried lawns of the test strains using a multiloop applicator such as that designed by Lidwell (1959) (Biddulf & Co, Manchester, UK). Instructions for phage typing with this machine have been described in detail by Parker (1972). All phage-typing plates are incubated overnight at 30°C.

3. *Examination of plates*

The plates are examined in oblique transmitted light, against a dark background, preferably with the aid of a ×5 magnifying lens. Lysis is recorded as follows.

$$\pm = 1\text{--}20 \text{ plaques}$$
$$+ = 21\text{--}50 \text{ plaques}$$
$$+ + = 51 \text{ plaques—semiconfluent lysis}$$
$$+ + + = \text{confluent lysis}$$
$$0 = \text{inhibition}$$

Strong lysis (+ + and + + + reactions) is generally easy to observe but the recognition of weaker lytic reactions and small plaques will only improve with experience. In addition to lytic reactions other phenomena can be observed, for example secondary growth on areas of confluent lysis is common. Inhibition reactions may be confused with confluent lysis or lysis with secondary overgrowth, these reactions are believed to be the result of phage adsorption and subsequent death of the cell without phage multiplication or lysis. The appearance resembles confluent lysis and is frequently observed when strong phage preparations are used.

V. Evaluation of the combined sero-phage typing system

The original set of 31 typing phages was reduced to 24, seven phages being eliminated either because of their restricted host range or because they were duplicated by others in the set (Stringer, 1980). Typing results using the set of 24 phages showed that numerous distinct phage patterns could be recognized within each serotype. Although a partial serotype association seemed to exist with certain phage patterns the majority could be observed within most serotypes. A two-stage typing system therefore was adopted in which phage typing is used to subdivide the serotypes.

The qualities demanded of a typing system are that it should be both reproducible and discriminatory, that is it should consistently identify epidemiologically related isolates as being the same and discriminate between strains that are different. A highly discriminatory typing system is one in which the number of distinct types is large and no one type predominates in an unselected population.

A. Reproducibility

The reproducibility of the serological and phage-typing system has been studied using epidemiologically related sets of isolates (Table IV) (Stringer, 1980); short-term isolates, that is those collected within 14 days; long-term isolates collected over a four-month period were compared. Only 2% of the short-term isolates differed from their index strain (first isolate received for typing) by one or more strong phage reactions. Some 12% of long-term isolates differed from their respective index strains by one or more strong phage reactions. This difference is probably the result of inherent variability of the typing method, because if a difference of one strong phage reaction was allowed, less than 3% of all isolates were found to differ from their index

TABLE IV

Differences in phage-typing patterns among epidemiologically related sets of Group B streptococci

	Number of		Number of (%) of strains differing from index strain by the indicated number of strong differences			
Criteria for inclusion	Sets	Strains	0	1	2	3
Same patient (or mother and baby); isolated within 14 days	39	134	131(97.8)[a]	2	0	1
			(99.3)[b]			
Same patient; isolated within four months (average)	30	170	150(88.2)[a] (88)	16	4	0
			(97.6)[b]			

[a] Percentage of correct results if no strong differences are permitted.
[b] Percentage of correct results if one strong difference is permitted.

Reprinted with permission from *Journal Medical Microbiology*.

strain. Therefore isolates were considered to be different only if they differed from each other by more than one strong phage reaction.

The same collection of strains was used to investigate the reproducibility of serotyping. Most of the discrepancies observed were either the loss or gain of a protein antigen. The variability in the presence of these antigens has been reported before (Pattison et al., 1955).

B. Discrimination

The discriminatory power of the two-stage typing system was investigated using epidemiologically independent strains isolated from one hospital in a six-month period (Stringer, 1980). By serotyping alone, seven different groups of strains could be recognized and three of these each comprised more than 20% of the total. If a difference of one strong phage reaction between strains was allowed, 94 different sero-phage types could be recognized. The largest single type comprised 6.3% of the total (this was phage pattern 12 on serotype III), but few others exceeded 3% in frequency. Thus the probability that two isolates would belong to the same sero-phage type by chance would seldom exceed 0.02 (Table V).

TABLE V
Discrimination among epidemiologically unrelated strains of Group B streptococci from one hospital

| | Number (%) of | | | Percentage frequency of the indicated more common sero-phage types | | |
Serotype	Strains examined	Strains untypable by phage	Distinct phage patterns[a]	1	2	3
1a	8 (3.9)	1(0.5)	5	1.0	—	—
1b	37(18.0)	5(2.4)	18	2.4	2.4	1.5
1c	55(26.7)	11(5.3)	21	3.4	3.4	2.9
II	42(20.4)	7(3.4)	20	3.9	1.9	1.0
III	42(20.4)	8(3.9)	17	6.3	1.5	1.0
X only	7 (3.4)	4(1.9)	3	—	—	—
R + NT	15 (7.3)	5(2.4)	10	—	—	—
Total	206	41	94			

[a] Difference established by > 1 strong phage reaction.

Reprinted with permission from Journal Medical Microbiology.

C. Storage and stability

The length of time between typing related isolates could influence the reproducibility of the system (Section IV.A). Many surveys involve long-term, multiple-site sampling of individual patients, therefore knowledge of the effect of time between typing, and of storage, is essential.

Repeated tests showed that approximately the same proportion (3%) of strains differed from their index strains by more than one strong reaction regardless of whether they were typed on the same day or over a longer period of time. However, when the phage-typing patterns of presumed individual strains obtained on first and second occasions (several months later) were compared, about 10% showed a two reaction difference. If a one reaction difference between strains was allowed, this would lead to a possible error in about 10% of long-term comparisons. This degree of error could be reduced to nil by widening the criterion for difference to more than two reactions, but only at the cost of considerable loss of discrimination (Sections I.A.1 and 2).

Serological typing does not appear to be affected by long-term storage (Stringer, 1980).

D. Interpretation and reporting of results

From the results of reproducibility experiments (Section V.A) isolates were considered to be unrelated if they differ by two or more strong phage reactions. In longitudinal studies, the time intervals between typing influence the reproducibility of the system and so this should be taken into consideration when interpreting results obtained over a long period of time. Blair and Williams (1961) have written a general guide for interpreting the results of *S. aureus* phage typing; the reader is referred to this for guidance. Although application of the term one strong reaction difference is advised, the experienced worker should consider all other factors influencing the organism under test. Some strains, for example, are much more variable than others in their phage-typing patterns, some grow more vigorously than others and experimental errors because of variation in the medium or phage preparations are possible additional causes of variability in typing results.

To report typing results it is usual to record only those phages that produce strong lysis of the culture, this is referred to as the typing pattern. When apparently related isolates are examined, all reactions that may comprise the phage pattern should be considered even if they are less than + +.

E. Sero-phage type associations

The results of the phage typing of isolates of Group B streptococci from clinical material showed that certain phages or phage patterns were associated more frequently with some serotypes than others. To investigate the extent of this phage–serotype relationship more than 200 epidemiologically unrelated strains were phage typed and serotyped and the results analysed for the percentage frequency of lysis of each phage within each serotype and the chi squared significance test was applied. The more common phage patterns were picked out by selecting frequencies greater than 30%. Phage pattern 17, 18, 19, 26, 28, 29, 30 was associated with serotype Ia and Ic strains, whereas pattern 12, 14, 16, 20 was associated with serotype Ib strains. Phage 12 lysed 60% of serotype III strains but this phage has a wide host range and was also associated with serotypes Ib and II. A series of chi squared tests confirmed significant associations between serotype Ib and phages 16 and 20, serotype Ic and phages 14, 26, 28 and 29 ($\chi > 10.83$, $P < 0.001$) and serotype III and phage 12 ($\chi > 6.63$, $P < 0.01$).

F. Virulent sero-phage types

The information accumulated to date in the UK suggests that there is little evidence of notably virulent bacterial strains identifiable as particular phage types among coagulase-positive staphylococci (Parker et al., 1974). Among Group B streptococci certain sero-phage types are more common than others, for example serotype III, phage type 12, accounts for more than 60% of Group B streptococci typed at Colindale but these are isolated from both superficial sites and systemic infections.

VI. Application

A. Longitudinal epidemiological studies

Surveys investigating the long-term carriage of Group B streptococci by pregnant women have shown that generally the same strain is carried throughout pregnancy (Ferrieri et al., 1977; Anthony and Okada, 1977; Anthony et al., 1978). Sanderson et al. (1979) used the combined sero-phage typing scheme to identify Group B streptococci isolated from women attending an antenatal clinic and from their babies at delivery. This survey attempted to establish the primary source of Group B streptococci in the female genital tract, with the aim of eradicating this organism either permanently or just prior to delivery. They found that the majority of women carried only a single strain and these could be isolated from various sites: in the ano-genital tract, from the rectum and from the skin of the perineum. They concluded that the alimentary canal was the major reservoir because consistently higher isolation rates were obtained from rectal swabs.

In a recent survey, Weindling *et al.* (1981) investigated the carrier status of pregnant women, of babies at delivery and during the neonatal period, as well as the siblings and fathers of those babies monitored for the first 12 weeks of life. There was little evidence of non-maternal acquisition during this period. A family-related strain could sometimes be isolated from multiple sites on members of a family. This family-related strain was identified by serotype and phage type and carriage, once established, appeared to be long-lasting.

B. Neonatal cross-infection investigations

Phage typing has been used to trace the spread of Group B streptococci within a variety of hospital environments. The more notable cases of nosocomial spread in neonatal units include those reported by Anthony *et al.* (1979), Boyer *et al.* (1980) and Band *et al.* (1981). Anthony and co-workers observed the spread of two different strains of Group B streptococci, which could be identified by serotype and phage type, in two infant nurseries. The source in both instances was found to be maternal and both strains spread rapidly throughout the nurseries although procedures aimed at preventing contact spread of the organism were rigidly adhered to by both nursing staff and parents. Boyer *et al.* (1980) reported a similar incident in 1980. Following the development of late-onset Group B infections in two prema-ture babies in a special care baby unit within a 24-h period, swabs were taken from the remaining babies in the unit, from their mothers and the nursery personnel. The isolates from the two infected babies were serotype III, phage pattern 7/11/12, but the relevant maternal cultures were negative. 32% of the nursery personnel carried Group B streptococci, but none of the seven type III isolates examined was the same phage type as the two infecting orga-nisms. These findings suggest that infant-to-infant spread can precede late-onset Group B infection in special care nurseries. The origin of the epidemic organism and the means of transmission from infant to infant however still remain obscure.

Band *et al.* (1981) have described probable nosocomial spread of Group B streptococci because late-onset disease developed in three infants after the introduction of an infant with early-onset disease into the nursery. Two of the infants developed late-onset disease before, and one after, being dis-charged from hospital. The epidemic strain was serotype III, phage type 27/31. A prospective study of nosocomial transmission, started three days after the first sick babies were diagnosed, showed that 11% of infants who were free from the organism at birth had acquired it by the time of discharge. Although type III strains were isolated from colonized babies none had a similar phage-typing pattern to the strains from the three sick babies.

VII. The future

A. International significance of phage typing and modification of the existing set of typing phages

The development and subsequent studies on phage typing of Group B streptococci have so far been carried out by a single laboratory (Division of Hospital Infection, Colindale), but this organism appears to be of world-wide interest. Experience with the staphylococcal phage-typing system showed that as that method was used more extensively throughout the world the composition of the basic phage typing set was changed. Phages that appeared to be of little value in some countries were replaced by others found to be useful in identifying strains specific to the country in question. An international set of typing phages for staphylococci now exists and the percentage typability and new phages are regularly monitored. Although comparable results in typability using the 24 Group B phages have been achieved with North European and American isolates, cultures from more distant places, for example Nigeria and Thailand, have proved more difficult to phage type. Skjold and Wannamaker (1976) have shown a geographical specificity of phages active on Group A M type 49 strains and a similar situation appears to exist with Group B streptococci. It would seem, therefore, that the composition of the basic set of typing phages might be changed in order to maintain high typability when used internationally.

The small numbers of bovine strains examined have been remarkably sensitive to the routine typing phages. A study using phages isolated from bovine strains might elucidate the epidemiology of bovine mastitis, and help to clarify the relationship between human and bovine strains.

References

Anthony, B. F. and Okada, D. M. (1977). *Annu. Rev. Med.* **28**, 355–369.
Anthony, B. F., Okada, D. M. and Hobel, C. J. (1978). *J. Infect. Dis.* **137**, 524–530.
Anthony, B. F., Okada, D. M. and Hobel, C. J. (1979). *J. Pediatr.* **95**, 431–436.
Baker, C. J. (1977). *J. Infect. Dis.* **136**, 137–152.
Baker, C. J. and Barrett, F. F. (1973). *J. Pediatr.* **83**, 919–925.
Baker, C. J. and Barrett, F. F. (1974). *J. Am. Med. Ass.* **230**, 1158–1160.

Band, J. D., Clegg, H. W., Hayes, P. S., Facklam, R. R., Stringer, J. and Dixon, R. E. (1981). *Am. J. Dis. Child.* **135**, 355–358.
Blair, J. E. and Williams, R. E. O. (1961). *Bull. W.H.O.* **24**, 771–784.
Boyer, K. M., Vogel, L. C., Gotoff, S. P., Gadzala, C. A., Stringer, J. and Maxted, W. R. (1980). *Am. J. Dis. Child.* **134**, 964–966.
Bradley, D. E. (1967). *Bacteriol. Rev.* **31**, 230–314.
Brailsford, M. D. and Hartman, P. A. (1968). *Can. J. Microbiol.* **14**, 397–402.
Brock, T. D. (1964). *J. Bacteriol.* **88**, 165–171.
Butter, M. N. W. and de Moor, C. E. (1967). *Antonie van Leeuwenhoek J. Microbiol. Serol.* **33**, 439–450.
Cinca, M., Baldovin-Agapi, C., Mihalcon, F., Belion, I. and Caffe, I. (1959). *Archs. Roum. Path. Exp. Microbiol.* **18**, 519.
Evans, A. C. (1933). *Public Health Rep.* **48**, 411–426.
Evans, A. C. (1934). *Public Health Rep.* **49**, 1386–1401.
Evans, A. C. (1940). *J. Bacteriol.* **40**, 215–222.
Evans, A. C. and Sockrider, E. M. (1942). *J. Bacteriol.* **44**, 211–214.
Ferrieri, P., Cleary, P. P. and Seeds, A. E. (1977). *J. Med. Microbiol.* **10**, 103–114.
Fisk, R. T. (1942). *J. Infect. Dis.* **71**, 153–160.
Franciosi, R. A., Knostman, J. D. and Zimmerman, R. A. (1973). *J. Pediatr.* **82**, 707–718.
Kjems, E. (1955). *Acta Pathol. Microbiol. Scand.* **36**, 433–440.
Kjems, E. (1958a). *Acta Pathol. Microbiol. Scand.* **42**, 56–66.
Kjems, E. (1958b). *Acta Pathol. Microbiol. Scand.* **44**, 429–439.
Kjems, E. (1960a). *Acta Pathol. Microbiol. Scand.* **49**, 199–204.
Kjems, E. (1960b). *Acta Pathol. Microbiol. Scand.* **49**, 205–212.
Lancefield, R. C. (1928). *J. Exp. Med.* **47**, 91–103.
Lancefield, R. C. and Freimer, E. H. (1966). *J. Hyg.* **64**, 191–203.
Leonova, A. G., Khatenever, M. L. and Christenkov, N. A. (1968). *Zh. Mikrobiol. Epidemiol. Immunobiol.* **45**, 104–108.
Lidwell, O. M. (1959). *Mon. Bull. Minist. Health Public Health Lab. Serv.* **18**, 49–52.
McKenna, J. M. (1952). *Rhode Island Med. J.* **35**, 601–603.
Maxted, W. R. (1955). *J. Gen. Microbiol.* **12**, 484–495.
Mihalcu, F. and Vereanu, A. (1978). *Archs. Roum. Pathol. Exp. Microbiol.* **37**, 217–222.
Paredes, A., Wong, P. and Yow, M. D. (1976). *J. Pediatr.* **89**, 191–193.
Paredes, A., Wong, P., Mason, E. O., Taber, L. H. and Barrett, F. F. (1977). *Pediatrics* **59**, 679–682.
Parker, M. T. (1972). *Methods in Microbiology* **7B**, 1–28.
Parker, M. T. (1977). *Postgrad. Med. J.* **53**, 598–606.
Parker, M. T. (1979). *J. Antimicrob. Chemother.* **5**, (Suppl. A), 27–37.
Parker, M. T. and Stringer, J. (1979). *In* "Pathogenic Streptococci" (M. T. Parker, Ed.), pp. 171–172. Reedbooks, Chertsey, Surrey.
Parker, M. T., Ashevov, E. H., Hewitt, J. H., Nakhla, L. S. and Brock, B. M. (1974). *Ann. N.Y. Acad. Sci.* **236**, 466–484.
Pass, M. A., Khare, S. and Dillon, H. C. (1980). *J. Pediatr.* **4**, 635–637.
Pattison, I. H., Matthews, P. R. J. and Maxted, W. R. (1955). *J. Pathol. Bacteriol.* **69**, 43–50.
Pitt, T. L. (1980). *J. Hosp. Infect.* **1**, 193–199.
Plecas, P. (1979). *In* "Pathogenic Streptococci" (M. T. Parker, Ed.), pp. 264–265. Reedbooks, Chertsey, Surrey.
Plecas, P. and Brandis, H. (1974). *J. Med. Microbiol.* **7**, 529–533.

Quirante, M. and Cassady, G. (1972). *Clin. Res.* **20**, 104.

Rippon, J. E. (1956). *J. Hyg.* **54**, 213–226.

Rotta, J. (1979). *Methods in Microbiology* **12**, 177–198.

Sanderson, P. J., Ross, J., Anderson, H. and Stringer, J. (1979). *In* "Pathogenic Streptococci" (M. T. Parker, Ed.), pp. 181–183. Reedbooks, Chertsey, Surrey.

Sharpe, M. E. and Shattock, P. M. F. (1952). *J. Gen. Microbiol.* **6**, 150–165.

Skjold, S. A. and Wannamaker, L. W. (1976). *J. Clin. Microbiol.* **4**, 232–238.

Spanier, J. G. and Timoney, J. F. (1977). *J. Gen. Virol.* **35**, 369–375.

Stringer, J. (1980). *J. Med. Microbiol.* **13**, 133–143.

Vereanu, A. and Mihalcu, F. (1979). *Archs. Roum. Pathol. Exp. Microbiol.* **38**, 265–272.

Vereanu, A., Mihalcu, F. and Ionescu, M. D. (1977). *Archs. Roum. Pathol. Exp. Microbiol.* **36**, 29–35.

Wannamaker, L. W., Skjold, S. and Maxted, W. R. (1970). *J. Infect. Dis.* **121**, 407–418.

Weindling, A. M., Hawkins, J. M., Coombes, M. A. and Stringer, J. (1981). *Brit. Med. J.* **283**, 1503–1505.

Wilkinson, H. W. (1978). *Annu. Rev. Microbiol.* **32**, 41–57.

Williams, R. E. O. and Rippon, J. E. (1952). *J. Hyg.* **50**, 320–353.

Wilson, G. S. and Atkinson, J. D. (1945). *Lancet* **1**, 647–649.

2

Phage Typing of *Brucella*

M. J. CORBEL

Ministry of Agriculture, Fisheries and Food, Central Veterinary Laboratory,
Weybridge, Surrey, UK

I.	Introduction	24
II.	Characteristics of *Brucella* species	26
	A. General properties and taxonomy.	26
	B. Cell structure	27
	C. Chemical composition	30
	D. Phage receptors	32
	E. Antigenic structure	33
	F. Genetics	34
	G. Variation	36
III.	Properties of *Brucella* phages	37
	A. Origins.	37
	B. Structure	38
	C. Plaque morphology	38
	D. Growth cycle	42
	E. Adsorption to bacterial cells.	44
	F. Resistance of phages	46
	G. Phage genetics	48
	H. Serological properties	49
IV.	General materials and methods	49
	A. Safety precautions	49
	B. Reference phages and seed strains.	50
	C. Phage propagating strains	50
	D. Media	52
V.	Isolation of *Brucella* phages.	53
	A. Wild strains	53
	B. Induction of lysogenic cultures	56
	C. Selection of host range mutants	56
VI.	Phage propagation	58
	A. Growth in liquid culture	58
	B. Growth in agar overlay cultures	59
	C. Growth on agar surface cultures	60
VII.	Phage standardization	60
	A. Determination of the routine test dilution (RTD) . .	60
	B. Determination of host range.	62
	C. Effects of culture medium composition on host range .	63

METHODS IN MICROBIOLOGY VOL. 16
ISBN 0-12-521516-9

VIII. Phage preservation 64
IX. Phage purification 65
X. Procedures for phage typing 65
XI. Conclusions. 71
      References 72

I. Introduction

Phage typing plays a central role in the routine identification of *Brucella* cultures. Historically, it has also made a major contribution to the differentiation of the species in the genus *Brucella*.

Originally, the three principal species *Brucella melitensis, B. abortus,* and *B. suis* were isolated independently from completely different sources, and no relationship was thought to exist between them. Subsequently Evans (1918) drew attention to the similarity of *Micrococcus melitensis* isolated from cases of Malta or Mediterranean fever and from the milk of infected goats, to *Bacillus abortus* (Bang's bacillus) strains isolated from bovine abortions and milk. Her conclusions were confirmed by Meyer and Shaw (1920) who proposed that the organisms should be included in a new genus: *Brucella.* Initially this consisted of two species *B. abortus* and *B. melitensis* but later the *B. abortus*-like strains isolated from porcine abortions and described by Traum (1914) were added as a third species: *B. suis* (Huddleson, 1929).

The three species were differentiated largely on the basis of their requirement for supplementary CO_2 on primary isolation, production of H_2S, ability to grow in the presence of selected dyes and their reaction in agglutination tests with absorbed antisera (Huddleson, 1931; Wilson and Miles, 1932). These procedures were generally effective for differentiating the three species on a basis consistent with the preferred natural host. Nevertheless it was recognized at an early stage that the differences between the species were quantitative rather than qualitative and that many inconsistencies occurred. As the studies were extended over a wider geographical area it became apparent that exceptions to the accepted standards were so numerous in some locations as to challenge the validity of the original definition of the species. This led to suggestions that the genus should be considered as a single species with an almost unlimited number of variants (Renoux, 1958; Moreira-Jacob, 1963).

The situation was resolved by the application of manometric methods for the measurement of oxidative metabolic profiles with selected substrates, and of phage sensitivity tests, to *Brucella* taxonomy. The results of these studies, summarized in the reports by Stableforth and Jones (1963) and Jones (1967), showed that *Brucella* strains isolated from sheep and goats, irrespective of their serological or cultural properties, generally had a phage sensitivity and oxidative metabolic pattern which was similar to that of classical *B. melitensis* recovered from these species in the Mediterranean. Similarly, strains isolated from cattle generally had the phage sensitivity and oxidative metabolic pattern of classical *B. abortus*. The same principle applied to *Brucella* strains of porcine origin, irrespective of their other properties or geographical origin. In general, the results of phage sensitivity and oxidative metabolism tests ran closely in parallel and for practical purposes could be considered equivalent (Jones, 1960; Meyer, 1961, 1962; Morgan, 1963).

Subsequently, similar criteria were applied as the new species *B. ovis* (Buddle, 1956), *B. neotomae* (Stoenner and Lackman, 1957) and *B. canis* (Carmichael and Bruner, 1968) were described. All of these studies were conducted with the Tbilisi (Tb) phage isolated in the USSR and described by Popkhadze and Abashidze (1957). Until recently these, or very similar isolates, were the only *Brucella* phages available for taxonomic purposes. Thus, although numerous phage strains were obtained from a variety of sources (Drożevkina, 1963), the majority if not all, were essentially similar to the Tb strain in general properties (Ostrovskaya, 1957, 1961; Morgan, 1963; Calderone and Pickett, 1965). Phages of this type are able to replicate efficiently only in smooth (S) or smooth–intermediate (SI) strains of *B. abortus*. Usually they replicate at low efficiency in S or SI strains of *B. neotomae*, but not in cultures of *B. suis, B. melitensis, B. ovis, B. canis* or rough (R) or mucoid (M) cultures. At very high concentrations, Tb and similar phages exert a bacterocin-like effect on S or SI cultures of *B. suis*, producing growth inhibition or "lysis from without". Phages of this type have been classified as Group 1 by Corbel and Thomas (1980).

More recently the range of *Brucella* phages has been considerably extended by the isolation of strains lytic for all *Brucella* species including non-smooth variants.

The Firenze (Fi) phages were isolated by Corbel and Thomas (1976a) and differ from those in Group 1 in that they are capable of replication in S or SI strains of *B. suis* although their efficiency of plating on biotypes 1, 2 and 3 of this species is much lower than in cultures of biotype 4 or those of *B. abortus* or *B. neotomae*. They have been classified as Group 2 by Corbel and Thomas (1980).

The third group of *Brucella* phages includes the M51 (also identified as M85) and S708 (believed synonymous with BM29) phages of Moreira-Jacob (1968), the Weybridge (Wb) phage (Morris and Corbel, 1973) and the MC/75 (Corbel and Thomas, 1976b) and D strains (Thomas and Corbel, 1977). These lyse S and SI strains of *B. abortus, B. neotomae* and *B. suis* with comparable efficiency, but are inactive on R and M form cultures including those of *B. canis* and *B. ovis*. They are not usually lytic for *B. melitensis* cultures, but may form plaques at low efficiency on some strains (Moreira-Jacob, 1968; Douglas and Elberg, 1976).

A single group of phages, the Berkeley (Bk) strains, form Group 4. These, of which three isolates, designated Bk_0, Bk_1 and Bk_2 are capable of lysing S phase cultures of *B. melitensis* (Douglas and Elberg, 1976, 1978). The Bk_2 strain is also lytic for S form cultures of *B. abortus*, *B. neotomae* and *B. suis*.

The phages in Group 5 are unique in that they are capable of lysing *Brucella* strains in the M and R phases. All are derived from phage R which was obtained from a mixture of smooth-specific phages by mutagenesis (Corbel, 1977a). These phages have proved to be genetically unstable and this has limited their application in phage typing but the R/C strain is relatively stable and is the most useful for this purpose. This will lyse non-smooth cultures of *B. abortus*, and *B. canis* and *B. ovis* with little or no effect on smooth cultures (Corbel and Thomas, 1980).

The five *Brucella* phage groups cover the entire range of species of *Brucella* and enable the majority of isolates to be identified without recourse to the more complex and hazardous procedures involved in oxidative metabolism tests.

II. Characteristics of *Brucella* species

A. General properties and taxonomy

The genus *Brucella* comprises six closely related species of small Gram-negative bacteria, namely *B. abortus, B. melitensis, B. suis, B. neotomae, B. ovis* and *B. canis*. They share several properties. All are non-motile cocci, coccobacilli or short rods, 0.5–0.7 μm in diameter by 0.5–1.5 μm in length, occurring singly, in pairs or short chains. They do not produce spores, flagella or capsules. They are not truly acid-fast but tend to resist decolorization by weak acids when stained by the Macchiavello or Ziehl-Neelsen methods. They are stricly aerobic and do not grow under anaerobic conditions, although many strains are carboxyphilic and require supplementary CO_2. Their metabolism is mainly oxidative and, with the exception of *B. neotomae*, they show little fermentative action on carbohydrates in conventional media. Their nutritional requirements are complex and multiple amino acids, thiamine, biotin, nicotinamide and magnesium are required for growth. Although some strains will grow on simple synthetic media, most will not. Growth is improved by the addition of blood, serum or tissue extracts but haemin (X factor) and nicotinamide adenine dinucleotide (V factor) are not essential.

Brucella strains are always catalase positive and most are oxidase positive. They possess a cytochrome-based electron transport system in which oxygen or nitrate may act as the terminal electron acceptor. Most strains produce nitrate reductase but *B. ovis* does not. The production of H_2S and hydrolysis of urea varies between strains but none produce indole, liquefy gelatin or lyse blood. The Voges–Proskauer and methyl red reactions are always negative and few strains utilize citrate or release *o*-nitrophenol from *o*-nitrophenol-β-D-galactoside (ONPG). Growth occurs in litmus milk, which either appears unchanged or may become alkaline. The optimum temperature is near 37°C but growth will occur in the range 20–40°C. The optimum pH for growth is between pH 6.6 and pH 7.4 and the optimum osmotic pressure is between 2 and 6 atmospheres.

The DNA base composition is between 56 and 58 (G + C) mole % DNA (Tm method). All species show >90% homology of nucleotide sequences in DNA hybridization tests. They contain characteristic fatty acids which produce a unique elution profile on gas-liquid chromatography (g.l.c.). The intracellular proteins are also characteristic and all species produce a similar pattern on disc electrophoresis of acid–phenol extracts. The intracellular antigens are mostly common to all strains irrespective of species, biotype or state of dissociation and are not shared by bacteria of other genera. The surface antigens differ between smooth and non-smooth strains. They include the A, M and R antigens, which although characteristic of the genus *Brucella* may cross-react serologically with unrelated organisms. *Brucella* strains are intracellular parasites which produce characteristic infections in a wide range of animals including man.

The species within the genus are distinguished on the basis of their oxidative metabolic profiles and their pattern of susceptibility to *Brucella* phages. These factors correlate closely with the host preference of each species (Table I).

Within each of the principal species, *B. abortus, B. melitensis,* and *B. suis,* biotypes can be distinguished on the basis of tests for CO_2 requirement, H_2S production, agglutination with monospecific A, M and R antisera and sensitivity to dyes at standard concentrations. Phage sensitivity tests do not distinguish between the biotypes except in the case of *B. suis* biotype 4 on the one hand and biotypes 1, 2 and 3 on the other (Table II).

B. Cell structure

The *Brucella* cell resembles that of other Gram-negative bacteria in general structure. The cell wall in electron micrographs appears as a multi-layered structure composed of an outer layer of lipopolysaccharide (LPS) protein about 9 nm thick with the polysaccharide chains exposed towards the

TABLE I

Classification of the genus *Brucella* into species

Brucella species	Lysis by phages at RTD							Preferred host	Oxidative metabolism measured by respirometry[a]												
	Tb	Wb	Fi	Bk$_2$	R	R/O	R/C		L-Alanine	L-Asparagine	L-Glutamic acid	L-Arabinose	D-Galactose	D-Ribose	D-Glucose	D-Xylose	L-Arginine	DL-Citrulline	DL-Ornithine	L-Lysine	L-Erythritol
B. abortus[b]	+	+	+	+	−	±	−	Cattle	+	+	+	+	+	+	+	±	−	−	−	−	+
B. melitensis[b]	−	−	−	+	−	−	−	Sheep, goats	+	+	+	−	−	−	+	−	−	−	−	−	+
B. suis$_{(1,2)}$[b]	−	+	±	+	−	−	−	Swine, hares	±	±	±	+	±	+	+	+	+	+	+	±	+
B. suis$_{(3,4)}$[b]	−	+	±	+	−	−	−	Swine, reindeer,	±	−	+	−	−	+	+	+	+	+	+	+	+
atypical	−	+	±	+	−	−	−	Rodents	−	+	+	+	−	+	+	+	+	−	+	+	+
B. neotomae	±	+	+	+	−	−	−	Desert woodrat	±	+	+	+	+	±	+	±	−	−	−	±	+
B. ovis	−	−	−	−	−	+	+	Sheep	±	+	+	−	−	−	−	−	−	−	−	−	−
B. canis	−	−	−	−	−	−	+	Dogs	±	−	+	±	±	+	+	−	+	+	+	+	±

[b] Smooth strains.

[a] +, $QO_2N \geq 50$; −, $QO_2N < 50$; ±, QO_2N variable.

TABLE II

Classification of *Brucella* biotypes

Species	Biotype	CO_2 required	H_2S produced	Growth on media[a] containing		Agglutination with monospecific antisera groups		
				Thionin	Basic fuchsin	A	M	R
B. abortus	1	$(+)^b$	+	−	+	+	−	−
	2	(+)	+	−	−	+	−	−
	3[c]	(+)	+	+	+	+	−	−
	4	(+)	+	−	+	−	+	−
	5	−	−	+	+	−	+	−
	6[c]	−	$(-)^b$	+	+	+	−	−
	7	−	(+)	+	+	+	+	−
	9	−	+	+	+	−	+	−
B. melitensis	1	−	−	+	+	−	+	−
	2	−	−	+	+	+	−	−
	3	−	−	+	+	+	+	−
B. suis	1	−	+	+	(−)	+	−	−
	2	−	−	+	−	+	−	−
	3	−	−	+	+	+	−	−
	4	−	−	+	(−)	+	+	−
	Atypical	−	−	+	−	−	+	−
B. neotomae		−	+	−[d]	−	+	−	−
B. ovis		+	−	+	(−)	−	−	+
B. canis		−	−	+	(−)	−	−	+

[a] Concentration = 1/50 000 (w/v).
[b] (+), most strains positive; (−), most strains negative.
[c] For more certain differentiation of biotypes 3 and 6 thionin at 1/25 000 (w/v) is used; type 3 = + ve, type 6 = − ve.
[d] Growth will occur in the presence of thionin at 1/150 000 (w/v).

surface. In non-smooth but not in smooth strains this carries numerous cation binding sites (Weber *et al.*, 1977, 1978). Beneath is an electron-dense layer between 3 and 5 nm thick and composed of heavily cross-linked muramic acid-containing peptidoglycan together with some lipoprotein. Interior to this is the periplasmic space. This is an electron-transparent layer which is 3–6 nm thick in S phase cells but may be up to 30 nm thick in non-smooth variants. This layer contains the periplasmic enzymes involved in cell wall metabolism during cell division (Dubray, 1972, 1976).

The cytoplasmic membrane is located beneath the periplasmic layer and has a typical triple layered lipoprotein membrane structure. Granular polyribosome complexes are usually located adjacent to this in the cytoplasm.

The cytoplasm is interspersed with small vacuoles and polysaccharide-containing granules. The nuclear vacuole can usually be located as a relatively large osmophobic body in the cytoplasm of cell preparations stained with OsO_4 (Dubray, 1976; Dubray and Plommet, 1976; Peschkov and Feodorov, 1978).

The phage receptors of *Brucella* cells have not yet been precisely defined in ultrastructural studies but they form part of the lipopolysaccharide–protein outer layer, at least in the case of smooth phase cells (Corbel, 1977b).

C. Chemical composition

The gross chemical composition of *Brucella* cells is also comparable with that of other Gram-negative species and varies somewhat with the conditions under which the organisms are grown. The approximate composition as a proportion of dry weight is 20–25% carbohydrate, 40–50% protein, 6–7% lipid, 1.5–3% DNA and 10–15% RNA.

The gross amino acid content includes aspartic acid, arginine, alanine, cysteine/cystine, glutamic acid, glycine, isoleucine, leucine, lysine, histidine, methionine, phenylalanine, proline, serine, threonine, tyrosine, valine and α, ε-diaminopimelic acid (Rasooly *et al.*, 1965; Parnas *et al.*, 1967; Kellerman *et al.*, 1970).

The carbohydrate composition includes glucose, galactose, mannose, ribose, deoxyribose, rhamnose, glucosamine, galactosamine, quinovosamine, 4-aminoarabinose, muramate and 2-keto-3-deoxyoctonate (KDO). The presence of xylose, arabinose, uronate and heptose is disputed by some authors (Tuszkiewicz *et al.*, 1966; Kellerman *et al.*, 1970; Kreutzer and Robertson, 1979; Moreno *et al.*, 1979).

The lipid composition of the *Brucella* cell includes aminophosphatides, branched and straight chain fatty acids, hydroxy fatty acids and neutral lipids (Wober *et al.*, 1964; Bobo and Eagon, 1968).

Detailed analysis of individual cell structures has been largely confined to cell wall components. The cell wall accounts for about 21% of the *Brucella* cell dry weight in smooth cultures and about 14% in non-smooth cultures. The smooth phase cell walls contain approximately 37% protein, 14% carbohydrate, 18% lipid, 0.46% muramate and 0.1% KDO. The cell walls of non-smooth organisms contain about 47.5% protein, 13% carbohydrate, 17% lipid, 0.4% muramate and 0.1% KDO (Kreutzer *et al.*, 1977). The cell walls contain the amino acids alanine, arginine, aspartic acid, cysteine/ cystine, glutamic acid, glycine, histidine, leucine, lysine, phenylalanine, proline, serine, threonine, tyrosine, valine and α, ε-diaminopimelic acid (Rasooly *et al.*, 1965; Kellerman *et al.*, 1970). Their carbohydrate content is reported to include glucose, galactose, mannose, fructose, aldodideoxyhex-ose, heptose, rhamnose, ribose, glucosamine, quinovosamine, 4-amino arabi-nose, muramic acid and KDO (Tuszkiewicz *et al.*, 1966; Kellerman *et al.*, 1970; Bowser *et al.*, 1974; Kreutzer and Robertson, 1979; Moreno *et al.*, 1979).

The lipid components of the cell wall account for approximately 18% of the dry weight. About two-thirds is comprised of loosely bound lipid and the remainder of firmly bound lipid. The loosely bound fraction in *B. abortus* contains 22.1% phospholipids and 76.1% free fatty acids and neutral lipids. The phospholipids include diphosphatidylglycerol compounds of cardiolipin type, phosphatidylethanolamine, lysophosphatidylethanolamine, phosphatidylcholine, and a number of uncharacterized compounds including an ornithine-containing phospholipid. Divalent cations including calcium, magnesium and zinc are associated with this fraction. The fatty acid components consist mainly of C_{18} saturated, monounsaturated and polyunsaturated fatty acids with smaller quantities of C_{16} saturated and monounsaturated acids, C_{19} saturated and cyclopropane fatty acids and traces of C_{12} and C_{14} saturated fatty acids. Hydroxy fatty acids have also been detected but these do not include β-hydroxymyristic acid (Wober *et al.*, 1964; Bobo and Eagon, 1968; Kreutzer and Robertson, 1979).

The firmly bound lipids of *Brucella* cells have not been extensively characterized but those of *B. abortus* include palmitic acid, oleic acid, *cis*-vaccenic acid and nonadecanoic acid (Kreutzer and Robertson, 1979).

The cell wall components have been resolved into separate lipopolysaccharide–protein and mucopeptide (murein)–lipoprotein fractions. Substantial differences have been found in the composition of these fractions from smooth and non-smooth cultures. Thus the lipopolysaccharide–protein complexes of smooth *Brucella* cells contain about 6.3% protein, 11.6% carbohydrate, 26.4% lipid, 0.62% KDO and 0.4% hexosamine. Those of non-smooth cells contain about 1.5% protein, 6.5% carbohydrate, 27% lipid, 0.74% KDO and 0.04% hexosamine. The carbohydrate fraction of both types contains glucose, galactose, mannose, dideoxyaldose, glucosamine, galactosamine and possibly quinovosamine and 4-aminoarabinose in differing proportions (Bowser *et al.*, 1974; Kreutzer *et al.*, 1979; Moreno *et al.*, 1979). The presence of heptose is disputed by different authors. Both types of lipopolysaccharide–protein complex are strongly hydrophobic and partition preferentially into the phenol phase on phenol–water or phenol–chloroform–petroleum ether extraction (Moreno *et al.*, 1979). In the case of smooth cultures the lipopolysaccharide–protein complex contains the A- and M-antigenic determinants in the carbohydrate component (Diaz *et al.*, 1968) and the phage receptors which are apparently associated with the protein component (Corbel, 1977b).

The mucopeptide–lipoprotein fraction accounts for 25% of the dry weight of the smooth *Brucella* cell and contains about 5.3% lipoprotein, 11.2% hexosamine, 4.4% carbohydrate and 2% muramic acid. The lipid fraction includes oleic, *cis*-vaccenic, palmitic and nonadecanoic acids and with its associated protein comprises 5.3% of the mucopeptide–lipoprotein complex. This complex accounts for 46% of the dry weight of the non-smooth *Brucella* cell and is composed of 29.2% lipoprotein, 8% hexosamine, 1% carbohydrate and 1.6% muramate. The lipid component is qualitatively similar in fatty acid composition to that of smooth *Brucella* cells, but together with its associated protein constitutes 29.2% of the mucopeptide–lipoprotein complex (Kreutzer and Robertson, 1979).

The mucopeptide fraction is susceptible to degradation by lysozyme and is presumably the target site for phage enzyme in both smooth and non-smooth *Brucella* cells.

D. Phage receptors

Phage receptors have been partially characterized for *B. neotomae* strain 5K33 (Corbel, 1977b). The nature of the receptors in other *Brucella* strains is not known but the indirect information available suggests that they share certain properties in common with those of *B. neotomae*. Thus the receptors for phages in Groups 1, 2, 3 and 4 are only found on cultures in the S, SI or I colonial phases. They appear to be absent from cultures in the M and R phases which lack the antigenic determinants characteristic of S and I cultures. The receptors are stable to heat and extraction with lipid solvents, including acetone and diethyl ether. They are labile to treatment with strong acids, alkalies and oxidizing agents and are largely inactivated on extraction by the standard Westphal phenol–water procedure. They may be extracted with dimethyl sulphoxide, although only part of the activity can be subsequently recovered in water-soluble form (Corbel and Thomas, 1980).

The phage receptor of *B. neotomae* 5K33 is associated with the lipopolysaccharide–protein agglutinogen complex, although it does not correspond to the major antigenic determinants. It seems that a similar situation exists in other smooth *Brucella* strains. Organisms which share antigenic determinants with smooth *Brucella* strains, including *Yersinia enterocolitica* 0:9, *Salmonella* serotypes of Kauffmann–White Group N $(0:30_{I,II}$-antigen), *Francisella tularensis*, *Vibrio comma* (Ogawa and Inaba) and *Pseudomonas maltophilia* 555, have not displayed receptor activity for our *Brucella* phages (Corbel and Phillip, 1972; Morris and Corbel, 1973; Morris *et al.*, 1973; Corbel and Thomas, 1976a,b; Thomas and Corbel, 1977). Nevertheless, there have been reports of adsorption of *Brucella* phages to other bacteria (Parnas, 1966; Parnas and Dominowska, 1966). The reason for these discrepancies is not known but may be related to differing experimental conditions.

The effects of various chemical treatments on the isolated phage receptor of *B. neotomae* 5K33 indicate that amino acid structures are probably involved in the interaction with the phage. Nevertheless these require the integrity of a carbohydrate component as the activity is substantially reduced by periodate oxidation (Corbel, 1977b). It seems clear that the phage receptors of different smooth *Brucella* cultures are not identical as the kinetics of adsorption vary considerably between species and even strains within a species. This may reflect structural differences between the receptor sites or it may be related to their accessibility and distribution over the cell surface.

The nature of the receptors for the phages in Group 5 has not been established. Nevertheless, it is quite clear that they are different from those involved in the adsorption of phages in Groups 1, 2, 3 and 4. Thus they are heat-labile and have not been demonstrated in heat-killed cell preparations, but they are detectable in live cell or acetone-killed suspensions. They are probably protein and distinct from the R-specific antigens of non-smooth *Brucella* strains (Corbel and Thomas, 1980).

E. Antigenic structure

Smooth *Brucella* strains of all species show complete cross-reactivity in agglutination tests with unabsorbed antisera. This cross-reactivity however does not include non-smooth cultures of the same species or the permanently non-smooth species *B. canis* and *B. ovis*. Otherwise, non-smooth *Brucella* strains show extensive cross-reactions with each other in tests with antisera to a non-smooth culture or the R antigen of *B. ovis*.

By cross-absorption of antisera to smooth cultures of *B. abortus* biotype 1 and *B. melitensis* biotype 1 the two major agglutinating antigens A and M can be demonstrated on the surface of these strains (Wilson and Miles, 1932). These antigens are not present on non-smooth *Brucella* cells. Similarly, the R antigen present on the surface of non-smooth cells does not occur on those in the smooth phase (Diaz and Bosseray, 1973).

Minor surface antigens common to *Brucella* strains have been reported including the B polysaccharide (Diaz and Levieux, 1972) and the *f* antigen of Freeman *et al.* (1970). The intracellular antigens of *Brucella* cells are, for the most part, common to all species and biotypes irrespective of colonial phase. Minor antigenic components apparently specific for certain strains or biotypes have been reported however (Nagy, 1967; Hatten and Brodeur, 1978). An attempt to classify the diffusible *Brucella* antigens has been made by Freeman *et al.* (1970) and a modification of their scheme is shown in Table III.

Although the distribution of the major surface antigens corresponds closely with the colonial phase and with the pattern of susceptibility to lysis by *Brucella* phages, none of the antigenic determinants so far described can be equated with the phage receptors. Thus the A and M antigens are common to all smooth strains of *B. abortus*, *B. melitensis*, *B. neotomae* and *B.*

TABLE III

The distribution of the major precipitating antigens between the smooth *Brucella* species

			Distribution			
Antigen[a]	Composition	Net charge[b]	*B. abortus*	*B. melitensis*	*B. neotomae*	*B. suis*
A	Protein	Anionic	+	+	+	+
B	Protein	Anionic	−	−	+	+
C	Protein	Anionic	+	+	+	+
D	Protein–carbohydrate	Anionic	−	−	+	+
E	Glycoprotein	Strongly anionic	+	+	+	+
F	Glycoprotein	Strongly anionic	+	+	+	+
G	Protein	Strongly anionic	−	−	+	+
LPS[c] (A,M)	Lipopolysaccharide –protein	Weakly anionic	+	+	+	+
2	Nucleoprotein	Cationic	+	+	+	+
3	Basic protein	Cationic	−	−	−	+
4	Protein–carbohydrate	Anionic	+	+	+	+
5	Protein or lipoprotein	Anionic	+	+	+	+
6	Protein	Anionic	−	+	−	−
7	Protein	None or weakly cationic	−	+	−	−

[a] Nomenclature modified from Freeman *et al.* (1970).
[b] In 0.05 mol l^{-1} of barbiturate buffer, pH 8.6.
[c] LPS, lipopolysaccharide; A,M are specific antigens.

suis yet these species differ considerably in both their susceptibility to lysis by phage and in their ability to adsorb phages. Similarly organisms of other genera, including the *Y. enterocolitica* 0:9, *Salmonella* 0:30 and *Escherichia coli* 0:157 serogroups, cross-react serologically with the surface antigens of smooth *Brucella* species but do not have receptor activity for *Brucella* phages (Corbel and Thomas, 1980). Thus serological tests with monospecific antisera are of value in predicting whether a culture will be susceptible to phages active on smooth or non-smooth *Brucella* strains but there is no further correlation between phage sensitivity and serology.

F. Genetics

DNA hybridization shows that the polynucleotide sequences of all *Brucella* species are very similar (Hoyer and McCullough, 1968a,b). Even *B. ovis*, the most aberrant member of the genus, still shows greater than 90% similarity of polynucleotide sequence with the DNA of the other species. This indicates a degree of genetic relatedness sufficient to justify considering all members of the genus as a single species. Nevertheless, there are adequate practical reasons for continuing to identify the species in relation to their preferred natural hosts.

Very few studies have been made of the genetics of the *Brucella* group. Transformation, conjugation and transduction have not been demonstrated in the genus. Sex factors and other plasmids have yet to be identified in *Brucella* and examination of a limited number of strains for extrachromosomal DNA has given negative results (Simon, 1979).

True bacteriocins have not been described for *Brucella* species. However, some of the *Brucella* phages, particularly those of the Tb group, behave as bacteriocins in their effect on *B. suis* cultures. Although numerous phages have been isolated from *Brucella* cultures, the question as to whether true lysogeny occurs in this genus has not been conclusively answered. Nevertheless, the behaviour of some cultures, for example those from which several of the Fi group of phages were isolated, is strongly suggestive of a lysogenic state (Corbel and Thomas, 1976a). Such cultures are smooth but phage resistant, and release phage particles which can be detected by plating onto other strains. The release of phages tends to be intermittent and unpredictable and is not permanently stopped by culture for several cell divisions, in the presence of anti-phage serum.

Some *Brucella* phages have been isolated after treatment of their cultures with mitomycin C or by prolonged incubation under conditions limiting growth (Morris and Corbel, 1973; Thomas and Corbel, 1977). This again is suggestive of lysogeny.

Most *Brucella* cultures shown to release phage have been in the non-smooth phase or in the process of dissociation to this. Such cultures have been described as phage carriers (Jones *et al.*, 1962). These can be cured of phage by growth in the presence of anti-phage serum. Occasionally phage maturation may be delayed in such cultures and the phage may persist for one or more cell divisions. Subsequently, however, phages are released promoting dissociation of the carrier culture. Resistance to phage in a *Brucella* strain of a species normally susceptible is usually associated with dissociation from the smooth to a non-smooth phase. This phenomenon occurs frequently *in vitro*, particularly when liquid media are used or cultures are grown on inadequately dried plates.

Much less frequently, smooth phage-resistant mutants may be produced which are otherwise indistinguishable from typical strains of the same species. Usually such mutants are resistant to all *Brucella* phage strains (Corbel and Morris, 1975; Harrington *et al.*, 1977). Such strains are very rarely isolated from clinical samples. They are also quite difficult to produce under laboratory conditions, exposure of smooth cultures to phage usually resulting in the selection of non-smooth resistant variants. Nevertheless, they are important in that such cultures can easily be misidentified if total reliance is placed upon phage typing without the use of supporting procedures.

G. Variation

Apart from smooth to non-smooth variation *Brucella* strains may undergo other modifications in structure which can influence phage susceptibility. This includes the development of cell wall defective forms. The latter are readily induced *in vitro* in culture media containing penicillin or glycine together with an osmotic stabilizer such as sucrose (Hines *et al.*, 1964; Roux and Sassine, 1971; Hatten, 1973) or by hormones such as diethylstilboestrol or testosterone (Meyer, 1976) or by cell cultures of immune macrophages (McGhee and Freeman, 1970). They may also be isolated from the blood or tissues of animals or man infected with *Brucella* strains (Nelson and Pickett, 1951; Corbel *et al.*, 1980).

Cell wall defective *Brucella* organisms are always pleomorphic, with irregular staining properties. Artificially induced spheroplasts will not grow on normal *Brucella* culture media unless these are supplemented to a high osmotic pressure.

The naturally occurring L forms are usually most exacting and require enriched media supplemented with thioglycollate (Nelson and Pickett, 1951) or a high proportion of horse serum (Corbel *et al.*, 1980). The colonial morphology may resemble that of mycoplasmas, including the production of film from phospholipids, or it may resemble that of minute *Brucella* colonies.

Usually cell wall defective forms do not form stable suspensions in saline solutions but behave as non-smooth cultures. They may have surface antigenic determinants common to both smooth and non-smooth *Brucella* strains. Usually they retain some susceptibility to *Brucella* phages (Nelson and Pickett, 1951; Roux and Sassine, 1971; Corbel *et al.*, 1980) but behave as partly dissociated cultures. Electron microscopy studies have indicated that the number of phage receptors is low (Roux and Sassine, 1971; Corbel *et al.*, 1980).

Apart from dissociation to non-smooth forms or vice versa, and the production of L forms, *Brucella* cultures show little variation in relation to phage susceptibility. This would seem compatible with observations that the species maintain their identity, although the biotypes may undergo change.

III. Properties of *Brucella* phages

A. Origins

The Tb phage strain was originally isolated from liquid manure from a cowshed in Tbilisi, Georgia, USSR (Popkhadze and Abashidze, 1957). The A422 strain was isolated from a *B. abortus* culture (Moreira-Jacob, 1968). Other phages of the same group have been isolated from stock *Brucella* cultures, fresh *Brucella* isolates, blood, urine, faeces and cerebrospinal fluid from man or from milk, blood, urine, faeces and aborted fetuses of cattle and other animals infected or vaccinated with *Brucella*. Strains have also been recovered from manure, soil and water (Droževkina, 1963; Parnas, 1963).

The M85 and S708 strains were isolated from cultures of *B. melitensis* and *B. suis*, respectively (Moreira-Jacob, 1968). The strain M51 (Morris *et al.*, 1973) is believed to be identical with M85 and the BM29 strain (Simon, 1979) is believed to be identical with S708.

The Wb phage strain was originally isolated from cultures of *B. suis* strain 1330 (Feesey and Parnas, 1975). The phage strain MC/75 was isolated from plaques produced when a batch of Tb phage stored at 4°C for five years in the presence of toluene was plated on *B. suis* strain 1330 (Corbel and Thomas, 1976). The phage strain D was isolated from a batch of Tb phage propagated on *B. abortus* strain 544 grown in the presence of mitomycin C and then plated on *B. suis* 1330 (Thomas and Corbel, 1977). The precise source of both of these phages is not clear. They may have arisen as mutants of the Tb phage or represent recombinants of Tb phage with defective phages carried by *B. suis* 1330 or by induction of temperate phages carried either by *B. abortus* 544 or *B. suis* 1330.

The Fi phages were isolated from a group of *B. abortus* cultures isolated in Italy. Four of the cultures were smooth and resistant to and occasionally shed phages. The fifth strain (75/13) was a typical phage carrier in that it was in the process of dissociation to a non-smooth state and was partially susceptible to lysis by its own phage (Corbel and Thomas, 1976a).

The Bk group of phages were derived from plaques produced when a Wb phage stock was plated on *B. melitensis* strain Isfahan (Douglas and Elberg, 1976). The three main variants Bk_0, Bk_1 and Bk_2 represent strains with differing host range, which were obtained by propagation on *B. melitensis* Isfahan for Bk_0 and Bk_1, and *B. melitensis* strain 16M for Bk_2 (Douglas and Elberg, 1978).

Apart from a single unconfirmed report (Hall, 1971), phages lytic for non-smooth *Brucella* strains have not been shown to occur naturally. Phage R was produced by mutagenesis of smooth-specific phages followed by selection on a non-smooth culture, *B. abortus* strain 45/20 (Corbel, 1977a).

Subsequently, host range variants active on smooth *Brucella* strains (R/S phage) and on non-smooth strains of *B. ovis* (R/O phage), *B. canis* (R/C phage) and *B. melitensis* (R/M phage) were selected by cultivation on their particular hosts.

B. Structure

All *Brucella* phages so far examined contain double-stranded DNA. The base composition has only been determined for 11 representatives of the Tb group (Calderone and Pickett, 1965). From the T_m values, which ranged from 88.6°C to 89.3°C, Calderone and Pickett (1965) determined the (G + C) mole % DNA ratio to be between 45.3 and 46.7. This is substantially lower than the range of 55–58 determined for the members of the host genus (Hoyer and McCullough, 1968a,b).

On ultracentrifugation in sucrose density gradients, the *Brucella* phages sediment to equilibrium in a zone ranging in density from 1.1 g ml^{-1} to 1.3 g ml^{-1} with most infectious particles found at densities between 1.18 g ml^{-1} and 1.22 g ml^{-1}. In buffered caesium sulphate density gradients the *Brucella* phages sediment to equilibrium in narrow zones corresponding to densities 1.2–1.4 g ml^{-1}. Frequently two discrete zones may be formed, with the bulk of the infectivity concentrated in the denser zone. The particles in the less dense zone, although morphologically complete, may show a reduced infectivity, possibly as a result of reduced DNA content (Corbel, 1979).

The morphology of the phages is consistent. All have a head and tail structure resembling the T-even phages of *Escherichia coli*. The head is hexagonal in outline and ranges in diameter from 55 nm to 80 nm but is in the range of 55 nm to 65 nm for the majority of strains. The tail length ranges from 14 nm to 33 nm but lies between 25 nm and 30 nm for most strains. Tail fibres are present at the distal end and are evidently responsible for attachment of the phage particle to the receptor sites on the bacterial cell surface (Figs 1 and 2).

C. Plaque morphology

The phages of Group 1 produce similar plaques on *B. abortus* cultures in the smooth colonial form. When a high phage input is used the plaques appear heterogeneous and give the impression of a mixture of small, turbid and large clear variants. These on cloning produce an apparent mixture of plaque types. This phenomenon has been attributed to asynchronous growth, and if a low phage input is used and the adsorption period limited by dilution or addition of antiserum, the plaques produced are large and clear and about 2–3 mm in diameter after 48 h incubation (McDuff *et al.*, 1962). A similar

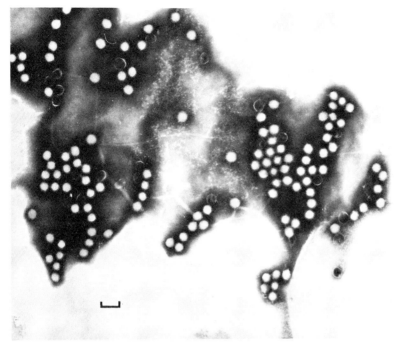

Fig. 1. Electron micrograph of concentrate of *Brucella* phage R purified by caesium sulphate density grade centrifugation. Bar = 100 nm. Phosphotungstic acid × 50 000.

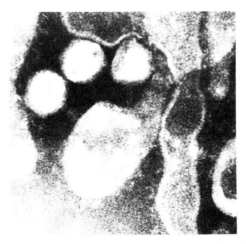

Fig. 2. Electron micrograph of *Brucella* phage MC/75 showing attachment to *Brucella* cell wall fragments via the tail fibres. Phosphotungstic acid × 200 000.

phenomenon occurs with the phages of the other groups (Fig. 3). However, true plaque variants have been detected in Tb phage batches (Jones *et al.*, 1968a). These are attributed to the production of enzyme-deficient phage mutants (Merz and Wilson, 1966; Jones *et al.*, 1968a). On *B. neotomae*, the phages of this group form small turbid, pinpoint plaques after 48 h incubation. True plaques are not produced on other *Brucella* species.

The phages in Group 2 produce plaques on *B. abortus* which are similar to those produced by Group 1 on this species. The plaques produced by Group 2 on *B. neotomae* are similar to those produced on *B. abortus*. Those produced on *B. suis* are small and late clearing and resemble the plaques formed by the phages in Group 1 on *B. neotomae*.

The phages in Group 3 vary in their plaque morphology both between strains and in the appearance of the plaques produced by individual strains on different host species. The Wb strain produces essentially similar plaques on *B. abortus, B. neotomae* and *B. suis* cultures. These are normally visible after 24 h incubation and appear to consist of two types, small turbid and large clear plaques. However, after 48 h incubation only large clear type plaques of 1–3 mm in diameter are present. The M51 and S708 strains also produce similar plaques on *B. abortus, B. neotomae* and *B. suis* strains. As in the case of the Wb strain, two types of plaque are present after 24 h incubation, small turbid plaques of 0.1–0.5 mm diameter and large clear plaques up to 2 mm in diameter. After 48 h incubation only large clear plaques of 2.5–3 mm in diameter are visible. The S708 plaques develop more rapidly than those produced by M51 and the majority are usually clear after 24 h incubation.

The MC/75 strain produces large clear plaques of 2–3 mm in diameter, after 24 h growth. They have a characteristic target appearance, that is they are circular with an outer opaque rim, immediately surrounded by an inner translucent zone of partially lysed cells, which encloses a wide central completely clear zone. The plaques produced on *B. suis* cultures are very similar in appearance but tend to be smaller, up to 2.5 mm in diameter. Those produced on *B. neotomae* are quite different, being 0.5–1.5 mm in diameter, more turbid and with a ragged outline. Phage D on the other hand produces large clear circular plaques of 3–4 mm in diameter, with a turbid periphery on *B. abortus* and *B. neotomae*. The plaques produced on *B. suis* are similar but no more than 2–3 mm in diameter. The R/S variant of phage R on initial isolation produces small, very turbid plaques up to 0.5 mm in diameter on *B. abortus*. At this stage the plaques produced on *B. neotomae* and *B. suis* are usually so small and turbid that they can be easily overlooked. On further passage, larger clearer plaques resembling those produced by the Wb–S708–M51 group are formed.

The Bk_2 strain, representing Group 4, produces plaques identical with

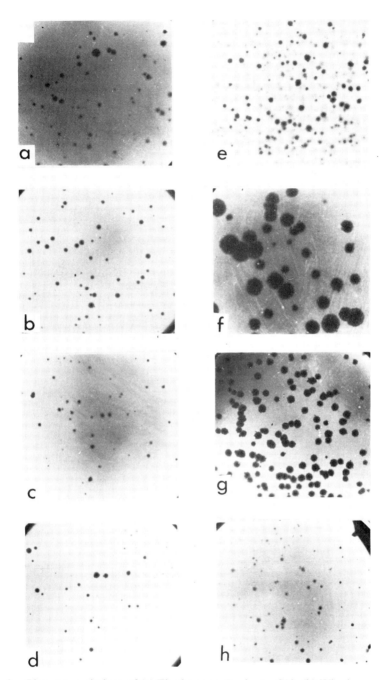

Fig. 3. Plaque morphology of (a) Tb phage on *B. abortus* S19, (b) Wb phage on *B. abortus* S19, (c) Fi phage on *B. abortus* S19, (d) Bk₂ phage on *B. abortus* S19, (e) phage R on *B. abortus* 45/20, (f) phage R/O on *B. ovis* 63/290, (g) phage R/C on *B. canis* RM 6/66, (h) phage R/M on *B. melitensis* B115. All plates were incubated for three days at 37°C.

those given by the Wb strain on *B. abortus, B. neotomae* and *B. suis*. On *B. melitensis* the plaque morphology is more variable. In the experience of the author, those formed on *B. melitensis* 16M are smaller, more turbid and take longer to develop than those produced on other strains of this species. Thus after 48–72 h incubation, the plaques on *B. melitensis* 16M are 0.5–1.0 mm in diameter and very turbid. Those produced on *B. melitensis* strains 63/10, Ether, Isfahan, H38 and Rev 1 are larger and clearer and 1–3 mm in diameter.

The phages in Group 5 characteristically produce very turbid, late clearing plaques on their rough hosts. Phage R produces large, irregular plaques of 2–3 mm in diameter with an outer turbid edge and an inner less turbid central zone after 48–72 h incubation on rough *B. abortus*. At 24 h the plaques are much smaller and so turbid so as to be very difficult to distinguish. The plaques produced by the R/O variant on *B. ovis* are similar to those produced by the R variant on *B. abortus* but tend to be larger and clearer in the central portion after 48 h incubation. If allowed to develop for 72 h or longer, both phages produce complete clearing of the central part of the plaques. This process is more rapid and complete at pH 6.5 than at pH 7.0 or higher. When fully developed the plaques of phage R/O on *B. ovis* may be up to 5 or 6 mm in diameter and completely clear except for a very narrow peripheral zone. The plaques produced by phage R/M on non-smooth *B. melitensis* strains tend to be very small and turbid. After 72 h or more they may have clear pinpoint centres. Phage R/C produces plaques on *B. canis, B. ovis* and non-smooth *B. abortus* cultures which initially resemble those produced by phage R but by 72 h they are completely clear and up to 4 mm in diameter. Plaque formation is much clearer on young *B. canis* cultures than on old mucoid growth. Examples of these plaque types are shown in Fig. 4.

D. Growth cycle

The latent period is longer than for the phages of more rapidly growing organisms and for most strains lies between 1 and 2 h. It is affected to some extent by the choice of propagating strain. The burst size is variable and ranges from 2 to 30 particles. This is determined to a major extent by the propagating strain. For example, for Fi phage 75/13 the latent period and burst size, respectively, are 1.5–2 h and 21 particles per infected cell with *B. abortus* 544 as propagating strain, but 2.5 h and 2 particles per infected cell with *B. suis* 1330 as host (Corbel and Thomas, 1980). This is reflected by the efficiency of plating, which can vary by several orders of magnitude on different host strains. Phage replication and the efficiency of plating is also influenced by the composition of the medium used for propagation of the *Brucella* host strain. This undoubtedly accounts for some of the variations in

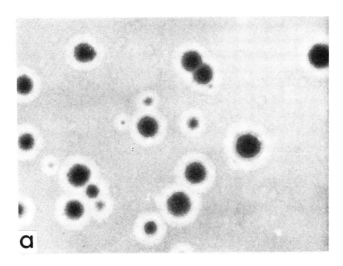

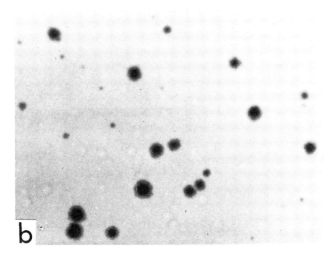

Fig. 4. Phage MC/75 on (a) *B. abortus* S19 and (b) *B. neotomae* 5K33, to show the effect of host species on plaque morphology.

the host range reported by different laboratories. According to Douglas (1978) the effect of the medium is to influence the growth rate of the host strain and this may determine its susceptibility to a specific phage. Such effects are usually less apparent when a nutritionally rich medium such as serum dextrose agar is used. The plating efficiency of the Bk group of phages may be improved by supplementing the culture medium with 0.001 to 0.005 mol l^{-1} glutamine (Douglas and Elberg, 1978). That of the R group is improved if the pH of the medium is lowered to 6.5. For other phage groups the effect of these modifications is minimal. However, the plating efficiency of all *Brucella* phages is substantially reduced at pH values below 6.0 or above 8.0. This is not attributable to a direct effect on the phages as they can tolerate much greater extremes of pH without loss of infectivity (Corbel and Thomas, 1980). All *Brucella* phages so far examined are relatively independent of divalent cations for activity.

E. Adsorption to bacterial cells

The phages in Groups 1, 2, 3 and 4 share the ability to adsorb to smooth *Brucella* cells as opposed to those in the R or M forms. In contrast, the phages in Group 5 adsorb efficiently to non-smooth *Brucella* cells and relatively poorly to those in the smooth phase (Table IV). The absorption patterns show quantitative differences between the phages in Groups 1, 2, 3 and 4. Thus, Group 1 phages absorb most efficiently to cells of *B. neotomae* and rather less efficiently to those of *B. abortus* and *B. suis*. Adsorption to *B. melitensis* cells is equivocal as the figures obtained are just outside the limits of experimental error.

The phages in Group 2 produce a pattern very similar to those in Group 1 except that the efficiency of adsorption to *B. suis* is similar to that obtained with *B. neotomae*.

The Group 3 phages fall into two categories in terms of the adsorption patterns produced by the various strains. Thus the Wb, S708 and M51 phages adsorb most efficiently to *B. neotomae* and somewhat less efficiently to *B. suis* and *B. abortus*, although to a greater extent than those in Groups 1 and 2. Adsorption to *B. melitensis* strains is marginal and to non-smooth strains, including *B. canis* and *B. ovis*, within the limits of experimental error. In contrast, phages D and MC/75 adsorb equally well to *B. neotomae* and *B. suis* and relatively poorly to *B. abortus*. Phage MC/75 adsorbs poorly to *B. melitensis* but the D strain adsorbs efficiently to the three biotypes of this *Brucella* species.

The only studied representative of Group 4, phage Bk_2, has an adsorption pattern resembling that of phage D.

All phages from Groups 1, 2, 3 and 4 adsorb to both living and killed cells of smooth *Brucella* strains. The efficiency of adsorption in general is higher with live cells rather than killed cells but the patterns of activity are qualitatively similar. The pattern of activity is similar with cells killed by heat or acetone.

The phages in Group 5 on the other hand show little tendency to adsorb to

TABLE IV
Adsorption of *Brucella* phages to acetone-killed bacterial cells

Adsorbing agent	Major surface antigens	Adsorption of phage strains as % p.f.u. remaining				
		Tb[b]	Wb[b]	Fi[b]	Bk$_2$[b]	R[c]
B. abortus biotype 1 strain 544	A	12	21	29	16	82
B. abortus biotype 1 strain 45/20[a]	R	93	90	99	90	42
B. abortus biotype 5 strain B3196	M	23	25	30	16	78
B. abortus biotype 7 strain 67/75	A,M	14	17	25	19	74
B. melitensis biotype 1 strain 16M	M	85	74	91	19	90
B. melitensis biotype 2 strain 63/9	A	82	82	101	22	92
B. melitensis biotype 3 strain Ether	A,M	90	79	104	25	84
B. melitensis biotype B115[a]	R	104	84	99	102	100
B. suis biotype 1 strain 1330	A	25	31	25	85	96
B. suis biotype 2 strain Thomsen	A	20	39.7	36	11.7	87
B. suis biotype 3 strain 686	A	22	33	32	ND[d]	ND
B. suis biotype 4 strain 40	A,M	18	11.8	7	2	ND
B. suis 76/4[a]	R	97	86	96	105	98
B. canis RM/66[a]	R	94	93	90	94	97
B. ovis 63/290[a]	R	100	95	97	ND	103
B. neotomae 5k33	A	0.5	10	0	2.0	94
Yersinia enterocolitica 0:1 NCTC 10461		99	96	95	96	ND
Y. enterocolitica 0:9 296/68		93	97	91	97	99
Acinetobacter sp.		99	100	93	100	95
Pseudomonas maltophilia 555		104	95	97	95	102
Salmonella urbana 0:30		98	99	99	99	90
0.15 mol l^{-1} NaCl		100	100	100	100	100

[a] A non-smooth (rough) strain.
[b] Plaque counts performed on *B. abortus* S19.
[c] Plaque counts performed on *B. abortus* 45/20.
[d] ND, not done.

heat-killed cells of either smooth or non-smooth *Brucella*. They do adsorb effectively to living cells however. Adsorption is most efficient with cells of *B. abortus* 45/20, rather less so with *B. ovis* and marginal with *B. canis* and non-smooth strains of *B. melitensis* and *B. suis*. Some adsorption to living cells of smooth *B. abortus* strains nevertheless does occur.

No adsorption occurs to either live or killed cells of organisms which share antigenic determinants in common with *Brucella* strains. These include *Escherichia coli* 0:157, *Francisella tularensis*, *Salmonella* serotypes of Kauffmann–White Group N, the Inaba and Ogawa serotypes of *Vibrio comma*, *Pseudomonas maltophilia* 555 and *Yersinia enterocolitica* serotype 0:9. Similarly no adsorption has been observed to cells of unrelated bacteria including *Escherichia coli*, *Pseudomonas aeruginosa*, *Salmonella typhimurium*, *Salmonella dublin*, *Pasteurella multocida*, *Bordetella pertussis*, *Staphylococcus aureus*, *Proteus* OX:2, *Proteus* OX:19, *Proteus* OX:K, *Proteus vulgaris*, *Streptococcus MG*, *Yersinia pseudotuberculosis* and *Y. enterocolitica* 0:1 and 0:3.

The kinetics of adsorption varies with the phage strain and with the cellular preparation used as substrate. In the case of Group 1, there is little variation between phage strains but considerable variation in their behaviour with different *Brucella* species. Adsorption is most rapid and most complete to cells of *B. neotomae* and *B. suis* biotype 4. Adsorption to *B. melitensis, B. canis* and *B. ovis* cells and to R or M phase cells generally is so slow as to be negligible. The rate of adsorption to *B. abortus* cells is influenced considerably by the host strain selected. It is relatively slow for *B. abortus S19 but faster for B. abortus 544* and the *B. abortus* biotype 5 strain B3196.

The kinetics of adsorption of the Group 2 phages are similar to those from Group 1 except that attachment to cells of *B. suis* strains was much more rapid and complete.

The phages in Group 3 give quantitatively variable responses but in general attach rapidly to cells of *B. abortus, B. neotomae* and *B. suis*.

The Bk$_2$ phage, representing Group 4, attaches rapidly to cells of *B. abortus, B. melitensis, B. neotomae* and *B. suis* but not to cells of non-smooth strains.

Because of the technical difficulties of performing kinetic adsorption experiments on live *Brucella* strains, little information is available on the reactions involved in attachment of the phages in Group 5. The available information indicates that for both phages R and R/O attachment to their preferred host is still incomplete after 30 min incubation.

F. Resistance of phages

In their relative sensitivity to physical and chemical inactivating agents the *Brucella* phages show considerable differences between groups and between strains within the groups (Table V). Nevertheless all strains share certain features. All are rapidly inactivated by heat at 100°C and by exposure to 0.5% (w/v) phenol or 0.1% (w/v) formaldehyde. The *Brucella* phages are generally resistant to toluene and diethyl ether, although vary in their sensitivity to chloroform. None of the phages are inactivated by non-ionic detergents of the Triton X-100 type but varied results are obtained with anionic detergents such as sodium deoxycholate. Cationic detergents such as cetyl trimethyl ammonium bromide completely inactivate all *Brucella* phages when used at concentrations of 1% (w/v). At lower concentrations, for example 0.2% (w/v), some phage strains are not completely inactivated.

The stability of phages to enzyme treatment also varies considerably between strains. Nearly all show considerable loss of infectivity after treatment with the broad specificity proteolytic enzymes such as pronase. Exposure to very high concentrations of RNAase and DNAase usually produces a small but significant reduction in plaque activity. The reason for

TABLE V

Sensitivity of *Brucella* phages to physical end chemical inactivation

Inactivating agent	Proportion of plaque forming activity remaining for phages[a]										
	Tb	A422	Wb	M51	S708	MC/75	D	Fi	Bk₂	R	R/O
0.15 mol l⁻¹ NaCl	100	100	100	100	100	100	100	100	100	100	100
Chloroform	10	2	74	11	57	5.9	12	1	79	15.5	22.6
Diethyl ether	90	81	97	93	102	106	98	98	110	101	91.4
Toluene	98	90	98	92	95	88	97	94	86	98	72.7
Alkyltrimethyl ammonium bromide 0.2% (w/v)	0	0	0	0	0	0.6	1.5	0	42.8	3.5	16.1
Sodium deoxycholate 1% (w/v)	92	94	66	64	46	50.9	96	100	122	99	79.8
Triton X-100 1% (w/v)	102	100	100	103	105	45.2	105	66	106	105	89.1
Chymotrypsin 100 µg ml⁻¹	5	4	17	17	18	42.4	41	95	102	34	78.6
Ficin 5 mg ml⁻¹	14	3	13	2	2	56.4	63	56	29	55	41.7
Pronase 10 mg ml⁻¹	ND[b]	ND	ND	ND	ND	56.6	56	42	33	25	29.2
Trypsin 10 mg ml⁻¹	10	17	20	7	2	101	76	42	54	60	31.8
DNAase 10 000 k units ml⁻¹	89	93	93	100	97	98	98	97	89	93	90.2
RNAase 5000 k units ml⁻¹	94	79	96	100	98	74	92	85	81	89	91
0.01 mol l⁻¹ DETC	98	99	101	98	100	99	98	98	101	100	105
0.01 mol l⁻¹ EDTA	98	106	97	95	105	97	93	99	81	92	100
0.02 mol l⁻¹ DTT	ND	ND	23	ND	ND	24.2	55	33	37	65	17.7
0.02 mol l⁻¹ NaIO4	0	0	0	0	0	0	3	0	0	0.5	0
60°C/1 h	3	9	19	9	9.5	3	7	14	ND	4.5	ND

[a] As % of BaCl control.
[b] ND, not done.

this is not known but may be related to the process of infection rather than a direct action on the phage particle itself. The reducing agent dithiothreitol (DTT) also produces a substantial decrease in the titre of most phage strains. This suggests that disulphide linkages are involved in the phage structure although their location is unknown. The chelating agents disodium ethylene-diaminetetracetic acid (EDTA) and sodium diethyl-dithiocarbamate do not interfere with plaque formation of *Brucella* phages, suggesting that divalent cations are not essential for infectivity.

The process of infection and plaque formation is much more susceptible to the influence of pH than is the stability of the phage particles themselves. Thus most of the phage strains show little reduction in titre following exposure to pH values spanning 2–3 pH units around the neutral point. Nevertheless, the efficiency of plating and the host range may be influenced by a pH variation of 0.2 pH unit.

All *Brucella* phages undergo a substantial reduction in titre following exposure to $0.02 \, \text{mol} \, l^{-1}$ of $NaIO_4$. While this may be interpreted as an indication of the presence of carbohydrate, or at least *cis-a*-glycol-containing structures in these phages, the effect is much reduced if the preparations are previously dialysed free of medium components. This suggests that the inactivating effect may be indirect, resulting from the production of formaldehyde by oxidation of medium components, particularly glucose.

All *Brucella* phage strains undergo a reduction in titre after exposure at 60°C for 1 h. The reduction is increased substantially if the preparations have been previously dialysed to remove low molecular weight components. The identity of the stabilizing factors present in crude lysates is not known but amino acids and possibly small peptides have been implicated as contributing factors. Even when the conditions are standardized and only dialysed preparations are used, the inactivation rate at 60°C varies considerably between strains.

G. Phage genetics

No DNA polynucleotide hybridization studies have been reported for the *Brucella* phages. Their base composition is substantially different from that of the host strains in terms of $(G+C)$ mole % DNA. However this does not mean that polynucleotide sequences of the phage and the host DNA do not overlap. Some degree of homology would be expected if lysogeny is to occur as has been supported by strong circumstantial evidence. The intracellular location of *Brucella* organisms growing *in vivo* makes it improbable that phages would survive unless able to persist in a non-lytic state within the bacterial cell. *Brucella* phages are strongly antigenic and are rapidly eliminated *in vivo*, with the production of neutralizing antibodies (Corbel and Morris, 1980).

Phage mutants are produced naturally as well as under the influence of mutagenic agents. They include plaque variants and mutants deficient in lytic enzymes (Merz and Wilson, 1966; Jones *et al.*, 1968b). The phage R group are extremely prone to spontaneous mutation, probably because they were originally produced as mutants under the action of *N*-methyl-*N*-nitro-*N*-nitrosoguanidine. This substance is known to produce multiple linked mutations at the point of replication (Cerda-Olmedo *et al.*, 1968).

H. Serological properties

Serological studies on the *Brucella* phages have been confined to neutralization tests. The antigenic relationship between phage strains has been determined by estimating the neutralization rate constants (K values) with absorbed homologous and heterologous antisera. The results of these studies have shown that some degree of antigenic relationship exists between all phages examined. The extent of this relationship does not correspond precisely to the classification of the phages in terms of host range, however. This is probably because the neutralization test detects antibodies directed towards the structures involved in attachment of the phage to the bacterial cell and does not give an overall picture of the antigenic structure of the phage particle.

According to cross-neutralization data, the phages in Group 1 are antigenically relatively homogeneous and are related to those in Group 2. The phages in Group 3 show a much greater variation in antigenic specificity although the Wb phage shows some serological relationship to Groups 4 and 5. It is difficult to compare the antigenic structure of the phages in Group 5 with those of the other groups as different indicator strains must be used. Nevertheless cross-neutralization occurs when antisera to the phages in Group 5 are used in tests against the smooth-specific phages of the other groups and vice versa.

Unabsorbed phage antisera show extensive neutralizing activity towards all *Brucella* phages. This is evidently because such antisera react with bacterial antigens adsorbed to the phage particles.

IV. General materials and methods

A. Safety precautions

Phage typing inevitably involves the handling of live virulent *Brucella* cultures and must be regarded as a potentially hazardous procedure. It is strongly recommended that, wherever possible, all stages of the process should be carried out in an efficient exhaust protective cabinet and that every precaution should be taken to avoid exposing the operator to *Brucella* infection. Specific recommendations for the safe execution of *Brucella* culture work have been made by Corbel *et al.* (1979a) and these should be followed wherever possible.

B. Reference phages and seed strains

The Tb phage is still the only *Brucella* phage strain universally recognized as a reference phage. Nevertheless the phages of the other groups have become more widely known and are in use at an increasing number of laboratories. A complete list, together with their propagating strains, is given in Table VI. As the phage groups are arranged on the basis of host range, usually a single strain representative of each will suffice for *Brucella* typing purposes. All of these phages and their recommended propagating strains are available from the FAO–WHO Collaborating Centre for Reference and Research on Brucellosis, Ministry of Agriculture, Fisheries and Food, Central Veterinary Laboratory, Weybridge, Surrey, UK.

C. Phage propagating strains

The *Brucella* strains recommended for propagating *Brucella* phages currently available are listed in Table VI. With the exceptions of *B. abortus* 45/20, *B. abortus* B1119R, *B. canis* Mex 51 and *B. ovis* 63/290, the propagating strains should be in the completely smooth colonial phase for optimum results. This is particularly essential for the phages in Group 4, which are very sensitive to dissociation in the host strain. In contrast, the propagating cultures for the phages in Group 5 should be completely rough. This is not a problem with the *B. canis* or *B. ovis* strains which only exist in the M or R phases but rough *B. abortus* cultures do show a tendency to produce S or I phase mutants. Thus each propagating culture should be checked before use to ensure that it is in the appropriate colonial phase. This is done by growing the strains on glycerol dextrose agar plates (*vide infra*) incubated in air or air + 10% (v/v) CO_2 at 37°C for four or five days. The colonial morphology of the cultures is then examined under oblique illumination as described by Henry (1933). Colonies in the S form are circular in outline with an entire edge and a smooth glistening surface. They appear blue or pale blueish and are soft and easily emulsified. R phase colonies are usually circular or sometimes irregular in outline but have a dry granular surface and appear dull white, yellowish-white or brownish-yellow. They are often friable or viscous and difficult to emulsify, forming clumps or granules in isotonic saline.

Colonies in the M phase tend to be larger than those in the S phase and usually circular in outline. They may be dome-shaped or slightly flattened

TABLE VI
Brucella phage propagating strains

Phage strain	Group	Propagating culture		
Tb		*B. abortus*	S19	(USSR)
A422	1	or		
22/XIV		*B. abortus*	544	
Fi 75/13	2	*B. abortus*	S19	(USDA)
Wb				
M51[a]				
S708[b]	3	*B. suis*	1330	
MC/75				
D				
Bk$_0$				
Bk$_1$	4	*B. melitensis*	Isfahan	
Bk$_2$				
R	5	*B. abortus*	B1119R	
		or		
		B. abortus	45/20	
R/O	5	*B. ovis*	63/290	
R/C	5	*B. canis*	Mex. 51	
R/M	5	*B. melitensis*	B115	

[a] Or strain M85.
[b] Or strain BM29.

with a shiny or slightly granular surface. They usually appear greyish-white in reflected light but the colour can vary. They have a slimy or sticky consistency and do not emulsify uniformly but form threads or clumps of bacterial growth.

The identification of these colonial phases may be facilitated by staining with crystal violet according to White and Wilson (1951). When examined under oblique illumination, S phase colonies appear yellow and transparent with a smooth shiny surface whereas those in the M or R phases are stained shades of red, blue or purple, and usually tend to be opaque with a granular or striated surface. This method is not suitable if subculture is to be attempted off the same plate.

As an additional test, small quantities of culture from single colonies should be emulsified on a clear glass slide with single drops of a freshly prepared 0.1% (w/v) solution of acriflavine in distilled water. S phase colonies form a uniform stable suspension whereas R phase colonies produce granular aggregates or clumps. M phase colonies form coarse clumps or threads. Colonies in the I phase may produce fine aggregations but SI phase colonies usually form a stable suspension.

Further tests for determining the dissociation status of *Brucella* cultures include double-staining with immunospecific conjugates prepared against R and S phase cultures (Brewer and Corbel, 1980), thermo-tube agglutination (Burnet, 1928; Wilson and Miles, 1975) and slide agglutination with antisera rendered monospecific for the A, M or R antigens. Full details of these methods are given by Corbel *et al.* (1979b).

D. Media

Phage dilutions, for the recovery of phage from solid medium and for propagation in liquid culture the most generally useful medium is Albimi brucella broth (ABB; Pfizer Diagnostics, New York, NY10017, USA). This has the following composition: mixed peptones 19.0 g, D(+) glucose 0.9 g, sodium bisulphite 0.1 g, sodium citrate 1.0 g, yeast autolysate 2 g, sodium chloride 5 g and distilled water 1 litre. It is prepared by dissolving 28 g of the dried powder in 1 litre of distilled water and autoclaving at 103.5 KPa (121°C) for 20 min.

For propagation in liquid medium, tryptone soya serum broth (TSSB) can be used. This is prepared by dissolving 30 g of tryptone soya broth powder (Oxoid, London) in 1 litre of distilled water and autoclaving at 103.5 KPa for 15 min. When cooled to 55–60 C, 50 ml of sterile horse serum previously heated at 56 C for 30 min is added. Alternatively, trypticase soy broth (TSB) (Baltimore Biological Laboratories, Baltimore, Maryland, USA) can be used.

For propagation on solid medium, serum dextrose agar (SDA) is recommended. This is prepared by dissolving 40 g of blood agar base (Oxoid, London) in 1 litre of distilled water and autoclaving at 103.5 KPa for 15 min. The agar base is allowed to cool to 55–60 C before adding 50 ml of sterile horse serum previously heated at 56 C for 30 min, then 50 ml of sterile 20% (w/v) D(+) glucose is added aseptically and mixed thoroughly before plating. This medium cannot be reheated. Alternatively, this medium may be prepared by suspending 40 g of trypticase soy agar (TSA) powder in 1 litre of distilled water, heating to boiling point to dissolve and then autoclaving at 103.5 KPa for 15 min. The molten agar is cooled at 50–60 C when 50 ml of sterile horse serum, previously heated at 56 C for 30 min, is added. Plates or slopes are poured immediately after addition of serum.

Glycerol dextrose agar (GDA) for the detection of dissociation in *Brucella* cultures may be prepared by adding glycerol and D(+) glucose to molten TSA or blood agar base cooled to about 60°C to give final concentrations of 2% (w/v) and 1% (w/v), respectively.

For overlay preparation ABB or TSB containing 0.4% (w/v) Oxoid No. 1 agar sterilized by autoclaving at 103.5 KPa for 15 min is used. Immediately before use this is melted by heating in a water bath at 100°C and then allowed to cool to 48°C. Volumes of 3 ml of melted medium are used to form overlays on 9-cm Petri dish layers of SDA.

For preparing dilutions of reagents, bacteria, phages or sera, or for dialysis, phosphate-buffered saline (PBS) is used. This is prepared by dissolving sodium chloride (8.5 g), disodium hydrogen phosphate (1.15 g) and potassium dihydrogen phosphate (0.2 g) in 1 litre of distilled water and autoclaving at 103.5 KPa for 10 min. The pH should be checked and should be between pH 7.2 and 7.4.

V. Isolation of *Brucella* phages

A. Wild strains

Phage strains may be recovered from *Brucella* cultures which spontaneously develop plaques or "pseudoplaques" (Moreira-Jacob, 1968; Corbel and Thomas, 1976a), from cultures which are in the phage carrier state (Jones *et al.*, 1962; Corbel and Thomas, 1976a) or from cultures which show no obvious signs of phage infection. They may also be recovered from materials other than cultures, including blood, faeces, liquid manure, aborted fetuses and sewer effluent (Popkhadze and Abashidze, 1957; Parnas, 1961, 1963). Many of the earliest *Brucella* phage isolates were obtained from these sources, although documentation is incomplete for some strains. The author has no experience of isolating *Brucella* phages from such material but the procedures have been described elsewhere (Droževkina, 1963; Parnas, 1963).

Several methods are available for isolating phages from cultures. In the first instance, as large a number of *Brucella* strains as possible should be examined. Preferably, these should include a proportion of fresh isolates as well as standard laboratory strains. The cultures should be grown up on SDA slopes incubated at 37°C for about 48 h or more in the presence of additional CO_2 for good growth.

The organisms should be suspended in ABB or TSB to give a final concentration of about 10^{10} organisms per millilitre. Volumes of 0.1–0.25 ml of these suspensions are mixed with 3 ml volumes of melted ABB + 0.4% (w/v) Oxoid No. 1 agar held at 48°C and poured onto the surface of 9-cm SDA plates. These are allowed to set and then incubated at 37°C in the presence of 10% (v/v) CO_2 and inspected every 18–24 h for up to seven days. They are first examined against a strong light, for example on an illuminated colony counter, for signs of plaques. If plaques are present, they are likely to be small (0.05–0.2 mm in diameter) and turbid. The plates should therefore be inspected under × 10 or × 15 magnification using a dissecting microscope fitted with a glass stage. Any plaques should either be cut out with a sterile Pasteur pipette or scraped with a sterile inoculating wire and the material transferred to 0.25 ml volumes of ABB or TSB. Drops of these can then be inoculated onto confluent lawn cultures of a selected range of indicator strains.

These are incubated as described above and inspected for propagation of plaques. If propagation occurs on any of the strains selected, single plaques are picked, the material suspended in broth and propagated on the same strain. After several passages, it is usually found that the plaques develop more rapidly and become larger and less turbid. More than one plaque type may become evident. These may or may not breed true to type on serial transfer. In each case, any phage isolate obtained should be cloned by picking single plaques, diluting in broth and plating out serial dilutions onto the indicator strain. The process is repeated by picking single plaques from the highest dilution showing phage activity followed by repeated transfers for purification purposes. At least six cycles of serial transfer from single plaques should be made before a phage can be considered cloned. Cloned isolates should then be fully characterized.

The spontaneous production of plaques on *Brucella* cultures is a rare event in the experience of the author. It was observed in the case of the Firenze cultures which yielded strains 75/6, 75/8, 75/9, 75/10, 75/12 and 75/13 (Corbel and Thomas, 1976a). It may be attributable to contamination of the cultures from an extraneous source or it may result from true carriage of phage either as a result of a lysogenic or pseudolysogenic state. Lysogenic cultures are likely to show anomalous phage sensitivity patterns, often being unusually resistant to phages which would normally lyse cultures of the same species and colonial phase. The behaviour of the Firenze cultures from which the five phages mentioned above were isolated was typical of this pattern (Corbel and Thomas, 1976a). Normally, however, lysogenic cultures are resistant to lysis by their own phages. The behaviour of these *Brucella* phages is similar to the phenomenon of cryptic lysogeny described in other bacterial species (Krizsanovich, 1973).

The existence of true lysogeny in the genus *Brucella* is a still not completely resolved question. Evidence does exist, however, for a phage carrier state (Jones *et al.*, 1962). In this situation, organisms are infected with phage which undergo replication in a lytic cycle but with delayed release of the phage progeny. This may be because changes in the wall structure of the infected cells make them more resistant to phage enzyme, or it may be attributable to defective enzyme production by the phage particles themselves. Usually phage carrier cultures are non-smooth and produce atypical sticky colonies which can undergo spontaneous lysis. If such a colony is streaked on agar and incubated, numerous plaques will appear. Phages are easily isolated from such carrier colonies. For detecting phage release from cultures, the Fredericq (1957) method may be used. In this case the suspected phage carrier is inoculated as a heavy streak or patch in the centre of a serum dextrose agar plate. This is incubated until heavy growth has occurred, usually within 24–36 h. The bacteria are then killed by exposure to diethyl ether (glass rather than plastic Petri dishes are preferred for this purpose) and the culture overlayed with an indicator strain suspended in ABB + 0.4% (w/v) Oxoid No. 1 agar. After incubation the plates are inspected for plaques, lysis and growth inhibition of the indicator strain in the immediate proximity of the suspected carrier culture. The procedure will detect bacteriocin activity as well as phage release. Usually discrete phage plaques can be seen at the periphery of the zone of growth inhibition or lysis. In the case of bacteriocin production, only local growth inhibition is produced. *B. abortus* Firenze strain 75/13 shows the characteristics of a phage carrier culture. In this instance, unlike truly lysogenic cultures, release of phage is "cured" by growth in the presence of anti-phage serum.

Brucella cultures may contain phage without apparently showing signs of lysis or any other direct evidence of infection. Such strains may be truly lysogenic. Phage release is detected by plating out the culture onto an indicator strain. Two procedures are recommended.

1. The test culture and the indicator culture are grown up separately on serum dextrose agar slopes for 24–48 h. The growths are suspended in ABB or TSB to concentrations of about 10^{10} organisms per millilitre. Equal volumes of the cultures are then mixed, diluted 1:100 in sterile broth and incubated on a shaker or magnetic stirrer at 37°C for 48 h. The mixtures are allowed to stand in the static condition at 37°C for about 8 h and then stored at 4°C for up to several months. At regular intervals samples of culture are removed and plated out on SDA. Plaques are cloned and propagated as described above.

2. The test culture and the indicator cultures are grown up as just described. The growths suspended in broth to a concentration of about 10^{10} organisms per millilitre are then mixed and volumes of about 0.25 ml are plated out on SDA and incubated. Alternatively the test culture can be inoculated as a confluent layer on the SDA, grown up for about 18–24 h and then overlaid with the indicator strain in ABB + 0.4% (w/v) Oxoid No. 1 agar. In either case it is advisable to use as many different indicator strains as possible. Any plaques are propagated and cloned as described above.

B. Induction of lysogenic cultures

Two methods generally used for phage induction are
 1. near-lethal ultraviolet irradiation;
 2. mitomycin C treatment.

Mitomycin C treatment may have been instrumental in the isolation of phage D (Thomas and Corbel, 1977). Using near-lethal ultraviolet irradiation the test culture is grown on serum dextrose agar slopes for 24–36 h. The growth is harvested in ABB and then washed twice by centrifugation in phosphate-buffered saline pH 7.2. The organisms are resuspended to give a concentration of 10^8 organisms per millilitre. The suspension is poured into a Petri dish to form a layer not more than 2–3 mm deep. This is placed on a sheet of aluminium foil and exposed to an ultraviolet radiation of 254 nm wavelength. The distance of the source and the duration of exposure have to be determined by experiment, but in general the conditions giving 90–99% kill of the bacteria will be most likely to achieve induction of lysogenic phage. The Petri dish should be agitated at intervals during irradiation. The treated culture is then plated out onto selected indicator strains using either the overlay method or inoculation of drops of material onto lawn cultures. Any plaques obtained are passaged and cloned as described previously.

Using the mitomycin C treatment a choice of methods is available for exposing cultures to mitomycin C. The simplest procedure is to plate the cultures onto SDA containing graded concentrations of mitomycin C in the range $0.01–1.0 \, \mu g \, ml^{-1}$. Concentrations much higher than that are likely to inhibit *Brucella* growth. The cultures are then incubated and after examination, overlaid with the indicator strain. After reincubation, the plates are inspected for plaques and any found are passaged and cloned by the standard procedures.

An alternative procedure is to inoculate the test culture at a concentration of about 10^9 logarithmic phase organisms per millilitre into ABB or TSB containing mitomycin C at concentrations between 0.01 and $100 \, \mu g \, ml^{-1}$. The mixtures are incubated on a shaker at 37°C and samples taken at 4, 8, 16 and 24 h. These are diluted to contain not more than $0.001 \, \mu g \, ml^{-1}$ and plated out onto indicator strains. Any plaques produced are processed and cloned according to the standard procedures.

C. Selection of host range mutants

Selection of host range mutants can be achieved by two processes
 1. selection of naturally produced mutants by the use of a suitable environment;
 2. deliberate production of mutants using mutagenic agents.

Both types of procedure have been successfully used with *Brucella* phages.

The simplest method for selecting naturally produced mutants is to mix the phage with the intended host strain in proportions which will give a multiplicity of infection of approximately 1. Sufficient of each must be plated out to ensure that a mutant occurring at a frequency of about 10^{-8} active phage particles will be detected at least 90% of the time. Thus the phage stock used should contain at least 10^{10} plaque-forming units per millilitre (p.f.u. ml^{-1}). The intended strain is grown up on SDA for about 48 h and suspended in ABB to give a concentration of about 10^{11} organisms per millilitre. This is added to ten volumes of phage suspension and the mixture is incubated at 37°C for 30 min to allow the phage to attach to the cells. The mixture is then plated out on SDA by the overlay method. After incubation for 24, 48, 72 and 96 h the cultures are inspected for plaques. Any found are picked off and adapted to the new host strain by further passage. Finally, the host range mutants are cloned by the standard procedure. Once adapted to the new host strain, such mutants are usually stable but it is advisable to use the new host as the regular propagating strain in future.

A number of agents can be used to produce phage mutants. These include ultraviolet and ionizing radiations, nitrous acid, ethyl methane sulphonate and N-methyl-N'-nitro-N-nitrosoguanidine. In the experience of the author, the latter reagent has given the most successful results with *Brucella* phages, and it is recommended. The selected host strain is grown up on SDA for 24–48 h at 37°C in air + 10% (v/v) CO_2. The cells are suspended in ABB to give a concentration of about 10^{10} viable units ml^{-1}. To about 100 ml volumes of this suspension, 10 ml volumes of the phage stock containing at least 5×10^8 p.f.u. ml^{-1} are added and the mixture warmed to 37°C. N-methyl-N-nitro-N-nitrosoguanidine is then added to a final concentration of 500 μg ml^{-1}. The mixture is agitated until the mutagen has dissolved (Caution: this substance is carcinogenic and must be handled accordingly). The phage–bacteria suspension is then kept static at 37°C for 1 h before being held at 4°C overnight. Volumes of 0.5 ml are then plated out on 9-cm SDA plates. As many replicate plates as possible should be set up. At intervals of about 24 h during incubation the plates are inspected for bacterial growth and plaque formation. Any abnormal colonies should be subcultured and examined for lysis. If spontaneous lysis is not seen then the material should be plated out on cultures of the selected host strain. Most of the cells in the original culture are killed by the mutagen treatment. Surviving cells may form colonies of an atypical appearance. These may contain phage carrier cells, and should be streaked out to permit plaque formation. Any plaques produced on the selected host strain should be passaged and cloned by the standard procedures. Mutants obtained by this method are usually less stable than those arising from natural mutations. It is advisable to preserve a large quantity of the low passage stock produced immediately after cloning for use as the future seed stock for propagation of the new phage strain.

VI. Phage propagation

When propagating *Brucella* phages it is essential to control a number of factors which may influence the final product. Foremost is the phage propagating strain. This should be selected for each phage by examining a number of host strains and choosing one which consistently gives a high yield. In the interests of safety a *Brucella* strain of low virulence should be chosen whenever possible. The propagating strain ideally should permit no variation in the properties of progeny phages from those of the original stock. This is facilitated by using a single large batch of seed stock for propagation, rather than a series of subcultures which may themselves have undergone changes. The seed stock should be preserved by freeze-drying or by storage in liquid nitrogen. To produce maximum phage yields it is essential to employ the optimum phage: bacterium ratio. This has to be determined for each phage and for each propagating strain. The conditions of incubation should also be standardized, since sub-optimal conditions may favour mutation and dissociation in the propagating strain. In particular, where liquid medium is used aeration should be well maintained and pH controlled. For propagation on solid media the accumulation of excessive surface moisture should be avoided and a well-buffered, nutritionally adequate medium used. Growth inhibiting agents must not be used in phage propagating media. The precise conditions which will give optimum phage yields have to be determined for each phage–host system. The methods described below may be adapted for each particular case.

Procedures
Three basic methods may be used for *Brucella* phage propagation. These are

- A. growth in liquid culture;
- B. growth in agar overlay cultures;
- C. growth on agar surface cultures.

A. Growth in liquid culture

This method is applicable for most phage strains but tends to give low titres. It is not suitable for phages in Group 5 and is not recommended for the Bk_2 phage. If a virulent *Brucella* strain is used for propagation, special care should be taken in handling large volumes of liquid culture. A suspension of a 24–36 h SDA slope culture of the propagating strain is prepared in ABB or TSB broth to give a concentration of about 10^{10} organisms per millilitre. A 1-ml volume of this is added aseptically to 150 ml of either ABB or TSB in a 1-litre Erlenmeyer flask. The broth should be prewarmed to 37°C. The suspension is incubated for 4–6 h at 37°C on a reciprocating shaker or magnetic stirrer to produce vigorous aeration. A 1-ml volume of phage suspension containing about 10^9 p.f.u. ml^{-1} is then added aseptically and the mixture reincubated for 48–72 h at 37°C with sufficient agitation to produce vigorous aeration. Foaming of the broth should be kept to a minimum, however. It is preferable to use a propagating strain which does not require supplementary CO_2 for growth, as this would have to be provided in the incubator atmosphere. After the incubation period is completed, the agitation is stopped and the culture left static for 2–3 h. It is then allowed to stand overnight at 4°C before centrifugation at 10 000 g for 15 min. The clear supernatant is collected, filtered through a 45-μm pore membrane filter and titrated for phage activity. The filtrate may then be saturated with toluene to help preserve bacterial sterility during subsequent manipulations.

B. Growth in agar overlay cultures

This method is recommended by Adams (1959) for producing high titre stocks of most phages. It is effective for most *Brucella* phages although simpler procedures can give equally good results. Volumes of 3 ml of ABB or TSB agar overlay medium are melted by heating in a water bath at 100°C and then allowed to cool to 48°C. To each 0.1 ml of a suspension of propagating strain containing about 10^{11} organisms per millilitre and 0.1 ml of a phage suspension containing 10^4–10^5 p.f.u. ml^{-1} are added and mixed thoroughly. The mixture is immediately poured onto the surface of an SDA plate (9-cm Petri dish) and allowed to set. This will usually be complete after 30 min at room temperature. It is important that the agar surface of the plates should be free of surplus moisture before use. One set of plates should be inverted and incubated in air, or air + 10% (v/v) CO_2, at 37°C. For phages in Group 1 incubation is continued for at least 36 h, whereas for those in Groups 2 and 3 incubation may be completed after 24 h. The phages in Groups 4 and 5 require incubation for 48–72 h if maximum yields are to be obtained. After incubation the plates should be inspected. The overlays should show extensive or complete lysis of the propagating strain and no colonies of contaminating bacteria or fungi should be present. The soft agar overlays are then scraped into a sterile container, about 3 ml of ABB added per layer and the material allowed to stand overnight at 4°C. The supernatant fluid is decanted from the macerated agar layers and centrifuged at 10 000 g for 10 min at 4°C. The deposited material is discarded and the supernatant collected. This is filtered through a 0.45-μm pore membrane filter and stored at 4°C until required. After the container has been opened the phage stock should be saturated with toluene to inhibit microbial contaminants.

C. Growth on agar surface cultures

According to Adams (1959) this method is less reliable than method B for producing high titre phage stocks. Nevertheless, it has given good results with all *Brucella* phage strains examined and has the advantage of simplicity. Volumes of 0.1 ml of propagating strain and phage strain, prepared as described for method B, are pipetted onto the surface of previously dried SDA plates. The phage–bacteria mixture is spread uniformly over the agar layers by means of L-shaped glass or stainless steel spreaders and allowed to dry. The plates are then inverted and incubated according to the schedules given for method B. After incubation the layers are inspected. Those showing confluent or near-confluent lysis without extraneous microbial contamination are flooded with 3 ml volumes of sterile ABB and left to stand on a level surface for 20 min at room temperature. The supernatant liquid is then decanted and processed exactly as described for method B.

VII. Phage standardization

A. Determination of the routine test dilution (RTD)

For the identification of *Brucella* cultures phage preparations standardized at the routine test dilution (RTD) are usually employed. For some purposes, particularly if dealing with cultures undergoing dissociation, preparations standardized at $RTD \times 10^2$ and $RTD \times 10^4$ can also be used. The RTD is defined as the highest dilution of the phage stock which will produce confluent lysis of a lawn inoculum of the propagating strain.

To determine the RTD, serial ten-fold dilutions of the phage stock in the range 10^{-1} to 10^{-10} are made in ABB. A separate pipette must be used for transferring each dilution to avoid carryover of phage. Cultures of the propagating strain are inoculated as a lawn by pipetting 0.1 ml of suspension containing from 10^{10} to 10^{11} organisms per millilitre onto previously dried SDA plates and spreading uniformly across the surface with an L-shaped glass or stainless steel spreader. The plates are allowed to dry on a level surface and then uniform drops of phage dilutions are pipetted onto the agar surface, ensuring that each remains discrete and well separated from its neighbours. The plate is left on a level surface for about 30 min, or until the phage suspension has soaked into the agar. It is then inverted and incubated at 37°C in air or air + 10% (v/v) CO_2 as appropriate.

The period of incubation required before the plates can be effectively examined for lysis varies between phages. For those in Group 1, 24–36 h is usually required, whereas for those in Groups 2 and 3, 24 h will usually suffice. For the phages in Groups 4 and 5, 48 h incubation may be needed before lysis is well defined, especially at higher dilutions.

The plates are best examined by transillumination. An illuminated colony counter is convenient for this purpose, although simply holding the plates up to the light will suffice. The areas of lysis will appear as zones of clearing in the bacterial lawn. The most concentrated phage suspension will produce complete clearing and the most dilute, no effect at all. The dilutions immediately below the RTD will produce zones of incomplete lysis and clusters of separate plaques. The last dilution to produce complete lysis is taken as the RTD (Fig. 5). It is advisable to set up two or three replicate plates for each phage titration. If a calibrated pipette, for example 50 dropper Pasteur or calibrated Eppendorf pipette with sterile disposable tips, is used then the absolute titre of the phage stock in p.f.u. ml^{-1} may be determined by counting the number of plaques produced by a unit volume of diluted phage. Counts should be confined to dilutions producing well-developed discrete plaques.

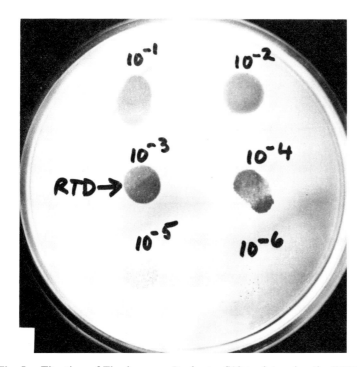

Fig. 5. Titration of Tb phage on *B. abortus* S19 to determine the RTD.

B. Determination of host range

Before a batch of phage is regarded as fully standardized for routine use its host range should be checked to ensure that it corresponds with that of the parent strain. The procedure to be used is essentially similar to that just described for the phage titration except that the reference strains of *B. abortus*, *B. melitensis*, *B. suis*, *B. neotomae*, *B. canis* and *B. ovis* are used as indicators instead of the propagating strain. It is advisable to include some rough strains of *B. abortus*, *B. melitensis* and *B. suis* in the host range tests in addition to the reference strains.

If variations in the host range are found they may result from cross-contamination of the phage stock by other phage strains, selection of a host–range mutant, mislabelling of a seed stock or variations in one of the indicator strains. All of these factors should be checked when possible. Phage batches with anomalous host range properties should not be used for routine culture typing. If they contain host–range mutants these may be of interest for further studies.

1. *Plaque formation and lysis by standard phage concentrations*

To determine patterns of lysis, phage standardized at RTD, $RTD \times 10^2$ and $RTD \times 10^4$ are used. As wide a range of host strains as possible should be tested. These should have been characterized by all of the methods available, including oxidative metabolism tests, examination for dissociation, CO_2 dependence, H_2S production, dye sensitivity and reaction with monospecific antisera.

Lawn-inoculated cultures of the host strains are prepared by spreading 0.1 ml volumes of fresh suspensions containing about 10^{10} viable organisms per millilitre across the surface of SDA layers. The plates are allowed to dry and then 0.02 ml volumes of the phage preparations are inoculated as discrete, well-separated drops. The positions of these are marked on the underside of the plates which are then incubated at 37°C in air + 10% (v/v) CO_2 (if available) for up to four days. The cultures are examined for zones of lysis or discrete plaques at one-day intervals. Usually phage preparations standardized at the concentrations stipulated will either produce confluent or near-confluent lysis, or no lysis at all, depending upon the host strain used. Nevertheless some phages, such as those in Group 3 may plate far less efficiently on some host strains than on others. In these cases discrete plaques may be produced in the one instance, whereas confluent lysis is produced in the other.

To examine plaque formation by phages, it is necessary to use preparations which have been diluted to about $RTD \times 10^{-2}$. If 0.1 ml volumes of these are plated out, between 10 and 100 plaques should be produced on the propagating strain. These will be sufficiently separated to allow full development. Other dilutions may be necessary for different host strains.

The phage and the indicator strain should be plated out by the overlay method as this allows the plaque morphology to be studied to best advantage. Plaque formation should be examined at intervals of 24 h for up to five days. Some phages, such as those in Groups 4 and 5, produce plaques which are initially very turbid and may not clear completely on some host strains. Consequently it is advisable to examine plates for plaque formation under low power magnification; for example with a dissecting microscope fitted with a glass stage provided with oblique illumination. The morphology of the plaques produced on the various host strains should be noted. In general, the more efficiently the phage replicates in a particular *Brucella* strain, the larger and clearer the plaques produced. Most *Brucella* phages apparently produce a range of plaque types if high phage concentrations are plated out. This is usually the result of asynchronous growth and a uniform plaque morphology is produced if the phage adsorption period is limited, either by dilution or by addition of anti-phage serum.

2. *Efficiency of plating on various host strains*

The host strains selected will usually include the reference strains for *B. abortus*, *B. melitensis* and *B. suis* biotypes 1 and *B. neotomae*, *B. canis* and *B. ovis*. Some additional non-smooth strains may also be included. Phage concentrations in the range $RTD \times 10^2$ to $RTD \times 10^{-4}$ are prepared and triplicate 0.1 ml volumes are plated by the overlay method onto the selected host strains. These are incubated under the normal cultural conditions until plaque development is at its maximum level. Careful plaque counts are then made with the aid of an illuminated colony counter. The efficiency of plating on any particular host strain is usually expressed as a percentage of the plaque count produced on the normal propagating strain.

C. Effects of culture medium composition on host range

This should be checked for each new phage–host system. The factors which are particularly likely to modify the plating efficiency of *Brucella* phages on various hosts are the pH of the medium, its nutritional adequacy for the host strain and the presence of low molecular weight components such as divalent or polyvalent cations or anions and amphoteric compounds such as amino acids. These factors are conveniently examined by using TSA agar as a basal medium and supplementing this or modifying the pH. Where defined medium is required, then Eagles' tissue culture medium (Eagle, 1959) may be used as the basic formula. Specific components may be added to or omitted from this at will. The medium is solidified by making it up at double strength, heating to 56°C in a water bath and adding an equal volume of 2% Oxoid No. 1 agar in distilled water, cooled to 56°C. Plates must be poured before the medium is allowed to set.

The *Brucella* phages hitherto examined are relatively independent of Mg^{2+} and Ca^{2+} concentration for their activity. Plaque formation may be influenced by comparatively small changes in pH, however. Thus phage R and its mutants form plaques most efficiently at pH 6.5. Phage Bk_2, on the other hand, plaques more efficiently on *B. melitensis* strains if the medium is supplemented with 4 mmol l^{-1} of glutamine (Douglas and Elberg, 1978). The other phage groups show quite small differences in plating efficiency at pH 6.5 compared with neutral pH.

VIII. Phage preservation

The *Brucella* phages as a group are relatively stable and preparations of most strains will remain infective for many years if stored in broth at 4°C. Loss of activity is accelerated by microbial contamination, probably as a result of production of proteolytic enzymes. Repeated freezing and thawing is also detrimental. Some are more susceptible to this than others but most strains undergo a reduction in titre of several log_{10} units with a single cycle of freezing and thawing. This method may be satisfactory nonetheless, if high titre phage stock is sealed in glass ampoules and snap frozen, for example in a solid CO_2–acetone freezing mixture, and stored at $-50°C$ or below.

Freeze-drying is also a satisfactory method of preservation if high titre phage stock is used. The process will usually produce at least a 90% reduction in titre but this is generally unimportant as once the material is dried and sealed in ampoules it will retain its infectivity for many years. Phages R and R/O are less easy to preserve by freeze-drying than most of the other strains; the smooth-specific mutants normally present in each batch tend to survive the process better than the rough-specific strain.

The most satisfactory preservatives for inhibiting microbial growth in phage stocks stored at 4°C are diethyl ether and toluene. Chloroform is to be avoided as it may inactivate some *Brucella* phages. Toluene is the most generally useful preservative as it is less volatile than ether and not subject to peroxide formation. Both of these solvents may be removed from phage stocks by dialysis or evaporation under reduced pressure. Phage stocks containing these chemicals should be stored in properly sealed containers in a spark-proof refrigerator. Phenols, cresols, merthiolate, sodium azide and dyes are not suitable preservatives for *Brucella* phages as they inactivate these and/or bacteria subsequently used for their propagation.

IX. Phage purification

For routine culture typing it is generally unnecessary to use highly purified phage preparations. For this purpose the filtered, clarified supernatant fluid prepared from crude lysates is usually adequate. Nevertheless, if further characterization of the phages is to be attempted it is usually necessary to use suspensions which are essentially free of bacterial components. *Brucella* phages may be purified by the following procedure.

Crude phage lysates are clarified by centrifugation at 10 000 **g** for 15 min at 4–6°C. The deposited material is discarded and the clear supernatant collected. This may then be dialysed against phosphate-buffered saline, pH 7.2, to remove low molecular weight bacterial and culture medium components. Further purification is required to separate the phage from high molecular weight soluble materials. Contaminating nucleic acids are removed by adding 1000 Kunitz units of DNAase (bovine pancreatic type 1, Sigma London, Poole) and 500 units of RNAase (Type 111-A, Sigma London, Poole) per 1 ml volume and incubating at 37°C for 1 h. This stage is preferably performed before the dialysis step. Final purification is achieved by concentrating the phage stock to about 10^{11} p.f.u. ml^{-1} by counter-dialysis against a concentrated solution (20 M) of polyethylene glycol in phosphate-buffered saline, pH 7.2. The concentrate is then layered on a preformed linear 10–60% (w/v) $CsSO_4$ density gradient and centrifuged at 100 000 **g** for 24–36 h at 8–10°C.

The phage will sediment to equilibrium density in the $CsSO_4$ gradient in the zone corresponding to a buoyant density of 1.2 g ml^{-1} to 1.4 g ml^{-1} as one or two discrete very narrow white or opalescent bands. The material in these may be recovered by direct aspiration with a syringe and needle or by means of a gradient fraction collector. The $CsSO_4$ is removed by dialysis against phosphate-buffered saline, pH 7.2. If further purification is required the density gradient centrifugation is repeated. Phage processed in this manner is essentially free of contaminating material derived from either the host cells or the culture medium. However, the infectivity of some phage strains may be reduced by exposure to high concentrations of $CsSO_4$.

X. Procedure for phage typing

As a preliminary step towards phage typing it is absolutely essential that the state of dissociation of the culture should be accurately determined. The importance of this cannot be too strongly emphasized as it is impossible to interpret accurately the results of phage sensitivity tests unless the colonial phase has been identified. Even a very low level of dissociation may modify the phage sensitivity pattern of a culture. This is particularly the case with the phages in Groups 4 and 5. More than one procedure must be used to check the dissociation status and it is recommended as a minimum that each culture should be examined by the following procedures:

1. oblique illumination of colonies grown on GDA (Henry, 1933);
2. slide agglutination tests with 0.1% (w/v) aqueous acriflavine;
3. slide agglutination tests with antiserum prepared against the group-specific R-antigen of *B. ovis* 63/290;
4. formation of a stable suspension in 0.15 mol l^{-1} of NaCl.

Additional tests which are of value include examination of colonies on GDA under reflected light after staining with ammonium oxalate–crystal violet (White and Wilson, 1951); staining with fluorochrome or enzyme-labelled antibody conjugates prepared against S and R forms of *Brucella* (Brewer and Corbel, 1980); thermo-agglutination or stability of cultures suspended in 0.15 mol l^{-1} of NaCl on heating at 80–100°C for 2 h (Wilson and Miles, 1975).

If the culture shows only a single colonial phase, representative colonies are picked off and subcultured on SDA slopes. If a mixture of smooth and non-smooth colonial forms are present, then smooth colonies are selected for subculture. The SDA slopes are incubated at 37°C in air + 10% (v/v) CO_2 24–48 h or until good growth is obtained. The bacteria are then suspended in ABB or TSB to give concentrations of about 10^{10} organisms per millilitre. Volumes of 0.1 ml are spread over the moisture-free surface of SDA plates with a swab or spreader. These are allowed to dry for about 1 h and are then inoculated with phage. The base of each plate should be marked at the point of inoculation of the phage. Standard drops of phage suspension, usually 0.02 ml, are applied with a 50-dropper pipette or a calibrated syringe and needle to the agar surface at the points indicated. The drops are allowed to absorb completely into the agar before the plates are inverted and incubated for 24 h. They are then inspected every 24 h for signs of lysis or plaques.

For routine typing purposes it is possible to employ only a single phage, for example the Tb strain, provided that this is used at concentrations of both RTD and RTD × 10^4. The test culture is inoculated across one-half of a plate and the RTD and RTD × 10^4 doses applied to separate sectors (Fig. 6). After incubation, the culture is examined for lysis as described in the preceding paragraph. This procedure has been used for many years and usually gives satisfactory results for the differentiation of the smooth *Brucella* species. It is less satisfactory when non-smooth cultures are to be examined and occasionally presents difficulty with rare atypical strains of *B. melitensis* which may be susceptible to "lysis from without" or growth inhibition by Group 1 *Brucella* phages.

The procedure currently used at this laboratory for routine typing employs one phage strain from each of Groups 1, 2, 3, 4 and 5 standardized at RTD.

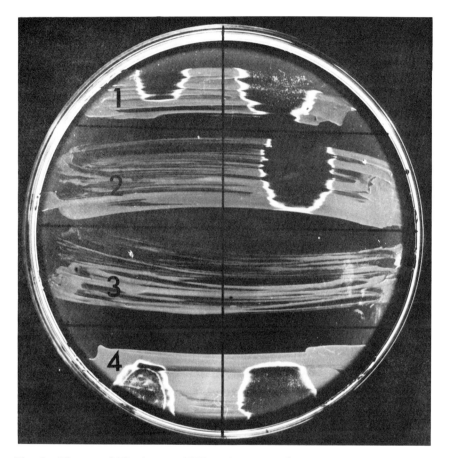

Fig. 6. The use of Tb phage at RTD and RTD × 10⁴ for routine typing of (1) *B. abortus* 544, (2) *B. suis* 1330, (3) *B. melitensis* 16M, (4) *B. neotomae* 5K33.

These are usually the Tb, Fi 75/13, Wb, Bk_2 and R/C strains. All five can be accommodated on a single plate as shown in Fig. 7. If only smooth cultures are under examination, phage R/C may be eliminated. However, this phage strain is very sensitive to dissociation in *B. abortus* and will form plaques on cultures which still react with smooth-specific phages. Therefore, it is useful to detect dissociation in SI or I form cultures, which may otherwise be difficult to demonstrate.

Screening cultures for species identity can usually be adequately carried out with phage preparations at RTD provided that typical *Brucella* isolates are being examined. If atypical cultures are under examination, particularly

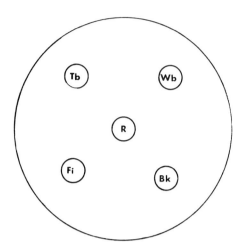

Fig. 7. The simultaneous use of five different phages at RTD concentration on a single plate.

those which may be partially dissociated, they should be retested with phage preparations standardized at $RTD \times 10^2$ and $RTD \times 10^4$. Not all of the phage strains are equally affected by dissociation. For example, the Wb strain will often still lyse cultures which have dissociated to an extent where they are resistant to the phages in Groups 1, 2 and 4. On the other hand, the Bk_2 phage is more sensitive to dissociation in *B. melitensis* than in *B. abortus* or *B. suis*. Using these phages at the recommended concentrations, the lytic patterns summarized in Table VII are obtained.

Phage typing may also be used to confirm the identity of *Brucella* isolates which are anomalous in other characteristics. For example, dye-resistant variants of *B. suis* biotype 1 or dye-sensitive strains of *B. melitensis* produce the phage sensitivity pattern typical of their respective species. This may be confirmed by using a large number of phage strains in Group 3 for example. At RTD these will readily lyse *B. suis* but not *B. melitensis*. Confirmatory evidence is provided by determining the efficiency of plating of these phages on the test strains. Comparative values for most of the phage strains in use are given in Table VIII. It is evident from this that the efficiency of plating of some phages varies not only between host species but also between biotypes of strains within the species. In most cases these differences are too small and subject to so much variation that they have limited significance. An exception to this is Fi phage and the *B. suis* biotypes. This phage strain plates with comparable efficiency on strains of *B. suis* biotypes 1, 2 and 3 but 5000–10 000 times more efficiently on those of *B. suis* biotype 4. This difference is only

TABLE VII
Lytic patterns of *Brucella* phages[a]

Phage strains	Group	Titre	*B. abortus* Smooth	*B. abortus* Non-smooth	*B. suis* Smooth	*B. suis* Non-smooth	*B. melitensis* Smooth	*B. melitensis* Non-smooth	*B. neotomae* Smooth	*B. neotomae* Non-smooth	*B. canis* Non-smooth	*B. ovis* Non-smooth
Tb, A422, 22/XIV	1	RTD	L	NL	NL	NL	NL	NL	NL or PL	NL	NL	NL
		RTD × 10^2	L	NL	PL	NL	NL	NL	PL	NL	NL	NL
		RTD × 10^4	L	NL	L	NL	NL	NL	L	NL	NL	NL
Fi 75/13	2	RTD	L	NL	PL	NL	NL	NL	L	NL	NL	NL
		RTD × 10^2	L	NL	L	NL	NL	NL	L	NL	NL	NL
		RTS × 10^4	L	NL	L	NL	NL	NL	L	NL	NL	NL
Wb, M51, S708, MC/75, D	3	RTD	L	NL	L	NL	NL[b]	NL	L	NL	NL	NL
		RTD × 10^2	L	NL	L	NL	NL[b]	NL	L	NL	NL	NL
		RTD × 10^4	L	NL	L	NL	NL[b]	NL	L	NL	NL	NL
Bk₂	4	RTD	L	NL	L	NL	L	NL	L	NL	NL	NL
		RTD × 10^2	L	NL	L	NL	L	NL	L	NL	NL	NL
		RTD × 10^4	L	NL	L	NL	L	NL	L	NL	NL	NL
R	5	RTD	NL	NL	V	NL	NL	NL	NL	NL	NL	NL
		RTD × 10^2	L	L	PL	NL	NL	NL	PL	NL	NL	NL
		RTD × 10^4	L	L	PL	V	NL	NL	PL	NL	NL	NL
R.O		RTD	L	NL	V	NL	NL	NL	PL	NL	NL	L
		RTD × 10^2	L	NL	V	NL	NL	NL	L	NL	NL	L
		RTD × 10^4	L	V	V	V	NL	V	L	NL	NL	L
R.C		RTD	NL	L	NL	NL	NL	NL	NL	NL	L	L
		RTD × 10^2	NL	L	NL	NL	NL	NL	NL	NL	L	L
		RTD × 10^4	NL	L	NL	NL	NL	NL	NL	NL	L	L

[a] L, confluent lysis;
PL, partial lysis, single plaques or growth inhibition;
NL, no lysis;
V, variable, some strains lysed;
RTD, routine test dilution.
[b] PL reported on some strains.

TABLE VIII
Relative efficiency of plating of *Brucella* phages on *Brucella* species

Brucella			Tb	Wb M51 S708	MC/75	D	Fi	Bk$_2$	R	R/C
					Plaque count (%) for phage strains[a]					
B. abortus	544	(A1)	100	160	60.6	117	100	640	7.3	0
	B3196	(A5)	116	140	78.7	130	100	660	8.6	0
	63/75	(A7)	100	180	72.7	130	200	780	7.6	0
B. suis	1330	(S1)	0	100	100	100	0.005	560	2.5	0
Thomsen		(S2)	0	53	69.6	121	0.002	560	2.1	0
	686	(S3)	0	37	75.7	117	0.002	580	1.9	0
	40/67	(S4)	0	130	91	152	9.3	560	2.3	0
B. melitensis	16M	(M1)	0	0	0	0	0	100	0	0
	Ether	(M2)	0	0	0	0	0	50	0	0
	63/9	(M3)	0	0	0	0	0	100	0	0
B. neotomae	5K 33		10.5	42	9	565	16.6	560	6.6	0
B. ovis	63/290		0	0	0	0	0	0	0	93.5
B. canis	RM6/66		0	0	0	0	0	0	0	100
B. abortus	45/20	(^{b})	0	0	0	0	0	0	100	83.6
B. suis	76/4	(^{b})	0	0	0	0	0	0	0	0
B. melitensis	B115	(^{b})	0	0	0	0	0	0	0	0

[a] The figures represent the % plaque count relative to that on the recommended propagating strain.
[b] Rough strain.

apparent if completely smooth cultures, either freshly isolated or after recent animal passage, are tested.

The typing of non-smooth isolates of *Brucella* can present difficulties. If phage R/C is used at RTD this will produce lysis of non-smooth strains of *B. abortus, B. ovis* and *B. canis* but will not differentiate between them. Inclusion of phages R and R/O at RTD in the typing scheme will enable these species to be distinguished. Phage R will produce lysis of non-smooth *B. abortus* but not the other species. Phage R/O will produce lysis of *B. ovis* but not *B. canis* or non-smooth *B. abortus* cultures. None of these phages will lyse non-smooth cultures of *B. melitensis, B. suis* or *B. neotomae* when used at RTD. They may occasionally produce partial lysis or inhibition of growth of these strains when used at RTD × 10^2 or RTD × 10^4. Phage R/O is particularly unstable genetically and each batch of this must be checked carefully for host range before use. Occasional batches will produce isolated plaques on *B. canis* or non-smooth *B. abortus* strains but the effect is usually not noticeable at RTD. The differentiation of non-smooth *Brucella* cultures with these phages is summarized in Table IX. Other phage R variants have been isolated which have produced plaques on non-smooth cultures of *B. meliten-*

TABLE IX

The differentiation of non-smooth strains of *Brucella* by phage typing[a]

Brucella species	Colonial phase	Sensitivity to lysis by phages					
		R		R/O		R/C	
		RTD	RTD × 10⁴	RTD	RTD × 10⁴	RTD	RTD × 10⁴
B. abortus	I, R or M	+	+	−	±	+	+
B. melitensis	I, R or M	−	−	−	±	−	−
B. suis	I, R or M	−	−	−	±	−	−
B. neotomae	I, R or M	−	−	−	−	−	−
B. ovis	R	−	−	+	+	+	+
B. canis	R or M	−	−	−	−	+	+

[a] +, lysis of all strains; ±, lysis of some strains; −, no strains lysed.

sis and *B. suis* but they have been too inconsistent in their properties to be of value for culture typing.

Phage R/C is the most stable of the R phage group and does not form plaques on completely smooth *Brucella* cultures. However, it is very sensitive to the presence of dissociation in *B. abortus* cultures and may be used for detection of this. On *B. abortus* cultures in the SI phase phage R/C will form very small turbid plaques. These tend to increase progressively in size and clarity as dissociation proceeds until fully rough cultures form typical large clear plaques with only a narrow turbid zone at the periphery. Plaque variation does not occur appreciably on cultures of *B. ovis* or *B. canis* strains provided that young actively growing cultures are used. On old cultures, particularly those of *B. canis*, plaque formation may be drastically reduced and even gross lysis by high concentrations of phage may be suppressed.

When interpreting the results of all phage sensitivity tests, it is absolutely essential that the state of dissociation of the host cultures is accurately known. Only then can the results of phage tests be meaningfully evaluated.

XI. Conclusions

Phage typing is a useful procedure for identification of *Brucella* cultures at the genus and species levels. Currently, with the exception of *B. suis*, it is not useful for differentiation at the subspecies level. The procedure is of epidemiological importance in that it enables species identification to be achieved more rapidly, safely and conveniently than by the only valid alternative method involving oxidative metabolism tests. It is possible that the range of *Brucella* phages will be extended in the future and will include isolates with a host range restricted to individual *Brucella* biotypes or strains. This would facilitate epidemiological studies for which the information has to be obtained at present by more complex procedures.

References

Adams, M. H. (1959). "Bacteriophages". Wiley (Interscience), New York.
Bobo, R. A. and Eagon, R. G. (1968). *Can. J. Microbiol.* **14**, 503–513.
Bowser, D. V., Wheat, R. W., Foster, J. W. and Leong, D. (1974). *Infect. Immun.* **9**, 772–774.
Brewer, R. A. and Corbel, M. J. (1980). *Br. Vet. J.* **136**, 484–487.
Buddle, M. B. (1956). *J. Hyg.* **54**, 351–364.
Burnet, E. (1928). *Arch. Inst. Pasteur Tunis* **17**, 128–146.
Calderone, J. G. and Pickett, M. J. (1965). *J. Gen. Microbiol.* **39**, 1–10.
Carmichael, L. E. and Bruner, D. W. (1968). *Cornell Vet.* **58**, 579–592.
Cerda-Olmedo, E., Hanawalt, P. C. and Guerola, N. (1968). *J. Mol. Biol.* **33**, 705–719.
Corbel, M. J. (1977a). *Ann. Sclavo.* **19**, 99–108.
Corbel, M. J. (1977b). *Ann. Sclavo.* **19**, 131–142.
Corbel, M. J. (1979). *J. Biol. Stand.* **7**, 349–360.
Corbel, M. J. and Morris, J. A. (1975). *Br. J. Exp. Pathol.* **56**, 1–7.
Corbel, M. J. and Morris, J. A. (1980). *Br. Vet. J.* **136**, 278–289.
Corbel, M. J. and Phillip, J. I. H. (1972). *Res. Vet. Sci.* **13**, 91–93.
Corbel, M. J. and Thomas, E. L. (1976a). *Dev. Biol. Stand.* **31**, 38–45.
Corbel, M. J. and Thomas, E. L. (1976b). *J. Biol. Stand.* **4**, 195–201.
Corbel, M. J. and Thomas, E. L. (1980). "The *Brucella* Phages: Their Properties, Characterisation and Applications". Ministry of Agriculture, Fisheries and Food, Pinner, Middlesex.
Corbel, M. J., Bracewell, C. D., Thomas, E. L. and Gill, K. P. W. (1979a). *In* "Identification Methods for Microbiologists" (F. A. Skinner and D. W. Lovelock, Eds), 2nd edn, pp. 71–122, The Society for Applied Bacteriology Technical Series No. 14. Academic Press, London and New York.
Corbel, M. J., Gill, K. P. W. and Redwood, D. W. (1979b). "Diagnostic Procedures for Non-smooth *Brucella* Strains". Ministry of Agriculture, Fisheries and Food, Pinner, Middlesex.
Corbel, M. J., Scott, A. C. and Ross, H. M. (1980). *J. Hyg.* **85**, 103–113.
Diaz, R. and Bosseray, N. (1973). *Ann. Rech. Vet.* **4**, 283–292.
Diaz, R. and Levieux, D. (1972). *C. R. Acad. Sci. Ser. D.* **274**, 1593–1596.
Diaz, R., Jones, L. M., Leong, D. and Wilson, J. B. (1968). *J. Bacteriol.* **96**, 893–901.
Douglas, J. T. (1978). *Diss. Abstr. Int. B 39*.
Douglas, J. T. and Elberg, S. S. (1976). *Infect. Immun.* **14**, 306–308.
Douglas, J. T. and Elberg, S. S. (1978). *Ann. Sclavo.* **20**, 681–691.
Droževkina, M. S. (1963). *Bull. W. H. O.* **29**, 43–57.
Dubray, G. (1972). *Ann. Inst. Pasteur* **123**, 171–193.
Dubray, G. (1976). *Ann. Microbiol. (Paris)* **127B**, 133–149.
Dubray, G. and Plommet, M. (1976). *Dev. Biol. Stand.* **631**, 68–91.
Eagle, H. (1959). *Science* **130**, 432–437.

Evans, A. C. (1918). *J. Infect. Dis.* **22**, 580–593.
Feesey, C. and Parnas, J. (1975). *Zentralbl. Vet. Med. Reihe B* **22**, 866–867.
Fredericq, P. (1957). *Annu. Rev. Microbiol.* **11**, 7–22.
Freeman, B. A., McGhee, J. R. and Baughn, R. E. (1970). *J. Infect. Dis.* **121**, 522–527.
Hall, W. H. (1971). *J. Infect. Dis.* **124**, 615–618.
Harrington, R. Jr, Bond, D. R. and Brown, G. M. (1977). *J. Clin. Microbiol.* **5**, 663–664.
Hatten, B. A. (1973). *Proc. Soc. Exp. Biol. Med. N.Y.* **142**, 909–914.
Hatten, B. A. and Brodeur, R. D. (1978). *Infect. Immun.* **22**, 956–962.
Henry, B. S. (1933). *J. Infect. Dis.* **52**, 376–402.
Hines, W. D., Freeman, B. A. and Pearson, G. R. (1964). *J. Bacteriol.* **87**, 1492–1498.
Hoyer, B. H. and McCullough, N. B. (1968a). *J. Bacteriol.* **95**, 444–448.
Hoyer, B. H. and McCullough, N. B. (1968b). *J. Bacteriol.* **96**, 1783–1790.
Huddleson, I. F. (1929). *Mich. State Coll. Agric. Exp. St. Technol. Bull. No. 100.*
Huddleson, I. F. (1931). *Am. J. Public Health* **21**, No. 5.
Jones, L. M. (1960). *Bull. W. H. O.* **23**, 130–133.
Jones, L. M. (1967). *Int. J. Syst. Bacteriol.* **17**, 371–375.
Jones, L. M., McDuff, C. R. and Wilson, J. B. (1967). *J. Bacteriol.* **83**, 860–866.
Jones, L. M., Merz, G. S. and Wilson, J. B. (1968a). *Appl. Microbiol.* **16**, 1179–1190.
Jones, L. M., Merz, G. S. and Wilson, J. B. (1968b). *Experientia* **24**, 20–22.
Kellerman, G. D., Foster, J. W. and Badakhsh, F. F. (1970). *Infect. Immun.* **2**, 237–243.
Kreutzer, D. L. and Robertson, D. C. (1979). *Infect. Immun.* **23**, 819–828.
Kreutzer, D. L., Scheffel, J. W., Draper, L. R. and Robertson, D. C. (1977). *Infect. Immun.* **15**, 842–845.
Kreutzer, D. L., Buller, C. S. and Robertson, D. C. (1979). *Infect. Immun.* **23**, 811–818.
Krizsanovich, K. (1973). *J. Gen. Virol.* **19**, 311–320.
McDuff, C. R., Jones, O. M. and Wilson, J. B. (1962). *J. Bacteriol.* **83**, 324–329.
McGhee, J. R. and Freeman, B. A. (1970). *Infect. Immun.* **2**, 244–249.
Merz, G. S. and Wilson, J. B. (1966). *J. Bacteriol.* **91**, 2356–2361.
Meyer, K. F. and Shaw, E. B. (1920). *J. Infect. Dis.* **27**, 173–184.
Meyer, M. E. (1961). *J. Bacteriol.* **82**, 950–953.
Meyer, M. E. (1962). *Bull. W. H. O.* **26**, 829–830.
Meyer, M. E. (1976). *Am. J. Vet. Res.* **37**, 207–210.
Meyer, M. E. and Morgan, W. J. B. (1962). *Bull. W. H. O.* **26**, 823–828.
Moreira-Jacob, M. (1963). *Nature (London)* **192**, 460.
Moreira-Jacob, M. (1968). *Nature (London)* **219**, 752–753.
Moreno, E., Pitt, M. W., Jones, L. M., Schurig, G. G. and Berman, D. T. (1979). *J. Bacteriol.* **138**, 361–369.
Morgan, W. J. B. (1963). *J. Gen. Microbiol.* **30**, 437–443.
Morris, J. A. and Corbel, M. J. (1973). *J. Gen. Virol.* **21**, 539–544.
Morris, J. A., Corbel, M. J. and Phillip, J. I. H. (1973). *J. Gen. Virol.* **20**, 63–73.
Nagy, L. K. (1967). *Immunology* **12**, 463–474.
Nelson, E. L. and Pickett, M. J. (1951). *J. Infect. Dis.* **89**, 226–232.
Ostrovskaya, N. N. (1957). *Izmenchivost Mikroorganizmov* **2**, 88.
Ostrovskaya, N. N. (1961). *Zh. Mikrobiol. Epidemiol. Immunobiol.* **6**, 70–78.
Parnas, J. (1961). *Zentralbl. Veterinaermed. Reihe B* **8**, 175–191.
Parnas, J. (1963). "*Brucella* Phages, Properties and Application". Bibliotheca Microbiologia Fasc. 3. Karger, Basel.

Parnas, J. (1966). *Rev. Immunol.* **30**, 55–68.

Parnas, J. and Dominowska, C. (1966). *Arch. Hyg.* **150**, 384–386.

Parnas, J., Pleszczynska, E., Mardarowicz, C. and Poplawski, S. (1967). *Z. Immunforsch.* **133**, 302–312.

Peschkov, J. and Feodorov, V. (1978). *Zentralbl. Bakteriol. Parasiterkd. Infektionskr. Hyg. Abt. Orig. Reihe A* **240**, 94–105.

Popkhadze, M. Z. and Abashidze, T. G. (1957). *Bakteriofagiya* **5**, 321–325. Gruzmedgiz, Tbilisi.

Rasooly, G., Olitzki, A. L. and Sulitzeanu, D. (1965). *Nature (London)* **207**, 1308–1309.

Renoux, G. (1958). *Ann. Inst. Pasteur* **94**, 179–206.

Roux, J. and Sassine, J. (1971). *Ann. Inst. Pasteur* **120**, 174–185.

Simon, F. (1979). "Contribution à l'étude des bacteries du genre *Brucella*: phage Tb, BM29 et Bk (aspects morphologiques serologiques et physiologiques)". Thesis, Paris.

Stableforth, A. W. and Jones, L. M. (1963). *Int. Bull. Bacteriol. Nomencl. Taxon.* **13**, 145–158.

Stoenner, H. G. and Lackman, D. B. (1957). *Am. J. Vet. Res.* **18**, 947–951.

Thomas, E. L. and Corbel, M. J. (1977). *Arch. Virol.* **54**, 259–261.

Traum, J. (1914). "Annual Report of the Chief Bureau of Animal Industry", p. 30. US Department of Agriculture.

Tuszkiewicz, M., Parnas, J., Pleszczynska, E., Fijalka, M., Winiarska, U. and Parnas, W. (1966). *Arch. Hyg.* **150**, 377–383.

Weber, A., Schiefer, H-G. and Krauss, H. (1977). *Zentralbl. Bakteriol. Parasitenkd. Infektionskr. Hyg. Abt. Orig. Reihe A* **239**, 365–374.

Weber, A., Schiefer, H-G. and Krauss, H. (1978). *Zentralbl. Veterinaermed. Reihe B* **25**, 324–328.

White, P. G. and Wilson, J. B. (1951). *J. Bacteriol.* **61**, 239–240.

Wilson, G. S. and Miles, A. A. (1932). *Br. J. Exp. Pathol.* **13**, 1–13.

Wilson, G. S. and Miles, A. A. (1975). *In* "Topley and Wilson's Principles of Bacteriology, Virology and Immunology", 6th edn, pp. 1052–1078. Edward Arnold, London.

Wober, W., Thiele, O. W. and Urbaschek, B. (1964). *Biochem. Biophys. Acta* **84**, 376–390.

3
Serological Characterization of Yeasts as an Aid in Identification and Classification

T. TSUCHIYA,[1] M. TAGUCHI,[1] Y. FUKAZAWA[2]
and T. SHINODA[3]

[1] *Department of Microbiology, Kanagawa Prefectural College of Medicinal Technology, Yokohama, Japan.* [2] *Department of Microbiology, Medical College of Yamanashi, Tamaho-mura, Yamanashi, Japan* and [3] *Department of Microbiology, Meiji College of Pharmacy, Tanashi, Tokyo, Japan*

I.	Introduction	76
II.	Classification and identification of yeasts	78
III.	Serological grouping of yeasts	78
	A. Serological approaches	78
	B. Antigenic structures	79
IV.	Methods of antigenic analyses	81
	A. Determination of serogroups of yeasts	81
	B. Identification of yeasts by serological characteristics	88
V.	Application of the serological grouping of yeasts	94
	A. Medically important species	94
	B. Frequency of yeast species identified by the serological method	97
	C. Detection of contaminants in brewing yeasts	100
	D. Serological diagnosis of systemic diseases due to yeasts	103
	E. Interrelationships among various grouping methods	104
	F. Significance of antigenic structures in yeast taxonomy	107
VI.	Immunochemical basis of the serological specificity of yeasts	112
VII.	Conclusions	113
	Appendix I	115
	References	118

Copyright © 1984 by Academic Press, London
All rights of reproduction in any form reserved

I. Introduction

Several species of the genus *Candida* are normal inhabitants of the oral cavity, the respiratory and intestinal tracts, and the vaginal cavity of man and animals. However, candidiasis sometimes breaks out because of the decreased resistance to infections in patients with diseases such as diabetes mellitus, leukaemia and malignant cancer. In addition, candidiasis has been found particularly in intensive and/or long-term treatments with antibiotics, corticosteroids and antimitotic agents.

The group of micro-organisms known as yeasts include microscopic fungi with predominantly unicellular forms and vegetative reproduction by budding, transverse division or both. Yeasts can generally produce ascospores in an ascus, originating either from a zygote or parthenogenetically from a single somatic cell, and they are further classified morphologically and physiologically. Therefore, they are regarded as a part of fungi among the lower plants being different from bacteria.

Several methods designed for the classification of yeasts have been published and among these the method universally used is the most elaborate one devised by Lodder and Kreger-van Rij (1952), who divided the yeasts into orders, families and genera depending upon their ability to form ascospores, and their morphology. The genera were then subdivided into species mainly on the basis of their biochemical characteristics. For a detailed taxonomic description of the various species of yeasts, the reader is referred to the second edition of Lodder (1970). In principle, the generic names used in this paper follow this edition, although various proposals have been presented by many investigators (Arx *et al.*, 1977). However, it is not easy to identify the yeasts precisely only on the basis of their morphological and biochemical characteristics (Barnett, 1960, 1961). Consequently, serological studies are important. Benham (1931, 1935), Yukawa and Ohta (1929) and Seeliger (1958, 1959, 1960, 1962a,b, 1968, 1974) reported serological properties and cross-reactions among various species of yeasts and yeast-like organisms. The differentiation was shown by agglutination, precipitation and complement fixation as well as by serum absorption. However, the serological characteristics of yeasts presented by several investigators have not yet been generally adopted for identification or classification.

The prerequisite for such tests is the establishment of antigenic structures or relationships among many species. Martin (1942), Pospíšil (1959), Biguet *et al.* (1961, 1962, 1965a,b), Campbell and Allan (1964), Kemp and Solotorovsky (1964) and Müller and co-workers (Müller *et al.*, 1964; Müller and Bürger, 1965; Müller, 1966; Müller and Hirsch, 1967) presented antigenic structures of some *Candida* and *Saccharomyces* species. Tsuchiya *et al.* (1955b, 1956a,b, 1957a) also established the antigenic structures of various species of pathogenic, manufacturing or often contaminated yeasts. They used absorbed antisera prepared for the rapid and reliable identification of various yeasts. Sweet and Kaufman (1970) prepared their own factor sera for medically important *Candida* species and indicated the possibility that many isolated strains could be identified.

It is the purpose here to review the methodology and application of a serogrouping system which can be applied to achieve rapid identification and reliable classification of yeasts.

TABLE I

Classification of yeasts and yeast-like fungi

Class *Ascomycetes (Protoascomycetes)*
  Order *Endomycetales*[a]
    Family 3. *Saccharomycetaceae*
      Subfamily 1. *Schizosaccharomycoideae*
      Genus *Schizosaccharomyces*
      Subfamily 2. *Nadsonioideae*
      Genera *Hanseniaspora, Nadsonia, Saccharomycodes, Wickerhamia*
      Subfamily 3. *Saccharomycoideae*
      Genera *Citeromyces, Debaryomyces, Dekkera, Endomycopsis, Hansenula, Kluyveromyces, Lodderomyces, Pachysolen, Pichia, Saccharomyces, Saccharomycopsis, Schwanniomyces, Wingea*
      Subfamily 4. *Lipomycetoideae*
      Genus *Lipomyces*
    Family 4. *Spermophthoraceae*
      Genera *Coccidiascus, Metschnikowia, Nematospora*

Class *Basidiomycetes*
  Order *Ustilaginales*
      Genera *Leucosporidium, Rhodosporidium*

Class *Fungi imperfecti (Deuteromycetes)*
  Order *Moniliales*
    Family 1. *Sporobolomycetaceae*
      Genera *Bullera, Sporidiobolus, Sporobolomyces*
    Family 2. *Cryptococcaceae (Torulopsidaceae)*
      Genera *Brettanomyces, Candida, Cryptococcus, Kloeckera, Oosporidium, Pityrosporum, Rhodotorula, Schizoblastosporion, Sterigmatomyces, Torulopsis, Trichosporon, Trigonopsis*

[a]Family 1. *Dipodascaceae* and family 2. *Endomycetaceae* of *Endomycetales* do not belong to yeasts.

II. Classification and identification of yeasts

According to the second edition of "The Yeasts" (Lodder, 1970), yeasts are divided into three main groups, i.e. *Ascomycetes, Heterobasidiomycetes* and *Deuteromycetes* or *Fungi imperfecti.* These were subdivided into 39 genera and 349 species. This classification is the most elaborate and widely accepted one, although a few other taxonomic arrangements have been published (Alex-opoulos, 1962; Ainsworth, 1963; Arx *et al.*, 1977). A simplified classification showing the major taxonomic groups based upon Lodder (1970), Kreger-van Rij (1969; 1970), is shown in Table I.

Identification of micro-organisms in this system follows morphological and biological properties such as cultural characteristics, microscopic findings, fermentation, and assimilation of carbohydrates, nitrate assimilation and urease production (Lodder, 1970; van der Walt, 1970; Phaff, 1970; Phaff and Ahearn, 1970; van Uden, 1970; Barnett and Pankhurst, 1974; Wicherham, 1951, 1970). However, it is not uncommon to find yeast isolates without typical spores or physiological and biological properties (Barnett, 1960; Seo, 1972). The procedures used for the identification of yeasts have, in many ways, been quite different from those employed in bacteriology. In addition, the methods for yeast identification are more time-consuming than those used in bacteriology because yeast-like fungi grow more slowly. Diagnostic sera for the yeast identification have not been investigated completely as in *Salmonella* and *Shigella* (Kauffmann, 1972; Edwards and Ewing, 1962). Rapid identifi-cation and differentiation of the yeasts is still required to identify pathogens and contaminants in industry, such as in wine making and brewing. The identification may be elaborate if traditional methods are followed (Campbell, 1968a; Nishikawa *et al.*, 1979a,b). Therefore, we have attempted to establish an antigenic schema for important yeast species using slide agglutination and absorbed antisera. Filamentous fungi with morphologically complicated structures are exempt from this system.

III. Serological grouping of yeasts

A. Serological approaches

Serological studies for differentiation and identification in mycology were started shortly after the discovery of antigen–antibody reactions in microbio-logy, especially on yeasts used in brewing, and pathogenic fungi such as those involved in thrush, sporotrichosis, blastomycosis and dermatophytosis (Preece, 1971; Oakley, 1971). In 1931 Benham differentiated four *Candida* species from *Hansenula anomala* and *Saccharomyces cerevisiae* by means of agglutination. Yukawa and Ohta (1928, 1929) and Yukawa *et al.* (1929) studied more than 20 species of *Saccharomyces, Willia (Hansenula), Debaryomyces* and *Saccharomycodes* using agglutination and complement fixation methods. They concluded that serological characteristics could be important in the classification of yeasts. Similar investigations were performed by Almon and Stovall (1934), Hines (1924), Stone and Garrod (1931), Kesten and Mott (1932), Lamb and Lamb (1935), Jonsen *et al.* (1953) and Winner (1955). Gordon (1958a,b, 1962) reported the differentiation of *Candida* species by fluorescent antibody staining.

On the other hand, in overcoming the technical problems involved in the differentiation of some of the deep-seated fungous infections, complement fixation and precipitation tests have proved to be useful for the diagnosis of fungous diseases such as North American blastomycosis, coccidioidomycosis and histoplasmosis (Kaufman and Kaplan, 1963; Kaufman, 1973, 1980a,b). However, serology seems to have gained little acceptance in mycology compared with bacteriology, probably due to the difficulty of preparing antisera. These investigations, however, have introduced more systematic approaches to fungous serology resulting in antigenic schemes of micro-organisms for classification.

B. Antigenic structures

Martin (1942) described the antigenic structures of four *Candida* species by the use of three antigenic factors (X, Y and Z) and their relative amounts as shown by agglutination and complement fixation reactions (Table II). Tsuchiya *et al.* (1955b) proposed antigenic structures of seven *Candida* species dividing them into thermostable and thermolabile antigens. Using slide agglutination with absorbed antisera, antigenic analyses of yeasts were also made with other *Candida* (Tsuchiya *et al.*, 1954, 1955a, 1956a,b, 1957a, 1958a,d, 1959, 1961a, 1967b; Kemp and Solotorovsky, 1964; Sweet and Kaufman, 1970) and several other species or varieties of the genera *Saccharomyces* (Tsuchiya *et al.*, 1957b,c, 1958c,e, 1961d, 1965b; Tsuchiya, 1959; Campbell, 1968a,b, 1970, 1971; Campbell *et al.*, 1964, 1966, 1968), *Hansenula* (Tsuchiya *et al.*, 1957e, 1958b, 1964a,b, 1967b), *Debaryomyces* (Tsuchiya *et al.*, 1960), *Hanseniaspora* and *Kloeckera* (Tsuchiya *et al.*, 1966; Tsuchiya and Imai, 1968; Tsuchiya and Fukazawa, 1969), *Pichia* (Ohtsuka and Tsuchiya, 1972; Kuroshima *et al.*, 1976), *Rhodotorula* (Tsuchiya *et al.*, 1957d), *Sporobolomyces* (Tsuchiya *et al.*, 1969), *Torulopsis* (Tsuchiya *et al.*, 1961b,c) and *Cryptococcus* (Tsuchiya et al., 1963a, 1965a; Fukazawa *et al.*, 1974; Shinoda *et al.*, 1980; Ikeda *et al.*, 1982).

Pospíšil (1959) presented the antigenic structures of five *Candida* species by agglutination and complement fixation and Biguet *et al.* (1961, 1962,

TABLE II

Antigenic structures of the medically important yeasts

Species	Martin (1942)	Tsuchiya et al. (1955–1969)	Pospíšil (1959)	Biguet et al. (1962)	Siluyanova (1964)	Müller et al. (1967)	Campbell (1968)	Sweet and Kaufman (1970)
C. albicans	X,4Y,3Z		a,1–4	1–7,8–12,**13,14**,15		I,II,III		1–3,**4**,5,10,11
C. albicans type A		1–5,6,7						1,2,5,10–**13**
C. albicans type B		1–5,7,**13b**						1–**3**,**4**,5,11
C. tropicalis	X,4Y	1–5,6	b,1–4	1–7,17,**18–22**		I,IV,V,IX		1,2,5,10–12
C. stellatoidea	X,2Y,2Z	1–5,**10**,32		1–7,8–12,15,16		I,II		
C. pseudotropicalis		1,**8,10**,28,31	c,1,5	1,30,**39**–**46**,47,48		IX,X		1,**7**
C. macedoniensis		1,**8,10**,28,31		1,30,**39**–**46**		IX,X		
S. fragilis		1,**8,10**,28,31					1,**3**	
S. cerevisiae		1–3,**10,14,18**,31					1,2,**3**	
S. ellipsoideus		1–3,**10,14,18**,31					1,**4**	
C. krusei		1,2,5,**11**	d,1,2,4	2,3,30,**31**–**38**	1,2,5,**11**,bB	VI,VII,VIII		1,**6**,11
C. mycoderma		1,2,5,**11**,12			1,2,5,**11**,12,22			
C. zeylanoides		1–**4,13,17**		1–5,17,**23**–**29**				
C. parapsilosis	X,4Z	1–3,5,13b,**13**–15				III		1,2,**8**,10–13
C. pulcherrima		1–3,5,**13**–15			1,2,5,**13**,22,C			
C. guilliermondii		1–4,9	e,1–3,5			IV		1–3,5,**9**,10,11
T. glabrata		1,4,**6**,**10**,34				I,III,IX		

1965a,b) showed the structures of seven *Candida* species using immunoelectrophoresis. They later revealed the antigenic structures of several species of aspergilli using the same methods (Biguet *et al.*, 1964). Campbell and Allan (1964), Campbell and Brudynski (1966), Campbell (1968a,b), Siluyanova (1964), Müller *et al.* (1964), Müller and Bürger (1965), Müller (1966), Müller and Hirsch (1967) and Sweet and Kaufman (1970) also proposed the structures of *Candida* species (Table II).

IV. Methods of antigenic analyses

A. Determination of serogroups of yeasts

1. *Preparation of antigens*

Yeast cells taken from colonies showing the typical characteristics of a certain species are inoculated on Sabouraud's agar plates containing 2% glucose, 1% peptone and 0.5% yeast extract (Difco), and cultured at 27°C for 48 h. The cells grown on the medium are harvested and washed with physiological saline solution, after heating at 100°C for 2 h. After washing three times by centrifuging with 0.5% formalinized saline solution (0.5% formalin in 0.85% NaCl), cell suspension is made in the same solution and the concentration is adjusted to McFarland scale No. 9 (Vera and Power, 1980). The cell suspension is ground in a glass grinder if necessary.

2. *Preparation of antisera*

Adult rabbits weighing about 2 kg and lacking antibodies for *C. albicans* are injected intravenously with 0.5, 1, 2, 4, 4 and 4 ml of cell suspension at four-day intervals. Blood is sampled from the ear vein and the sera are tested for agglutinin titre against the homologous organisms. When the titre of a given antiserum is 1:640 or higher, the rabbits are bled to death seven days after the last injection. The highest titre of agglutinin is 1:2560 in the majority of the *Candida* species, but less than 1:320 in some other species such as in *Rhodotorula*. In cases where the titre does not reach the desired level, it is better to immunize another animal, because continued injections generally will not improve the response.

3. *Absorption tests*

The antiserum to be absorbed is diluted to obtain an agglutinin titre of approximately 1:256, although the titre formerly used was 1:128. Antisera are generally diluted ten times with physiological saline solution, because it was necessary to perform the experiments sometimes using both high and low titred antisera concomitantly for antigenic analyses of several yeasts. A 2-ml sample of diluted serum and 1 ml of packed wet cells are mixed and incubated at 37 C for 2 h with agitation, and left standing overnight at 4°C. The suspension is then centrifuged, and the supernatant is tested against the cells used for absorption by the slide agglutination. Absorptions have to be performed repeatedly to remove antibodies reacting to the antigen. The absorbed antisera are stored in a refrigerator at 4 C. The absorbed antiserum may retain its agglutinating ability for over a year when stored under refrigeration. Sodium azide is added as a preservative to the absorbed antiserum at a final concentration of 0.1%. Merthiolate (0.01%) can also be used as a preservative.

4. *Antigenic analyses of* Candida

Antigenic analyses of the genus *Candida* were carried out with seven species: *C. albicans*, *C. tropicalis*, *C. pseudotropicalis*, *C. krusei*, *C. parapsilosis*, *C. guilliermondii* and *C. stellatoidea* (Tsuchiya *et al.*, 1955b). For antigenic analyses of the species, experiments were performed first on the thermostable antigens and then on the thermolabile ones. The thermostable antigens are indicated by Arabic numerals and thermolabile antigens by the letters of the alphabet. The antiserum was absorbed with a strain belonging to a species different from that used for immunization. There are 42 possible combinations, one species for immunization and the other for absorption (Table III). All of the reciprocally absorbed antisera were examined by slide agglutination against heated cells of each of the seven species.

The first experiment was on the combination of *C. albicans* and *C. tropicalis*. After absorption with heated cells of *C. tropicalis*, anti-heated *C. albicans* serum gave a positive slide agglutination against heated *C. albicans*, but it did not agglutinate the other six strains of different species. However, when anti-heated *C. tropicalis* serum was absorbed with heated *C. albicans*, no positive agglutination was shown against any of the seven species. From the results obtained by the reciprocal absorptions, it can be concluded that *C. albicans* possesses an independent thermostable antigen which is not present in any of the other six *Candida* species, and that *C. tropicalis* has no thermostable antigens other than those present in *C. albicans*. The unique antigen of *C. albicans* was designated antigen 7.

In experiment 2-A and 2-B (Table III), the anti-heated *C. albicans* serum previously absorbed with heated cells of *C. pseudotropicalis* did not agglutinate heated cells of *C. pseudotropicalis*, but the absorbed antiserum still showed positive reactions against the other six species. On the other hand, the anti-heated *C. pseudotropicalis* serum which had been absorbed with

TABLE III

Antigenic analyses of seven species of *Candida*

Experiment No.	Antiserum for heated cells	Heated cells for absorption	Thermo-stable antigen assumed	Heated cells used as antigens for agglutination							Thermostable antigens for remained antibodies after absorption
				C. albicans	*C. tropicalis*	*C. pseudotropicalis*	*C. krusei*	*C. parapsilosis*	*C. guilliermondii*	*C. stellatoidea*	
1-A	*C. albicans*	*C. tropicalis*	7	+	−	−	−	−	−	−	7
1-B	*C. tropicalis*	*C. albicans*		−	−	−	−	−	−	−	
2-A	*C. albicans*	*C. pseudotr.*[a]	2	+	+	−	+	+	+	+	2,3,4,5,6,7,
2-B	*C. pseudotr.*	*C. albicans*	8	−	−	+	−	−	−	−	8
3-A	*C. albicans*	*C. krusei*	3	+	+	−	−	+	+	−	3,4,6,7
3-B	*C. krusei*	*C. albicans*		+	−	−	−	−	−	−	
4-A	*C. albicans*	*C. paraps.*	4	+	+	−	−	−	+	+	4,6,7
4-B	*C. paraps.*	*C. albicans*		−	−	−	−	−	−	−	
5-A	*C. albicans*	*C. guillierm.*	5	+	+	−	+	+	−	+	5,6,7
5-B	*C. guillierm.*	*C. albicans*	9	−	−	−	−	−	+	−	9
6-A	*C. albicans*	*C. stellat.*	6	+	+	−	−	−	−	−	6,7
6-B	*C. stellat.*	*C. albicans*		−	−	−	−	−	−	−	
7-A	*C. tropicalis*	*C. pseudotr.*		+	+	+	+	+	+	+	2,3,4,5,6
7-B	*C. pseudotr.*	*C. tropicalis*		+	−	+	−	−	−	−	8
8-A	*C. tropicalis*	*C. krusei*		+	+	−	−	+	+	+	3,4,6
8-B	*C. krusei*	*C. tropicalis*		−	−	−	−	−	−	−	
9-A	*C. tropicalis*	*C. paraps.*		+	+	−	−	−	+	+	4,6
9-B	*C. paraps.*	*C. tropicalis*		−	−	−	−	−	−	−	
20-A	*C. paraps.*	*C. stellat.*		−	−	−	−	−	−	−	
20-B	*C. stellat.*	*C. paraps.*		+	+	−	−	−	+	+	4
21-A	*C. guillierm.*	*C. stellat.*		−	−	−	−	−	+	−	9
21-B	*C. stellat.*	*C. guillierm.*		+	+	+	+	+	−	+	5

[a] *C. pseudotr.*: *C. pseudotropicalis*; *C. paraps.*: *C. parapsilosis*; *C. guilliderm.*: *C. guilliermondii*; *C. stellat.*: *C. stellatoidea*.

heated cells of *C. albicans* agglutinated only the heated cells of *C. pseudotro-picalis*. This indicates that *C. albicans* has a thermostable antigen which is lacking in *C. pseudotropicalis*, and that *C. pseudotropicalis* possesses another thermostable antigen which is absent in *C. albicans*. The former was designated as antigen 2 and the latter as antigen 8. Antigen 2 is distributed among the other five species, whereas antigen 8 is unique for *C. pseudotropi-calis*. Similar experiments (3-A to 21-B) were performed to establish the antigenic structures of thermostable antigens of the seven *Candida* species.

Finally, unabsorbed antisera from rabbits immunized with heated cells of each species were observed to show agglutination reactions against heated cells of every species. The common thermostable antigen present in every species was designated antigen 1. Antigenic structures of each species are shown in Table VI.

Concerning the thermolabile antigens, antiserum against non-heated, formalinized antigen was tested by slide agglutination against fresh orga-nisms of each species after the absorption with heated cells of homologous species. Only three species, *C. pseudotropicalis, C. krusei* and *C. parapsilosis*, were found to possess thermolabile antigens. These were designated a, b and c, respectively. However, antisera for antigens 2, 3, 7, a, b and c are not used at present for practical identification because of their slightly lower stability or decreased specificity.

5. *Further antigenic analyses of* Saccharomyces

(a) *Examination of known antigens.*   Antigenic analyses of various species were carried out subsequently using anascosporogenous and ascosporoge-nous yeast species cross-reacting with non-absorbed antiserum to *C. albi-cans*. Each species was tested first by the slide agglutination using monospeci-fic or absorbed antisera (factor sera) prepared on the basis of antigenic structures for simplified analyses. When one of 15 species of micro-organisms is used for immunization and another for absorption, the total number of possible combinations for absorption experiments comes to 15×14 sets for the establishment of antigenic structures.

In the slide agglutinations using 24 factor sera for antigens 2 to 25, the heated cells of *S. oviformis* and *S. steineri* were agglutinated by factor sera for the thermostable antigens 2, 3, 10, 14 and 18 (Table IV) (Tsuchiya *et al.*, 1961d). Similarly, *S. rouxii, S. carlsbergensis* and *S. mellis* had the thermos-table antigens 8 and 10, and *S. chevalieri* and *S. italicus* the thermostable antigens 2, 3, 8, 10, 14 and 18.

(b) *Confirmation of antigenic structures and detection of new antigens.*   In experiments 1-A and 1-B, anti-heated *S. oviformis* serum absorbed with

TABLE IV

Slide agglutination with factor sera of seven *Saccharomyces* species

No. of factor serum	Heated cells used as antigens for agglutination						
	S. ov.[a]	*S. st.*	*S. rou.*	*S. car.*	*S. mel.*	*S. che.*	*S. it.*
2	+	+	−	−	−	+	+
3	+	+	−	−	−	+	+
4	−	−	−	−	−	−	−
5	−	−	−	−	−	−	−
6	−	−	−	−	−	−	−
7	−	−	−	−	−	−	−
8	−	−	+	+	+	+	+
9	−	−	−	−	−	−	−
10	+	+	+	+	+	+	+
11	−	−	−	−	−	−	−
12	−	−	−	−	−	−	−
13	−	−	−	−	−	−	−
14	+	+	−	−	−	+	+
15	−	−	−	−	−	−	−
16	−	−	−	−	−	−	−
17	−	−	−	−	−	−	−
18	+	+	−	−	−	+	+
19	−	−	−	−	−	−	−
20	−	−	−	−	−	−	−
21	−	−	−	−	−	−	−
22	−	−	−	−	−	−	−
23	−	−	−	−	−	−	−
24	−	−	−	−	−	−	−
25	−	−	−	−	−	−	−

[a] *S. ov.*: *S. oviformis*; *S. st.*: *S. steineri*; *S. rou.*: *S. rouxii*; *S. car.*: *S. carlsbergensis*; *S. mel.*: *S. mellis*; *S. che.*: *S. chevalieri*; *S. it.*: *S. italicus*.

heated cells of *S. steineri* as well as anti-heated *S. steineri* serum absorbed with heated cells of *S. oviformis* gave negative agglutination reactions against every antigen of all the species (Table V). The observations in the reciprocal absorption tests among *S. oviformis*, *S. steineri* and *S. cerevisiae* were similar, and the antiserum for one species was exhausted completely after absorption with heated cells of the other species. Therefore, all of these species are identical to each other so far as thermostable antigens are concerned. Such serological identities were also found among other species.

Subsequently, attempts to detect new antigens were made by further successive absorption with antigenically related species. In experiment 16, anti-heated *S. steineri* serum absorbed with heated cells of *C. albicans*

TABLE V
Antigenic analyses of seven species of *Saccharomyces*

Experiment No.	Antiserum for heated cells or antiserum absorbed in Experiment No.	Heated cells for absorption	Heated cells used as antigens for agglutination											Thermostable antigens for remained antibodies after absorption
			S. ov.[a]	S. st.	S. rou.	S. car.	S. mel.	S. che.	S. it	S. cer.	C. alb.	C. pt.	C. pp.	
1-A	S. ov.	S. st.	−	−	−	−	−	−	−	−	−	−	−	
1-B	S. st.	S. ov.	−	−	−	−	−	−	−	−	−	−	−	
3-A	S. st.	S. cer.	−	−	−	−	−	−	−	−	−	−	−	
3-B	S. cer.	S. st.	−	−	−	−	−	−	−	−	−	−	−	
5-A	S. rou.	C. pt.	+	+	+	+	+	−	−	−	−	−	−	32
5-B	C. pt.	S. rou.	+	−	−	−	+	+	+	+	−	+	−	31
6-A	S. mel.	S. rou.	−	−	−	−	−	−	−	−	−	−	−	(31)
6-B	S. rou.	S. mel.	−	−	−	−	−	−	−	−	−	−	−	
13	S. ov.	C. alb.	+	+	−	−	−	+	+	+	−	−	+	(10)14,18(31)
14	As (13)	C. pp.	+	+	−	−	−	+	+	+	−	−	−	18
15	As (14)	S. cer.	−	−	−	−	−	−	−	−	−	−	−	
16	S. st.	C. alb.	+	+	−	−	−	+	+	+	−	+	+	10,14,18,31
17	As (16)	C. pt.	+	+	−	−	−	+	+	+	−	−	+	14,18
18	As (17)	C. pp.	+	+	−	−	−	+	+	+	−	−	−	18
19	As (18)	S. cer.	−	−	−	−	−	−	−	−	−	−	−	
25	S. che.	C. alb.	+	+	−	−	−	+	+	+	−	−	−	(8)(10)(14)18(31)
26	As (25)	S. cer.	−	−	−	−	−	−	−	−	−	−	−	

[a] S. ov.: S. oviformis; S. st.: S. steineri; S. rou.: S. rouxii; S. car.: S. carlsbergensis; S. mel.: S. mellis; S. che.: S. chevalieri; S. it.: S. italicus; S. cer.: S. cerevisiae; C. alb.: C. albicans type A; C. pt.: C. pseudotropicalis; C. pp.: C. parapsilosis.

showed positive agglutination reactions against five *Saccharomyces* species and two *Candida* species. The reactions are interpreted on the basis of residual antibodies homologous to antigens 10, 14, 18 and 31. It must be added here that several other strains were used as antigens in slide agglutinations to ascertain the existence of some antigens in addition to the species shown in Table V. The absorbed antiserum was reabsorbed with heated cells of *C. pseudotropicalis* and then with those of *C. parapsilosis*, and still the antiserum gave positive reactions against five *Saccharomyces* species. In experiment 18, the antiserum was further absorbed with heated cells of *S. cerevisiae* possessing antigen 18, and finally, the positive reactions turned negative. Therefore, it can be concluded that *S. steineri* has an antigenic structure consisting of antigens 2, 3, 10, 14, 18 and 31. None of the new antigens was found in this case.

All species possessing antigen 10 were found to have either antigen 31 or 32 as shown in experiments 5-A and 5-B, suggesting that antigens 31 and 32 are subunits of antigen 10. In addition, in experiment 6-A, anti-heated *S. mellis* serum absorbed with heated cells of *S. rouxii* showed negative agglutination reactions against every antigen, notwithstanding the fact that several *Saccharomyces* and *Candida* species possess antigen 31. In this case, it can be assumed that the immunogenicity of antigen 31 of *S. mellis* is very weak, although its agglutinability is "normal". Such an antigen is shown in parenthesis in Table VI. The thermolabile antigen of *S. steineri* possesses antigen e.

It would be convenient if the antigenic structures of yeasts were simple. Kauffmann (1954) has already reported that the determination of antigenic structures of *Salmonella* is necessary only for diagnostic purposes. However, the division or establishment of new antigens occasionally occurs, when yeast strains used for antigenic analyses are changed or increased in number.

TABLE VI

Antigenic structures of *Candida* and *Saccharomyces* species

Species	Antigens	Species	Antigens
C. albicans type A	1,4,5,6,(7)	*S. cerevisiae*	1,10,18,31
C. albicans type B	1,4,5,(7)13b	*S. oviformis*	1,10,18,31
C. tropicalis	1,4,5,6	*S. steineri*	1,10,18,31
C. stellatoidea	1,4,5	*S. chevalieri*	1,(8)(10)18,31
C. guilliermondii	1,4,9	*S. italicus*	1,8,10,(18)28,31
C. krusei	1,5,11	*S. mellis*	1,(8)(10)(28)(31)(32)
C. parapsilosis	1,5,13,13b	*S. rouxii*	1,(8)(10)28,32
C. pseudotropicalis	1,8	*S. carlsbergensis*	1,(8)(10)28,32
T. glabrata	1,4,6,34		

Antigenic structures of various yeasts established to date are summarized in Appendix I with their $G + C$ contents and their coenzyme Q patterns for reference purposes.

B. Identification of yeasts by serological characteristics

1. *Serological identification of yeasts*

(a) *Isolation.* Sabouraud's agar, which contains 2% glucose, 0.5% yeast extract, 1% peptone and 1.5% agar, is recommended as an appropriate isolation medium. It is generally used in the form of slants or agar plates, depending on whether antibacterial agents, such as streptomycin (50 mg l^{-1}) or chloramphenicol (50 mg l^{-1}) and cycloheximide (500 mg l^{-1}), are present or not. Littman's oxgall agar with streptomycin and crystal violet is also used for isolating yeasts, since oxgall agar restricts the growth of mould contaminants. In addition, a variety of agar media, which contain salts of metals such as bismuth or molybdenum, are commercially available and aid in the identification of yeasts by their colour reactions. The inoculated media are incubated at 27°C for 48 h or at 37°C for 18–24 h. The yeast colonies are further subcultured on Sabouraud's agar and an examination for various properties is performed after a check for pure growth has been obtained. Most human pathogenic yeasts will form pasty, opaque colonies at 37°C or room temperature on the usual agar media.

(b) *Preparation of factor sera.* Factor sera containing antibodies homologous to thermostable antigens are prepared on the basis of antigenic structures. The factor sera may be monospecific or not (Table VI). For example, the factor serum for antigen 9 of *C. guilliermondii* is monospecific, but the factor for antigen 4 also contains anti-6 and -7. In addition, there may be several sets of species combinations in factor sera for thermostable antigens. However, only the set of combinations between sera and absorption with specific strains shown in Table VII is used for preparation of specific factor sera. Factor sera for thermolabile antigens are relatively less important for rapid identification, but they can be prepared if necessary for differentiation.

(c) *Slide agglutination.* One or two drops of the diagnostic antisera are placed within a circular marking on an ordinary glass slide (75 × 25 mm). A sufficient amount of organisms from agar culture to be tested is mixed with each drop of the antisera to produce a homogeneous suspension (Fig. 1). A loopful of the same culture is mixed with a drop of 0.85% saline solution as a control. After the culture has been well stirred, the slide is tilted back and

Group of factor sera	Factor sera	Antiserum	Organisms for absorption
Group specific	A 1[a]	C. albicans A (1060)	None
	6	C. albicans A	C. stellatoidea (0692)
	8	C. pseudotr. (0586[b])	C. albicans A
	9	C. guillierm. (0679)	C. albicans A
	11	C. krusei (0584)	C. albicans A
	13	C. parapsilosis (M[c]1015)	C. albicans A, C. albicans B (10108)
	16	H. anomala (0707)	C. parapsilosis (0640)
	24	S. rosei (0428)	C. albicans A, C. parapsilosis
Subgroup specific	7	C. albicans A	C. tropicalis (1070)
	10	S. bisporus (0723)	C. albicans A
	12	C. mycoderma (0734)	C. krusei (0584)
	17	C. utilis (0619)	H. anomala
	18	S. cerevisiae (0718)	C. alb. A, C. paraps., C. pseudotr.
	20	C. pelliculosa	C. alb. A, C. utilis, S. cerevisiae
	40	Kl. antillarum (0669)	C. albicans A, C. guilliermondii
Fairly specific or species specific	4	C. albicans A	C. parapsilosis
	5	C. albicans A	C. guilliermondii
	13, 13b	C. parapsilosis (M 1015)	C. albicans A
	19	C. rugosa (0591)	C. albicans A, C. krusei
	21	H. saturnus (0117)	C. albicans A, C. utilis, H. anomala
	23	S. unisporus (0724)	C. albicans A
	25	C. melinii (0747)	C. alb. A, C. utilis, H. anomala, C. pseudotr
	34	T. glabrata (0005)	C. albicans A, C. pseudotropicalis
Group specific	S 1[a]	Schiz. pombe (0358)	None
	R 1	Rh. glutinis (0559)	None
	R 4	Rh. minuta (0387)	None
	C 1	Cr. neoformans (0608)	None

[a] A 1, S 1, R 1, R 4 and C 1 mean antigen 1 of C. albicans, Schiz. pombe, Rh. glutinis, Rh. minuta and Cr. neoformans, respectively. Strain number is as per IFO (Institute for Fermentation, Osaka).

[b] C. pseudotr.: C. pseudotropicalis; C. guillierm.: C. guilliermondii; C. alb.: C. albicans A; C. paraps.: C. parapsilosis; C. pellicul.: C. pelliculosa.

[c] M.: Meiji College of Pharmacy.

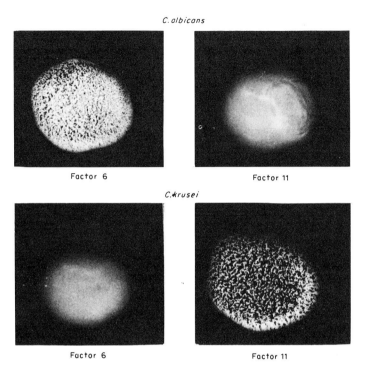

Fig. 1　Slide agglutination reactions of *Candida* species using factor sera 6 and 11.

forth for thirty seconds to two minutes. Agglutination against light, preferably with the naked eye, is recorded.

Recently Shinoda *et al.* (1981) reported a modified slide agglutination method. Briefly, cultures were suspended in saline to give a concentration of 2×10^8 cells per millilitre. Slides (75×50 mm) marked with 12 circles were used. Approximately 0.05 ml of each of the factor sera and control physiological saline was added to the circles on the slides, and approximately 0.05 ml of a cell suspension was then added to each of them. The reactants were mixed on a platform rotary shaker (A. H. Thomas Co., Philadelphia, Pennsylvania) at 125 ± 25 r/min for 2–3 min. For isolates forming rough colonies, agglutination should be performed with specimens prepared by heating the isolated cells at 100°C for 1 h followed by ultrasonic vibration (Taguchi *et al.*, 1979). The indirect fluorescent antibody method is recommended for cases of spontaneous agglutination in saline. This occurs in about 10–15% of the yeast isolates, especially in *C. krusei* and *C. tropicalis*.

(d) *Commercial sources of factor sera.*　Fukazawa *et al.* (1968a,b) and

Fukazawa and Tsuchiya (1966, 1969) designed a scheme of slide agglutination using factor sera, whereby it is possible to make rapid and accurate identification of medically important yeasts including *C. albicans* type B and *T. glabrata*.

Accordingly, a commercial kit called "Candida Check" has been produced by Iatron Laboratories, Tokyo 101, Japan. The kit consists of antisera for antigens 1, 4, 5, 6, 8, 9, 11, 13, 13b and 34, and is widely used in clinical laboratories in Japan (Fig. 2). In addition, two rapid biochemical tests, sucrose disc and sucrose strip tests, are recommended. The sucrose (SY) disc test discriminates between *C. albicans* type A (negative reaction) and *C. tropicalis* (positive yellow reaction). The sucrose (SA) strip test is used to discriminate between *C. albicans* type B (yellow or orange reaction) and *C. stellatoidea* (pink reaction). Shinoda *et al.* (1981) compared "Candida Check", the API-20C kit and the conventional modified Wickerham's technique, in terms of rapid and accurate identification of medically important *Candida* species, and revealed that "Candida Check" gave a 95% accurate result. Diagnostic reagents for several species of *Cryptococcus*, *Saccharomyces* and *Rhodotorula* are not present in the currently available kits but they can be prepared on the basis of their antigenic structures.

2. *Improvement in the preparation of diagnostic sera*

(a) *Preparation of antisera.*   Antisera for *Candida, Torulopsis, Saccharomyces* and other yeast species have been prepared by immunizing rabbits with heated cells, and antigenic analyses of many species are carried out with antisera titres of 1:128. It was generally difficult to ascertain the desirable titres of antisera against *Rhodotorula* and *Cryptococcus* (Tsuchiya *et al.*, 1957d, 1963a). Shinoda *et al.* (1980) performed antigenic analyses of *Cr. neoformans* and the serologically related *Candida* species at a titre of 1:256, using heat-killed cell antigen (100°C, 1 h) or bovine γ-globulin (BCG)-conjugated cell antigens (Fukazawa *et al.*, 1972b). BGG-conjugated cell antigens were prepared as follows: *Cr. neoformans* grown on Sabouraud's medium was harvested, washed with physiological saline after heating at 100°C for 1 h. Heat-killed cells (1 ml) were suspended in physiological saline solution (5 ml) at 4°C. A solution of 2,4,6-trichloro-*s*-triazine (10 mg) in dimethylformamide (1.0 ml) was then added to the cooled solution. The mixture was stirred at 4°C for 1 h, to which the solution of BGG (200 mg) in physiological saline solution (5 ml) was added. Then the mixture was stirred at 25°C for 1 h, kept at 4°C for 18 h and dialysed against distilled water for 30 h. The cell-BGG conjugate was suspended in formalinized saline solution at a concentration equivalent to McFarland scale No. 9 (ca 3.8×10^7 cells per millilitre) (Vera and Power, 1980). Rabbits were given intravenous injections

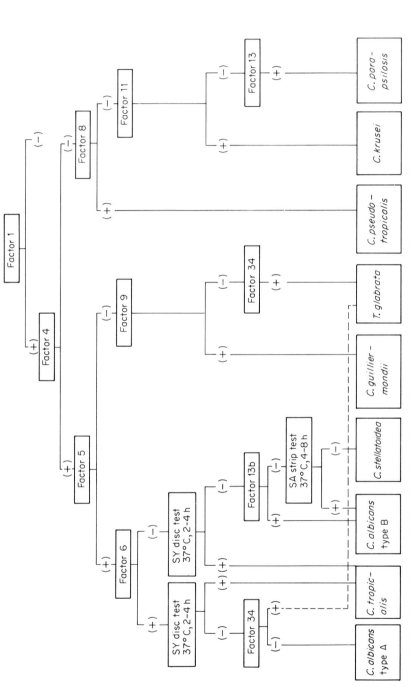

Fig. 2. Flow chart for the identification of medically important yeast species by slide agglutination using factor sera.

with the cell-BGG conjugate of 1, 2, 4, 4, 4, 4 and 4 ml at four-day intervals. Seven days after the last injection, blood was sampled from ear veins and the homologous agglutination titre determined. When the titre had reached a desired level of over 1:1280, the rabbits were bled to death.

(b) *Specificities of antisera.* Electrophoretically isolated γ-globulin of immune serum has been reported to be heterogeneous (Askonas *et al.*, 1960; LoSpalluto *et al.*, 1962), and most antibody activities are recovered in 7S (IgG) and 19S (IgM) fractions in man (Treffers *et al.*, 1942), rabbit (Steros and Taliaferro, 1959) and many other species of animals (Fahey, 1960, 1961; Fahey and Horbett, 1959). As revealed by Fukazawa *et al.* (1968a, 1969) and Tsuchiya *et al.* (1967a) with *Shigella flexineri* and *Salmonella* O-agglutinins, the immune rabbit IgG showed a much higher specificity than the IgM for *Candida*, although both IgG and IgM fractions react with *C. albicans* in both agglutination and immunofluorescence (Fukazawa *et al.*, 1968a, 1968b; Fukazawa and Tsuchiya, 1966, 1969). Furthermore, the IgG fraction is known to be more stable than IgM or whole serum for lyophilization. Thus IgG is useful for serotyping *C. albicans* and other *Candida* species by slide agglutination.

(c) *Preparation of refined antibody fractions.* For the assays of immune sera, the separation of IgG and IgM fractions was carried out (Fukazawa *et al.*, 1968a; Sober *et al.*, 1956). The immunoglobulin fractionation was performed by salting-out with half-saturated ammonium sulphate, followed by either fractionation on gel filtration, ion-exchange column chromatography, or both. Although there are various procedures for fractionation of immunoglobulins, here we describe only one.

The salting-out procedures are carried out at 2–4°C. Immune sera at 4°C are mixed with an equal volume of saturated ammonium sulphate solution, with the latter added dropwise to 50% saturation. The mixture is allowed to stand for 30 min or longer to obtain maximal precipitation. It is then centrifuged at $1000 \times g$ for 10 min and the precipitate is dissolved in the original volume of phosphate buffer saline (PBS). The precipitation step is repeated three times, and the residual ammonium sulphate is removed by dialysis against saline or PBS.

Two millilitres of the resulting immunoglobulin solution adjusted to a concentration of 1–2% protein are applied to a column (25×450 mm) containing 200 ml of Sephadex G-200 (Pharmacia, Uppsala, Sweden) equilibrated with 0.2 mol l^{-1} of Tris (hydroxymethyl) aminomethane (Tris)–0.5 mol l^{-1} of glycine-HCl buffer (pH 8.0) containing 0.2 mol l^{-1} of NaCl. Elution of immunoglobulin solution with the same buffer yielded two peaks, labelled PI and PII according to elution time. The PII (second) fraction contains the IgG antibody and the PI (first) fraction contains the IgM antibody.

Another method of separating the immunoglobulins is to run them on a diethylaminoethyl(DEAE)-Sephadex A-50 column (chromatography). After activation with 0.1 N NaOH, the gel is treated with $0.02 \, \text{mol} \, l^{-1}$ of phosphate buffer (pH 6.6) until equilibrium is attained. The equilibrated ion-exchanger is packed into a chromatographic column ($15 \times 450 \, \text{mm}$) and 5–10 ml of immunoglobulin solution containing 1–2% protein are applied to the column. After the sample has been applied to the column, stepwise elution is performed with the phosphate buffer containing $0.1 \, \text{mol} \, l^{-1}$ of NaCl as a starting eluant (B1), followed by buffers containing increasing NaCl concentrations, first $0.17 \, \text{mol} \, l^{-1}$ (B2) and then $0.30 \, \text{mol} \, l^{-1}$ (B3). Three protein fractions (FI to FIII) are obtained when buffers B1 to B3 are employed. The FI and FIII fractions contain IgG and IgM, respectively. The IgG fractions prepared by the two methods described above have been proven to be electrophoretically pure.

Factor sera can be prepared from the IgG fractions by the absorption methods based upon antigenic structures, since specificity of IgG fraction is higher than that of IgM. In addition, the IgG fraction can be made by the rivanol–ammonium sulphate procedures which allow large-scale production (Heide and Schwick, 1978; Hořejší and Smetano, 1956).

V. Application of serological grouping of yeasts

A. Medically important species

1. Candida

Various yeast species including *Candida, Cryptococcus, Rhodotorula* and *Saccharomyces* are regularly isolated from man and animals, and opportunistic *Candida* infections (Rogers and Balish, 1980) have increased remarkably since the advent of broad-spectrum antibiotics and corticosteroids. The seven *Candida* species described above as well as *C. lusitaniae* and *C. viswanathii* are emphasized as being potential pathogens by Holzschu *et al.* (1979) and Emmons (1972). *C. albicans* and *Torulopsis glabrata* (*C. glabrata*), (Wickerham, 1957; Yarrow and Meyer, 1978) are the first and second most frequently isolated species (Hurley, 1967; Winner and Hurley, 1964; Rieth, 1958; Skinner and Fletcher, 1960; Tsuchiya *et al.*, 1963b). Hasenclever and Mitchell (1960, 1961a,b,c) proposed the serologically different types A and B of *C. albicans*, and that the former was antigenically identical to *C. tropicalis*, and the latter to *C. stellatoidea*. Fukazawa *et al.* (1972a) found that *C. albicans* type B and *C. parapsilosis* have antigen 13b in common.

TABLE VIII

Principal characteristics of yeasts frequently isolated from clinical specimens[a,b]

Species	Main antigenic structures	Chlamydospore	Pseudomycelium	Urease test	Assimilation K.[a]	Gl.	Ga.	Su.	Ma.	La.	Ce.	Tr.	Me.	Ra.	Xy.	In.	Fermentation Gl.	Ga.	Su.	Ma.	La.
C. albicans type A	1,4,5,6,(7)	+	+	−	−	+	+	+	+	−	−	+	−	−	+	−	+	+	−	+	−
C. albicans type B	1,4,5,(7)13b	+	+	−	−	+	+	+	+	−	−	+	−	−	+	−	+	+	−	+	−
C. tropicalis	1,4,5,6	−	+	−	−	+	+	+	+	−	+*	+	−	−	+	−	+	+	+	+	−
C. viswanathii		−	+	−	−	+	+	+	+	−	+	+	−	−	+	−	+	+	+	+	−
C. stellatoidea	1,4,5	R	+	−	−	+	+	−	+	−	−	+	−	−	+	−	+	+	−	+	−
C. pseudotropicalis	1,8	−	+	−	−	+	+	+	−	+	+	−	−	+	+*	−	+	−	+	−	+
C. krusei	1,5,11	−	+	−	−	+	−	−	−	−	−	−	−	−	−	−	+	−	−	−	−
C. rugosa	1,4,11,19	−	+	−	−	+	+	−	−	−	−	−	−	−	+	−	−	−	−	−	−
C. parapsilosis	1,5,13,13b	−	+	−	−	+	+	+	+	−	+	+	−	−	+	−	+	+*	−	−	−
C. guilliermondii	1,4,9	−	+	−	−	+	+	+	+	−	+	+	+	+	+	−	+	+*	+	−	−
T. glabrata	1,4,5,6,34	−	−	−	−	+	−	−	−	−	−	+	−	−	−	−	+	−	−	−	−
T. famata	1,4,9	−	−	−	−	+	+	+	+	+	+	+	+	+	+	−	+*	+*	+*	−	−
S. cerevisiae	1,10,18,31	−	−	−	−	+	+	+	+	−	−	+	−	+	−	−	+	+	+*	+	−
D. hansenii	1,4,9	−	−	−	−	+	+	+	+	+*	+	+	−	+	+	−	+	+*	+*	+	−
Cr. neoformans type A	1,2,3,7	−	R	+	−	+	+*	+	+	−	+	+	+	+	+	+	−	−	−	−	−
Cr. diffluens	1,2,3	−	−	+	−	+	+	+	+	+	+	+	+	+	+	+	−	−	−	−	−
Cr. laurentii		−	−	+	+	+	+	+	+	+	+	+	+	+	+	+	−	−	−	−	−
Rh. glutinis	1,2,5	−	−	+	+	+	+*	+	+	−	+	+	−	+	+	−	−	−	−	−	−
Rh. rubra	1,2(3)	−	−	+	−	+	+*	+	+	−	+*	+	−	+	+	−	−	−	−	−	−

[a] K.: potassium nitrate (KNO₃); Gl.: glucose; Ga.: galactose; Su.: sucrose; Ma.: maltose; La.: lactose; Ce.: cellobiose; Tr.: trehalose; Me.: melibiose; Ra.: raffinose; Xy.: xylose; In.: inositol; +*: +variable; R: rare.
[b] Compare with Table IX.

Recently, the serology of various yeasts has become important for identification and epidemiological tracing. Serological identification using ten factor sera is proved to be a rapid and accurate identification for eight medically important *Candida* species (Shinoda *et al.*, 1981).

Biological examination is generally made according to various procedures laid out in the following papers (Lodder, 1970; Lodder *et al.*, 1958; Vanbreuseghem *et al.*, 1978; Barnett and Pankhurst, 1974; Rieth, 1958, 1978; Gentles and La Touche, 1969; Ahcarn *et al.*, 1977; Silva-Hutner and Cooper, 1980). Recently, a rapid biochemical identification procedure for clinical isolates has been developed (API 20 C, Analytab Products, Plainview, New York) (Buesching *et al.*, 1979).

2. Cryptococcus *and the other yeast genera*

Most yeast species are not pathogenic for man with the exception of some of the species of the genera *Cryptococcus* and *Candida*. However, various species of yeasts present in foodstuffs are ingested by man. They are sometimes found in fermenting fruits, grape must (*S. cerevisiae*) and sausage (*Debaryomyces*). *Rhodotorula* species have occasionally been isolated from the mouth or can be transiently present in the blood stream because of prolonged intravenous therapy. *Cr. neoformans* is frequently isolated from pigeon droppings (Emmons, 1955, 1972) and it causes meningitis in a compromised host (Hart *et al.*, 1969; Emmons *et al.*, 1977). Since no serological cross-reaction has been found among *C. albicans, Cr. neoformans, Rh. glutinis, Rh. minuta* and *Schizosaccharomyces pombe* serogroups (Tsuchiya *et al.*, 1965a, 1974; Tsuchiya and Fukazawa, 1966; Seeliger, 1957, 1958), it is necessary to have such sera, especially anti-*Cr. neoformans* serum (Table IX).

As for the serological differences between strains of *Cr. neoformans*, Evans (1949, 1950) and Evans and Kessel (1951) divided *Cr. neoformans* into three serotypes A, B and C by agglutination, precipitation and capsular reaction. Later, Wilson *et al.* (1968) added serotype D. Therefore, it is recognized that *Cr. neoformans* has four serotypes. On the other hand, Tsuchiya *et al.* (1963a, 1965a) reported a serological rapid identification method of *Cr. neoformans* and antigenic structures of several *Cryptococcus* species, although the immune sera were prepared by immunization with micro-organisms possessing very thin capsules unlike those of ordinary *Cr. neoformans* strains. Recently, high titered antiserum for *Cr. neoformans* was prepared by Shinoda *et al.* (1980) using male New Zealand white rabbits immunized with BGG-conjugated cell antigens. Then Ikeda *et al.* (1982) established antigenic patterns of *Cr. neoformans* serotypes and serologically related *Cryptococcus* and *Candida* species (Appendix I). Hopefully, reagents of factor sera prepared on the basis of antigenic patterns will appear on the

TABLE IX

Serological groups of yeasts

Antiserum	Antigens				
	C. albicans	Schiz. pombe	Rh. glutinis	Rh. minuta	Cr. neoformans
C. albicans	+	−	−	−	−
Schiz. pombe	−	+	−	−	−
Rh. glutinis	−	−	+	−	−
Rh. minuta	−	−	−	+	−
Cr. neoformans	−	−	−	−	+
	P. fermentans	C. lipolytica	Rh. aurantiaca	Rh. pallida	Cr. diffluens
	H. anomala		Rh. rubra	Sp. gracilis	Cr. albidus
	D. hansenii		Sp. roseus		C. humicola
	Kl. magna		Sp. odorus		C. curvata
	S. cerevisiae		Sp. pararoseus		T. aeria
	Hs. valbyensis				Tr. cutaneum
	Sm. ludwigii				
	Schw. occidental				
	Nem. coryli				

market in the near future. *Cryptococcus* and *Rhodotorula* species are biochemically characterized by urease activities (Seeliger, 1956; Seneca *et al.*, 1963), inositol assimilation (Phaff and Spencer, 1966) and extracellular DNAase (Sen and Komagata, 1979), and their DNA has been shown to have a high $G + C$ content: 49.0–59.0 $(G + C)$ mole % DNA in *Cryptococcus* and 50.2–67.8 $(G + C)$ mole % DNA in *Rhodotorula* compared with almost all the urease-negative yeast species (Nakase and Komagata, 1971c,f). However, the immunogenicity of these species for rabbits is generally very low. In addition, different antigenic structures of the same species have been established by several authors, probably because of differences in the methods, strains or in titres of antisera used. Within the same species, however, each antigen proposed by different authors and indicated by different symbols appears to correspond fairly well (Table II).

B. Frequency of yeast species identified by the serological method

1. *Isolates from man*

Yeast-like fungi are usually present as saprophytes, and yeasts frequently present in, for example, the mouth, intestinal, respiratory or vaginal tracts and mucosa may or may not cause disorders. The distribution of yeast species at various sites in the human body has been reported by a number of authors, revealing the isolation of more varied species than those from the clinical materials. Rieth (1958), Vörös-Felkai and Novák (1961), Mackenzie (1961) and Stenderup and Pedersen (1962) reported the isolation of yeasts from patients with mycotic infection and from healthy persons (Table X), following conventional methods of identification such as the examination of morphological and biological properties.

On the other hand, Tsuchiya et al. (1959, 1963b), Sweet and Kaufman (1970) and several other investigators first performed slide agglutinations to identify yeasts by absorbed antisera and followed them by biological tests to confirm the serological identification. Taguchi et al. (1972), Shinoda et al. (1978) and others (Kanno and Suzuki, 1972; Seo, 1972; Inoue et al., 1973; Yamazaki et al., 1976; Senju et al., 1978) have recommended the "Candida Check" kit containing factor sera for antigens 1, 4, 5, 6, 8, 9, 11, 13, 13b and 34 for the identification.

Almost all the yeast isolates belonged to the genus Candida, and the most common species was C. albicans, representing 33.3–68.0% of the isolates from clinical human sources. Similarly, the other Candida isolates were identified as C. tropicalis, C. krusei, C. parapsilosis, C. pseudotropicalis, C. guilliermondii and C. stellatoidea by slide agglutination with factor sera. Also T. glabrata was frequently identified by factor serum 34. Among C. albicans, 81.5–88.9% belonged to serotype A and 11.1–18.5% to type B, when tested by factor sera 6 and 13b. Strains of Cryptococcus and Rhodotorula were occasionally found, but did not react with factor serum 1 of the C. albicans serogroup.

The identification of the genera Candida and Torulopsis by slide agglutination appears to be both practical and accurate, while identification on the basis of biological properties alone is often difficult (Seo, 1972; Kanno and Suzuki, 1972).

2. Isolates from animals and chickens

Yeasts are widespread in nature, and varied species can be isolated from different species of animal. Although Ainsworth (1963), van Uden et al. (1958) and Kuttin and Beemer (1975) used the conventional methods for identification of several yeasts from animal sources, Tsuchiya et al. (1962) and Taguchi et al. (1979) classified yeast isolates from animals and chickens by slide agglutination. In the first step, unabsorbed group antisera against heat-killed micro-organisms of C. albicans, Cr. neoformans, Rh. glutinis, Rh. minuta and Schizosaccharomyces pombe groups were used, and then the isolates showing positive agglutination against anti-C. albicans serum were further subjected to slide agglutination using the "Candida Check" kit or

TABLE X

Frequency of yeast species isolated from man

Species	Rieth (1958)	Mackenzie (1961)	Stenderup and Pedersen (1962)	Tsuchiya et al. (1963)	Sweet and Kaufman (1970)	Taguchi et al. (1972)	Seo (1972)	Shinoda et al. (1978)	Senju et al. (1978)
C. albicans	1111	141	659	1078					
C. albicans type **A**					29	255	585	81	950
C. albicans type **B**					13	44	73	5	112
C. tropicalis	203	12	16	122	17	17	69	7	151
C. krusei	31	3	19	71	11	34	26	7	12
C. parapsilosis	342	9	19	32	15	6	3	12	25
C. pseudotropicalis	27		12		7		8		
C. guilliermondii	170	1	1	37	12	3	11	14	17
C. stellatoidea	5				8				
Other Candida spp.	225	8	17	4					
T. glabrata	84	31	137	129		80	261	7	163
T. famata	172	1	2					1	
Other Torulopsis spp.	430	1	12	2					
Tr. cutaneum	78	1	3	13				2	
Other Trichosporon spp.	63	1	2	1					
Cr. neoformans		1		1					
Other Cryptococcus spp.	24	6	2					1	
Rh. mucilaginosa	44	16	22	5					
Rh. rubra	27	1		1					
Other Rhodotorula spp.	91							5	
S. cerevisiae	22	4	20	1					
Other Saccharomyces spp.	60	5	5	4					
Other species	123	5	20	12		21	23	1	
Total	3332	247	968	1515	112	460	1059	143	1430

absorbed antisera as in the identification of medically important yeasts from human cases. Tsuchiya et al. (1965a) noted that each of five yeast serogroups was antigenically distinct (Appendix I, Table IX). Antigenic analyses were performed on yeast species which each belonged to the five serogroups not showing mutual cross-reactions. Accordingly, as seen in Appendix I, antigen 5 which appears in both C. albicans and Rh. glutinis is different, although symbolized by the same Arabic numeral (Tsuchiya et al., 1963a, 1965a, 1974).

Most isolates from animals and chickens were also isolates of Candida (Table XI). With the exception of dogs and chickens, C. albicans was less common in animals than in man (van Uden et al., 1958). All strains of C. albicans isolated from animals and chickens belonged to serotype A, and none to type B. C. krusei was the most common species among isolates from dogs, pigs and cattle.

Torulopsis glabrata was not isolated as frequently from animals and chickens as from man. As reported by Seeliger and Schröter (1963), Trichosporon cutaneum and Cryptococcus have antigen E in common, and many isolates from cattle and chicken could be serotyped quickly as Tr. cutaneum using antiserum from Cr. neoformans (Taguchi et al., 1979).

C. Detection of contaminants in brewing yeasts

1. Slide agglutination

The contamination by wild yeasts in the brewing of beer, wine and other microbial beverages must be carefully checked and controlled to detect contamination as soon as possible. For the detection of wild yeasts in the brewing field, various selective media have been proposed; lysine media (Walters and Thiselton, 1953) or crystal violet agar (Kato, 1967) are most widely used. Although wild isolates have been identified according to van der Walt (1970), such conventional procedures might take two weeks or longer.

Nishikawa et al. (1979a,b) collected strains of wild contaminants in beer, wine, other beverages and baker's yeast together with authentic strains from the bottom and top yeasts, and grouped them first into ten serogroups A to J on the basis of agglutination reactions using the "Candida Check" kit described in Section IV.B.1(d). They examined the contaminants further, for example for growth on lysine and crystal violet agar and development of a visible haze in bottled beer, when stored at room temperature for 20 days. Harmful contaminants belonged to the serological group D, grew on crystal violet medium, but not on lysine agar. The same authors prepared a reagent for rapid serological detection of yeast contaminants by a mixture of sera

TABLE XI

Frequency of yeast species isolated from animals and chickens

Species	van Uden et al. (1958)					Tsuchiya et al. (1962)		Taguchi et al. (1979)		
	Horses	Sheep	Goats	Pigs	Cattle	Chickens	Cattle	Dogs	Pigs	Chickens
C. albicans	11	21	2	23		205	5	45	5	71
C. albicans type **A**										
C. albicans type **B**										
C. tropicalis	11	3	2	15	45		30	3	1	6
C. krusei	21	6	4	37	33	3		5	38	
C. parapsilosis	9	3		2	3			2		5
C. pseudotropicalis								1	13	
C. guilliermondii	4						1	5	2	4
C. sloofii	6			121						
Other *Candida* spp.	2			12	11	7		10	15	14
T. glabrata	3		2	8	4					
Other *Torulopsis* spp.	1			1				16	25	40
Tr. cutaneum	55			1	5	24	25	3	1	7
Other *Trichosporon* spp.	1						2		1	
Cr. neoformans								4	1	
Other *Cryptococcus* spp.					1			18	5	
Rh. minuta								8	7	
Other *Rhodotorula* spp.		1						5	2	
S. cerevisiae	8		4	22	12					
Other *Saccharomyces* spp.	9		1	38	8					
Other species	5		1	9	9					
Total	146	35	16	289	131	238	63	124	116	147

against each of *Saccharomyces bisporus* IFO 0723 and *S. pastorianus* BSRI YW 1–2 absorbed with *S. carlsbergensis* BSRI YB 1–2 cells. The ensuing reagent which was used for quality control in the yeast-related industries gave results compatible with immunofluorescence.

2. *Fluorescent antibody technique*

Several investigators have studied the fluorescent antibody technique for identification of *Candida* and other yeasts, since Coons and Kaplan introduced the technique in 1950 (Gordon, 1958a,b, 1962; Gordon *et al.*, 1967; Kaufman and Kaplan, 1963; Everand *et al.*, 1957; Wood *et al.*, 1965; Gerencser, 1979; Cherry, 1980). The reaction can be seen by using antibody conjugated with a fluorescent dye. The immunofluorescence technique provides laboratory workers with a valuable adjunct to conventional diagnostic tests, since it is far more rapid and sensitive than other methods for identifying yeasts. It makes possible direct identification of both viable and non-viable micro-organisms in cultures as well as clinical materials such as pus, exudate, blood, tissue impression smears and spinal fluid. Thus, the use of immunofluorescence for the identification of micro-organisms has become quite common in the medical fields and also in brewing microbiology. In brewing, the Analysis Committee of the Institute of Brewing, Nutfield, England (1973) recommended immunofluorescence for the detection of wild yeasts in the brewing process, since the technique is simpler than the plating methods and has been shown to be applicable to routine quality control. For details of the technique, the reader is referred to Narin (1976), Riggs *et al.* (1958), Goldman (1968), Hebert *et al.* (1967) and Kawamura (1977).

For indirect immunofluorescence, dried smears of thoroughly washed yeast cells are fixed in acetone for 10 min or by gentle heating (Richards and Cowland, 1967; Haikara and Enari, 1975) and treated with non-labelled anti-yeast rabbit serum showing an appropriate titre for 30–45 min. Then antiserum is removed by washing in phosphate buffer for 15 min, and the yeast smears are treated for 30–45 min with fluorescein-conjugated anti-rabbit IgG goat or swine serum (Wellcome Research Laboratories, Beckenham, England; Dakopatts, Copenhagen, Denmark; BBL, Cockeysville, Maryland; Difco Laboratories, Detroit, Michigan). After final washing with phosphate buffer for 15 min, the smears are mounted in phosphate-buffered (10%, pH 7.2) glycerol and sealed with a cover-slip and paraffin wax. The above treatments with antisera are performed in a moist chamber. Indirect immunofluorescence is said to be more sensitive than direct immunofluorescence, although non-specific reactions may appear frequently, but it is more widely used, probably because diagnostic reagents are available, commercially.

D. Serological diagnosis of systemic diseases due to yeasts

While slide agglutination using absorbed antisera has proved to be useful for the identification of yeast isolates, it is also possible for sera from patients to be diagnosed using specific antigens, because the isolation of organisms is sometimes difficult and diagnosis is important for chemotherapy. For a detailed account of methods see Palmer *et al.* (1977) and Kaufman (1980a,b).

Positive serological reactions at titres of 1:8 or over are mostly significant (Table XII). In candidiasis, however, some patients do not exhibit any significant immune response at the initial stage of infection. Therefore, it must be borne in mind that negative serology itself cannot exclude the possibility of a mycotic infection, and it is necessary to make some other examination, particularly in *Cryptococcus* infections

TABLE XII
Serological test for Candidiasis and Cryptococcosis

Disease	Tests	Diagnostic value
Candidiasis	Immunodiffusion (ID)[a] and Counterelectrophoresis (CEP)[a]	88% of systemic infections detected. Specific for *Candida* spp. antibodies.
	Latex agglutination (LA)[a]	1:8 titre or greater, titre movement, 80% of systemic cases detected.
Cryptococcosis	Latex agglutination (LA)[a]	Antigen is present in sera or cerebrospinal fluid (CSF) of pulmonary and/or meningeal cases. 1:8 titres or greater are considered as strong evidence of active infection.
	Tube agglutination (TA)[a] and Indirect fluorescent-antibody (IFA)[a]	Antibodies are present in sera of pulmonary and/or meningeal cases. Antibodies may occur in early or localized infection.

[a] Kaufman (1980a,b); Palmer *et al.* (1977).

E. Interrelationships among various grouping methods

1. *Chemical properties and DNA base sequence relatedness of yeasts*

In the process of classification, morphology, growth patterns and biochemical properties reflecting intermediate metabolism are first taken into consideration. Morphological criteria are not always sufficient for the classification of minute and simple organisms whereas there is no discrepancy in the classification of higher plants based on such criteria. Antigenic properties are considered to be important for the classification of bacteria, but they are still not really considered for the classification of yeasts. The $(G+C)$ mole % DNA has been proposed as a major criterion for the classification of yeasts (Nakase and Komagata, 1966, 1968, 1969, 1970a,b, 1971a–g, 1972; Nakase *et al.*, 1972; Stenderup and Bak, 1968) and new species were at times proposed on the basis of the differences in $(G+C)$ mole % of DNA of several strains in the same species either with or without physiological properties, when differences of 2.0% or more were observed (Nakase *et al.*, 1972; Nakase, 1972). The studies of these authors covered almost all the species of yeasts (Storck and Alexopoulos, 1970), attracting much attention to the close relationship between serological characteristics and the $(G+C)$ mole % of DNA. In bacteriology, this is also an important criterion.

Yamada and Kondo (1972a,b, 1973) and Yamada *et al.* (1973a,b, 1976a,b) reported that the coenzyme Q system was useful for the classification of yeasts and together with the $G+C$ content and antigenic structure was valuable for definition of yeast species. The types of ubiquinone, although five categories, seem to be a significant generic parameter. The properties of the chemical constituents of yeasts are more stable than the biological properties, and consequently are considered to be important for classification.

On the other hand, DNA base sequence relationships were investigated among phenotypically similar yeasts to determine whether the existing taxonomic criteria actually reflect the evolutionary affinities within the group (Price *et al.*, 1978; Fuson *et al.*, 1979). Although the strains used were limited nearly to only type strains, the results indicated that the methodology generally used for delimiting yeast species was inadequate for defining such natural taxa as *Torulaspora delbrueckii*. It is desirable to carry out further comparative studies with several strains of the same species using this method.

2. *Physicochemical properties*

Proton magnetic resonance (p.m.r.) spectroscopy was first used as a taxonomic tool by Gorin and Spencer (1968) to distinguish the chemical structure of mannans in *Trichosporon* species. The p.m.r. spectra of mannans were specific for each yeast species and divided the yeasts into 21 groups and 126 subgroups except for the genus *Ceratocystis* (Gorin and Spencer, 1970, 1972; Spencer and Gorin, 1968, 1969a,b,c, 1970, 1971; Spencer *et al.*, 1970a,b).

The findings of antigenic analyses are consistent with the p.m.r. spectra (Gorin and Spencer, 1970), since the antigenic properties of the cells depend mainly on mannans or mannose-containing polysaccharides in the outer layer of the cell wall. In our recent studies on *Saccharomyces, Pichia, Candida* and *Torulopsis* species, the p.m.r. spectra of mannans in the H-1 region are characterized by the number of signals and their intensities (Fukazawa *et al.*, 1972a, 1975, 1980a,b; Kuroshima *et al.*, 1976; Shinoda, 1972). Here, we described only seven *Candida* species and *T. glabrata*. The p.m.r. spectra of mannans of *C. albicans* type A and *C. tropicalis* were very similar. That of *C. guilliermondii* was closely related to the two other species, except for a lack of one weak signal (Fig. 3). In addition, p.m.r. spectra of *C. albicans* type B, *C. stellatoidea* and *C. parapsilosis* were also similar. *C. parapsilosis* was fairly analogous to them, although *C. parapsilosis* had another minor signal at 5.37 ppm. *C. krusei* and *C. pseudotropicalis* demonstrated clearly spectral patterns different from those of the above five species, both of them possessing a main signal at 5.08 ppm and a minor one at either 5.21 or 5.28 ppm. The spectrum of *T. glabrata* had three signals which were partially similar to those of *C. albicans* type A and *C. tropicalis*. After all, the relationships among p.m.r. spectra of nine *Candida* and *Torulopsis* species and their antigenic patterns corresponded well with one another (Table XIII). We are now using ppm units for the chemical shift instead of $\tau(\tau = 10.00 - ppm)$.

Serological grouping of yeasts would have reasonable correlations with their classification by chemical characteristics, since microbial antigenicities depend on the stereoscopic molecular structure of determinant groups in the cell wall. Therefore, serological and physicochemical characteristics correspond, so far as the strains used are identical, and almost all the results are considered to reflect DNA homology. Thus, it is considered that *S. exiguus* (p.m.r. spectral group 5d) and *S. dairensis* (group 6f) should be grouped into 5d, and *H. californica* (11a), *H. beijerinkii* (11d), *H. mrakii* (11d), *H. saturnus* (10a) and *H. saturnus* subsp. *subsufficiens* (11e) should be brought together in groups 11a and 11e, on the basis of their p.m.r. spectra and antigenic structures as in the case of *Torulaspora delbrueckii* shown in the List of Cultures, Centraalbureau voor Schimmelcultures, 1978 (Tsuchiya and Taguchi, 1980; Spencer and Gorin, 1969b; Wickerham, 1951, 1970; Wickerham and Burton, 1962).

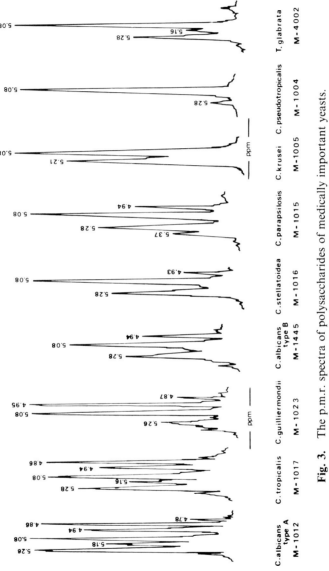

Fig. 3. The p.m.r. spectra of polysaccharides of medically important yeasts.

C.albicans type A M-1012
C.tropicalis M-1017
C.guilliermondii M-1023
C.albicans type B M-1445
C.stellatoidea M-1016
C.parapsilosis M-1015
C.krusei M-1005
C.pseudotropicalis M-1004
T.glabrata M-4002

TABLE XIII

The p.m.r. signals from mannans of seven *Candida* species and *T. glabrata*

Strains		Signals (ppm)[a]						Main antigenic patterns
		A	B	C	D	E	F	
C. albicans type A	M-1012		5.26	5.18	5.08	4.94	4.86	4.78	1,4,5,6
C. tropicalis	M-1017		5.28	5.16	5.08	4.94	4.86		1,4,5,6
C. guilliermondii	M-1023		5.26		5.08	4.95	4.87		1,4,9
C. albicans type B	M-1445		5.28		5.08	4.94			1,5,13b
C. stellatoidea	M-1016		5.28		5.08	4.93			1,4,5
C. parapsilosis	M-1015	5.37	5.28		5.08	4.94			1,13,13b
C. krusei	M-1005			5.21	5.08				1,11
C. pseudotropicalis	M-1004		5.28		5.08				1,8
T. glabrata	M-4002		5.28	5.16	5.08				1,4,6,34

[a] A, B, C, D, E and F indicate peaks of p.m.r. spectra.

F. Significance of antigenic structures in yeast taxonomy

1. Rapid identification

Conventional identification methods of yeasts are relatively time-consuming, while rapid diagnosis is usually desirable. Some strains of *C. albicans* do not produce chlamydospores, especially at the beginning of their isolation (Taguchi *et al.*, 1972), and biologically atypical isolates are not rare (Seo, 1972). Meanwhile, antigenic structures of yeast species are now very useful for the rapid identification required in the medical and industrial fields (Nishikawa *et al.*, 1979a,b) and also in inspection of the single cell protein (SCP) manufacturing. Actually any bacteriologist can prepare specific diagnostic antisera by himself.

Fukazawa *et al.* (1968a) demonstrated highly specific and stable IgG antibody fractions for *Candida* species based upon antigenic structures. The "Candida Check" kit containing ten kinds of IgG factor antibodies is now available and widely used in Japan, and it has proved to be superior to the usual method with respect to speed, accuracy and simplicity. In addition, there is a plan to distribute additional diagnostic sera such as those against *Cr. neoformans* and *S. cerevisiae*.

TABLE XIV

Species names in the lists of culture collections

Species names Lodder (1970)	CBS[a] (1978)	ATCC[b] (1978)	NCYC[c] (1978)	IFO[d] (1978)
D. coudertii	D. coudertii	Tsp. hansenii[e]	—	D. coudertii
D. hansenii	D. hansenii	Tsp. hansenii	D. hansenii	D. hansenii
P. etchellsii	P. etchellsii	Tsp. etchellsii	P. etchellsii	—
P. pseudopolymorpha	P. pseudopolymorpha	Tsp. pseudopolymorpha	P. pseudopolymorpha	P. pseudopolymorpha
S. bailii	Zs. bailii	S. bailii	S. bailii	S. bailii
S. bisporus	Zs. bisporus	S. bisporus	S. bisporus	S. bisporus
S. delbrueckii	Tsp. delbrueckii	Tsp. delbrueckii	S. delbrueckii	S. delbrueckii
S. diastaticus	S. cerevisiae	S. diastaticus	S. diastaticus	S. diastaticus
S. fermentati	Tsp. delbrueckii	Tsp. delbrueckii[f]	S. fermentati	S. fermentati
S. florentinus	Zs. florentinus	Tsp. florentinus	S. florentinus	S. florentinus
S. heterogenicus	S. cerevisiae	S. heterogenicus	S. heterogenicus	S. heterogenicus
S. inconspicuus	Tsp. delbrueckii	—[g]	S. inconspicuus	S. inconspicuus
S. kloeckerianus	Zs. globosus	—	—	S. kloeckerianus
S. pretoriensis	Tsp. pretoriensis	Tsp. pretoriensis	S. pretoriensis	S. pretoriensis
S. rosei	Tsp. delbrueckii	S. rosei	S. rosei	S. rosei
S. rouxii	Zs. rouxii	S. rouxii	S. rouxii	S. rouxii
S. saitoanus	Tsp. delbrueckii	Tsp. mongolia	—	S. saitoanus
S. uvarum	S. cerevisiae	S. uvarum	S. uvarum	S. uvarum
S. vafer	Tsp. delbrueckii	Tsp. vafer	—	S. vafer
T. colliculosa	Tsp. delbrueckii	T. colliculosa	T. colliculosa	T. colliculosa
T. sphaerica	K. lactis	T. sphaerica	K. lactis	K. lactis

[a] Centraalbureau voor Schimmelcultures, Delft, Netherlands.

[b] American Type Culture Collection, Rockville, USA.

[c] National Collection of Yeast Cultures, Brewing Industry Research Foundation, Nutfield, England.

[a] Institute for Fermentation, Osaka, Japan.

[e] Tsp. coudertii ATCC (1980).

[f] Tsp. fermentati ATCC (1980).

[g] Tsp. inconspicuus ATCC (1980).

2. Unification of species

The relationships among yeast groupings shown by various methods are of practical importance, leading to a greater interest in the chemotaxonomy of yeasts. The p.m.r. spectra of polysaccharides of many *Saccharomyces* and other strains have revealed that the patterns can be mostly correlated with antigenic structures already established (Tsuchiya *et al.*, 1958c, 1965b; Tsuchiya and Taguchi, 1980; Gorin and Spencer, 1970). However, scarcely any amendments of the various classification schemes have been proposed, even when antigenic differences were indicated in the same species or the same antigenic structures were found among different species, because morphological and biological properties are regarded as being of prime importance (Tsuchiya *et al.*, 1958e).

Price *et al.* (1978) proposed that *S. delbrueckii, S. saitoanus, S. inconspicuus, S. fermentati, S. rosei* and *S. vafer* should be considered as one species, *S. delbrueckii*, based on their DNA homology. Subsequently, Yarrow (1970, 1971) brought together these species as well as *Torulopsis stellata* subsp. *cambresieri* and *T. colliculosa* into the new species *Torulaspora delbrueckii* on the basis of his own serological and hybridization studies and probably our serological findings (Appendix I), even though differences exist in eight physiological characteristics between *S. saitoanus* and *S. rosei* (Fig. 4) (van der Walt, 1970). In addition, it is considered that *S. dairensis* should be included in *S. exiguus*. Similarly, *S. uvarum, S. heterogenicus, S. florentinus, S. diastaticus, S. oviformis* and *S. logos* should be combined with *S. bayanus* as described previously (Tsuchiya *et al.*, 1974). Fukazawa *et al.* (1980a) proposed grouping *S. cerevisiae* and *S. uvarum* into single species with three serotypes.

3. Separation of species

Fukazawa *et al.* (1975) proposed dividing *C. sake* into the four species, *C. maltosa, C. natalensis, S. cerevisiae*-like species and *C. sake* on the basis of serological and physicochemical findings. Furthermore, *S. exiguus* CBS 3019, CBS 5647, CBS 5648, IFO 0270, IFO 0215 and IFO 0286 would be transferred either to *S. cerevisiae* or *S. unisporus* according to both p.m.r. spectra of cell wall mannans and serological findings (Tsuchiya and Taguchi, 1980, Tsukiji, 1983), although Gorin and Spencer (1970) considered some of them to be different serotypes within *S. exiguus* (Fig. 5). It was also observed that *S. chevalieri* CBS 420, CBS 1576, CBS 1591, CBS 3077 and CBS 3078 were different from the type strain CBS 400 as well as strains CBS 403, CBS 405 and IFO 0210. Besides, *D. subglobosus* and *T. candida* should regain their species names (Tsuchiya *et al.*, 1965a) because they are different from *D. hansenii* as Price *et al.* (1978) has shown by DNA homology.

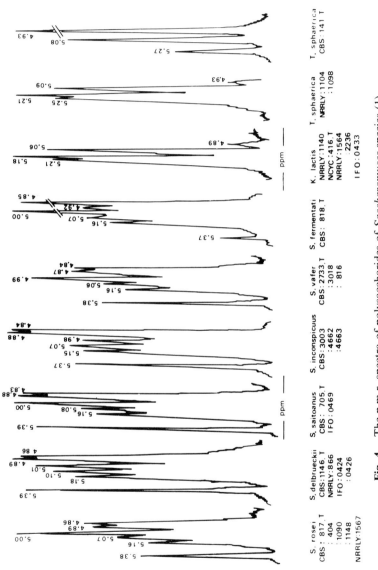

Fig. 4. The p.m.r. spectra of polysaccharides of *Saccharomyces* species (1).

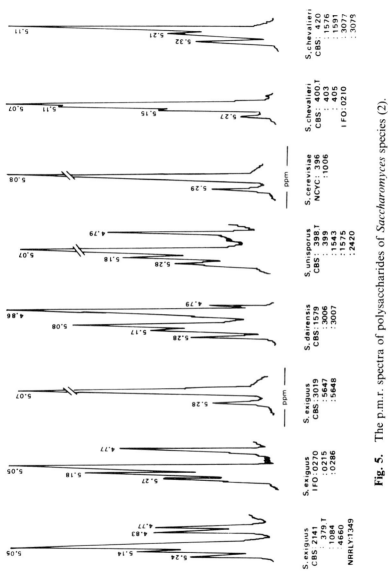

Fig. 5. The p.m.r. spectra of polysaccharides of *Saccharomyces* species (2).

4. *Change of genus*

Torulopsis sphaerica CBS 141 (type strain and a strain of *Kluyveromyces lactis*) is serologically and physicochemically different from *K. lactis*, although the former is regarded as an imperfect form of the latter (van Uden, 1970; Gorin and Spencer, 1970; Tsuchiya and Taguchi, 1980). However, *T. sphaerica* NRRL Y-1104 and NRRL Y-1098 are completely the same as *K. lactis* in serology and physicochemistry (Fig. 4). Therefore, it is necessary to change either the description of relationships between the two species or the definition of the type strain *T. sphaerica*. Also it may be possible to change the type strain of *T. sphaerica* as was formerly done in the case of *S. delbrueckii* CBS 398 (type strain of *S. unisporus* at present). It is considered that such a definition might have arisen by chance.

VI. Immunochemical basis of the serological specificity of yeasts

The chemical structures of some yeast mannans have been demonstrated by many authors (Raschke and Ballou, 1972; Ballou *et al.*, 1974; Suzuki *et al.*, 1968; Mitchell and Hasenclever, 1970) and relationships between immunochemical structures and antigenic factors of yeast polysaccharides have not yet been fully elucidated. By the various reactions or phenomena in immunochemical analyses, such as p.m.r. spectra of yeast polysaccharides (Gorin and Spencer, 1968, 1970; Spencer and Gorin, 1969c, 1971), precipitation inhibition tests, fragmentation of acetolysis polysaccharides and methylation analysis of acetolysis fragments have been thoroughly studied (Kocourek and Ballou, 1969; Hakomori, 1964; Hellerqvist *et al.*, 1968; Björndal *et al.*, 1970). Recently, Fukazawa *et al.* (1980b) reported the immunochemical structures of the Tsuchiya and Fukazawa factors, mainly factor 18 specific for *S. cerevisiae* serotype I (Fukazawa *et al.*, 1980a), factor 8 specific for *C. pseudotropicalis*, factor 10 common among several *Saccharomyces* species and factor 6 specific for *C. albicans* type A.

As for the antigenic factors 10 and 18, it was suggested that the determinant group of antigenic factor 18 was a terminal α-1,3 linked mannose unit of tri-, tetra- and penta-saccharides of *S. cerevisiae* type I mannan, on the basis of immunochemical analyses of *S. cerevisiae* type I and type II mannans. The determinant of antigenic factor 10 was suggested to be a mannotriose composed of two α-1,2 linkages. Subsequently, the immunochemical analyses of acetolysis oligosaccharide fragments from *C. pseudotropicalis* mannan suggested that α-1,2 linked mannobiose is most responsible for the specificity of antigenic factor 8, which is present in high amounts in *C. pseudotropicalis* mannan (Fig. 6).

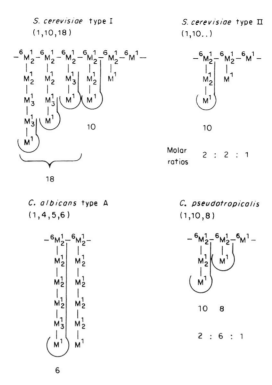

Fig. 6. Presumptive structures of the immunochemical determinants of antigenic factors of yeasts.

Finally, immunochemical analyses of antigenic factor 6 for *C. albicans* type A was performed in relation to antigens of *C. albicans* type B. The results suggested that the determinant of type B mannan was shorter than that of type A mannan, and that the serological difference between the two serotypes was attributable to the different structures of mannohexaose. Possible models of immunochemical determinants of antigenic factors 6, 8, 10 and 18 are shown in Fig. 6 (Fukazawa *et al.*, 1980b).

VII. Conclusions

For the rapid and reliable identification of yeast species serological investigations have become as important in clinical, biological and industrial laboratories as in bacteriology. Since yeast species have specific antigenic structures, they can be identified by slide agglutination. The diagnostic antisera can be prepared by each researcher in the laboratory or commercially available reagents can be used, at least for the nine medically important *Candida* species. On the other hand, some yeast contaminants in beer can be detected by slide agglutination with absorbed antisera (Nishikawa *et al.*, 1979a,b). Results to this effect have been confirmed by indirect fluorescent antibody techniques. However, reagents for agglutination of yeasts of a primary industrial concern are not yet commercially available, probably because of the lack of demand from factories. Furthermore, various immune reactions with sera from patients can be performed for clinical diagnosis of the systemic opportunistic infections.

Classically yeasts have been classified into genera based on their morphological characteristics and sexual reproduction as well as into species on the basis of their biochemical properties. However, these bases of classification are not necessarily stable. Since antigenic structures of yeasts have been shown to correspond well with the patterns of p.m.r. spectra of the cell wall polysaccharides, the p.m.r. spectra are also considered to be useful as an aid to the classification of yeasts (Gorin and Spencer, 1970; Tsuchiya and Taguchi, 1980). In addition, serologically classified yeast groups are found to correspond fairly well to those grouped by DNA homology or by $(G+C)$ mole % DNA. Therefore, in considering the transfer of some yeast strains from a certain species to another because of biological properties, it would be necessary to perform at least two additional tests using p.m.r. spectra of the cell wall polysaccharides, serological characteristics and DNA homology, although $(G+C)$ mole % DNA and coenzyme Q are also very important. Also, it might be better to refer to other characteristics which have been adopted in bacteriology.

Finally, it would seem reasonable to propose the organization of a committee on yeasts, with a nomenclature subcommittee including members of chemo- and sero-taxonomical fields as in bacteriology, since there exist some problems regarding reference to genera or species even of strains of culture collections, which have a very important role in this era of molecular biology.

Acknowledgement

The authors would like to thank K. Fujiwara, Institute of Medical Science, Tokyo University, for his helpful discussion and sincere advice during the preparation of the manuscript. We also wish to acknowledge all our colleagues who helped with the research on the antigenic analyses and the related problems of yeasts.

Appendix I

Antigenic structures, coenzyme Q and $(G+C)$ mole % DNA of various yeast species[a]

Group	Species	Coenzyme Q	(G + C) mole % DNA	Antigenic structure
	P. vini	Q-9	36.8	1,4,9,14
	P. etchellsii	Q-9	38.5	1,4,9(14)
	T. famata	Q-9	36.8–37.3	1,4(9)(14)
	D. kloeckeri	Q-9	36.8	1,4(9)(14)
	D. marama	Q-9	36.6	1,4,9(14)
	D. hansenii	Q-9	36.6–37.3	1,4(9)14
V	D. coudertii	Q-9	37.4[b]	1,4,9,14
	Schw. occidentalis	Q-9	34.6	1,4,9,14,j
	D. subglobosus	Q-9	36.6–37.1	1,4,9,14,33
	Cit. matritensis	Q-9	44.4	1,4,9(14)
	P. haplophila	Q-9	39.0	1,4,9
	C. guilliermondii	Q-9	44.1–44.4	1,4,9
	C. tenuis	Q-9	43.2	1,5,9
	C. stellatoidea	Q-9	34.6	1,4,5,10,32
	C. sake	Q-9	38.8	1,4,5,6
	C. tropicalis	Q-9	33.9–34.9	1,4,5,6
	C. subtropicalis	Q-9	36.3–36.8	1,4(5)(6)
	C. claussenii	Q-9	35.4	1,4,5,6(7)
	C. albicans type A	Q-9	34.9	1,4,5,6(7)
	C. albicans type B	Q-9	34.9	1,4,5(7)13b
	C. intermedia	Q-9	43.7–44.4	1,4,5,6,24,38
I	T. glabrata	Q-6	38.5–39.5	1,4,6,10,34,k
	S. unisporus	Q-6	32.4–32.7	1,4,5,6(10)23
	S. bisporus	Q-6	43.7–44.1	1,4,10,26
	S. exiguus	Q-6	33.2–33.7	1,4(5)6,10,26,32
	S. dairensis	Q-6	37	1,4(5)6,10,26,32
	T. holmii	Q-6	34.1	1(4)(5)6,10,26,32
	Kl. africana	Q-6	37.6–38.0	1,5,6(7)40
	Kl. corticis	Q-6	37.3–37.8	1,5,6(7)40
	Hs. vineae	Q-6	37.3–38.8	1,5,6(7)40
	Hs. osmophila	Q-6	37.3–37.8	1,5,6(7)40
	Kl. antillarum	Q-6	33.7	1,5,6(7)40
	Kl. javanica	Q-6	34.1–34.6	1,5,6(7)8,10,28,40
	Kl. lafarii	Q-6	33.9	1,5,6(7)8,10,28,40
	Kl. apiculata	Q-6	27.1–31.7	1,8,10,28, l
	Hs. uvarum	Q-6	31.2–31.5	1,8,10,28, l
	Kl. japonica	Q-6	27.1	1,8,10,28
	Hs. valbyensis	Q-6	26.8–31.7	1,8,10,28, l
	K. marxianus	Q-6	41.0	1,8,10,28,31, a
	C. macedoniensis	Q-6	39.8	1,8,10,28,31 (a)
	K. fragilis	Q-6	40.0–40.7	1,8,10,28,31, a
	C. pseudotropicalis	Q-6	39.0–40.5	1,8(10)28,31, a
	S. fructuum	Q-6	39.3–39.5	1,8,10,28,31,14
II	S. microellipsodes	Q-6	39.3–39.5	1,8,10,28,31,14
	S. bailii	Q-6	40.5	1,8(10)28,32
	S. rouxii	Q-6	40.0	1(8)(10)28,32

Group	Species	Coenzyme Q	(G+C) mole % DNA	Antigenic structure
	S. carlsbergensis	Q-6	38.5–39.5	1(8)(10)28,32
	S. pastorianus	Q-6	39.5	1,8,(10)28,32
	K. veronae	Q-6	44.4–45.9	1,8,10,28(32)
	T. sphaerica	Q-6	39.0	1,10,28,30
	S. italicus	Q-6	38.3–39.3	1,8,10(18)28,31
	S. chevalieri	Q-6	38.8–39.0	1(8)(10)18,31
	S. heterogenicus	Q-6	39.0	1,10,18,31
	S. diastaticus	Q-6	38.8–39.8	1,10,18,31
	S. oviformis	Q-6	39.3–40.0	1,10,18,31
II	S. bayanus	Q-6	38.8–40.2	1,10,18,31
	S. logos	Q-6	39.0–39.3	1,10,18,31
	S. uvarum	Q-6	39.8	1,10,18,31
	S. florentinus	Q-6	40.5	1,10,18,31
	S. cerevisiae	Q-6	38.8–40.2	1,10,18,31, *a, e*
	S. ellipsoideus	Q-6	40.2	1,10,18,31, *a, e*
	C. robusta	Q-6	39.0	1,10,18,31, *a, e*
	S. willianus	Q-6	39.0	1,10,18,31 (*a*)(*e*)
	S. steineri	Q-6	39.5	1,10,18,31, *e*
	K. lactis	Q-6	39.3–40.0	1,10,18,31,42
	K. drosophilarum	Q-6	40.0	1,18,31
	W. fluorescens	Q-9	35.0	1,10,18,31,43
	T. gropengiesseri	Q-9	57.1	1,10,35,46
	T. magnoliae	Q-9	60.0	1,10,35
	T. anomala	Q-9	45.4–46.6	1,10,35,36
	T. versatilis	Q-9	45.6	1(10)35,36
	S. rosei	Q-6	42.9–43.2	1,4,24
	S. delbrueckii	Q-6	42.9–43.2	1,4,24
	T. cambresieri	Q-6	43.7	1,4,24
	S. vafer	Q-6	43.8[b]	1,4,24
	S. saitoanus	Q-6	42.9–43.2	1,4,24
VII	S. inconspicuus	Q-6	43.7[b]	1,4,24
	S. fermentati	Q-6	42.9–43.4	1,4,24
	T. colliculosa	Q-6	42.7	1,4,24
	S. pretoriensis	Q-6	45.6	1,24,27, *i*
	S. kloeckerianus	Q-6	42.7–42.9	1,10,24,27, *i*
	H. silvicola	Q-7	34.1–35.1	1(15)(16)20
	H. subpelliculosa	Q-7	33.9	1,15(16)20
	H. schneggii	Q-7	35.6–36.3	1(15)16(17)20
	H. anomala	Q-7	35.9–36.6	1,15,16(17)20
	C. pelliculosa	Q-7	36.1	1,15,16,20
	H. ciferrii	Q-7	32.2	1,15,16(17)20
	H. fabianii	Q-7	44.4–45.6	1(15)(16)17,20(21)
	H. petersonii	Q-7	43.9–44.1	1,15(16)17,20,21, *p*
	H. mrakii	Q-7	42.0	1,15,16(17)20,21
	H. saturnus	Q-7	42.4–43.7	1(15)16(17)20,21
VI	H. saturnus subsp. subsufficiens	Q-7	42.9–43.2	1,15,16(17)20,21
	H. bimundalis	Q-7	40.0–40.2	1(15)(16)17,20,21
	P. bovis	Q-7	39.8	1,15,16,20,21

Group	Species	Coenzyme Q	(G+C) mole % DNA	Antigenic structure
	H. californica	Q-7	43.9–44.1	1(15)16(17)20,21,22
	H. beijerinckii	Q-7	42.7	1,15,16(17)20,21,22
	H. jadinii	Q-7	43.2	1,(16)17, (c)
VI	C. utilis	Q-7	44.4–45.4	1,(16)17, c
	H. minuta	Q-7	46.8–47.3	1,5,45
	H. canadensis	Q-7	39.3–40.0	1,5,17,18,25,31,48
	H. wingei	Q-7	39.0	1,5,17,18,25,31,48
	H. beckii	Q-7	36.3–36.8	1,5,17,18,25,31,48
	C. krusei	Q-7	38.8–39.3	1,5(11) b
	P. krusei	Q-7	38.5	1,5,11, b
	P. fluxuum	Q-7	32.2	1,5,11, m
	T. pinus	Q-7	37.3	1,5,11(17)25
	C. melinii	Q-7	39.5	1,5,10(17)25
	P. toletana	Q-7	39.3	1,5,11,17,49
	P. fermentans	Q-7	42.2	1,5,11
	H. angusta	Q-7	48.0	1(5)11(17)
	C. trigonopsoides	Q-7	35.6–36.6	1,5,11,12
	C. mycoderma	Q-7	38.5	1,5,11,12
III	P. kudriavzevii	Q-7	38.5	1,5,11,12, b
	P. terricola	Q-7	36.6–36.8	1,5,11,12,18,31
	P. pijiperi	Q-7	40.7	1,5,11,12,20,21,37, c
	C. solani	Q-7	40	1,5,11,12,20,21,37, c
	P. membranaefaciens	Q-7	41.5–42.4	1,5,11,12
	M. reukaufii	NT	44.2	1(5)(11)(12) b, f
	T. inconspicua	Q-8	35.6	1,5(11)(12)
	H. capsulata	Q-8	46.8–47.1	1,4,5,11,12,47
	P. pastoris	Q-8	40.2–41.0	1,4,17,24,25
	P. farinosa	Q-9	38.8–39.5	1,4,29
	C. catenulata	Q-9	53.2–53.4	1,5,11, b
	C. brumptii	Q-9	54.4	1,19
	C. rugosa	Q-9	50.2–50.7	1,4,11,19, g
	H. holstii	Q-8	36.8	1(13)(15)
	T. ernobii	Q-8	36.1	1,5,13(15) c
	D. castellii	Q-9	34.4	1(5)13(15) a
IV	P. pseudopolymorpha	Q-9	35.7	1,4,5,13,15, c
	C. parapsilosis	Q-9	39.3–40.0	1,5,13,13b(15) c
	M. pulcherrima	Q-9	45.4–47.1	1,5,13(15) d
	C. zeylanoides	Q-9	54.7–55.9	1,4,13,17,(a)c,h
Rhodotorula group	Sp. salmonicolor	Q-10	63.0–64.5	1,9,8
	Sp. odorus	Q-10	65.0	1,9,8
	Sp. pararoseus	Q-10	51.5–60.5	1,11
	Rh. glutinis	Q-10	60.2–67	1,2,5,8
	Rh. aurantiaca	Q-10	55.4	1,2,5,8
	Rh. mucilaginosa	Q-10	60.2–60.7	1,3,8
	Rh. rubra	Q-10	60.0–60.7	1,2(3)8
	Sp. gracilis	Q-10		(4)(6)
	Rh. pallida	Q-10	50.0	4,6,10
	Rh. minuta	Q-10	51.0	4,7

Group	Species	Coenzyme Q	(G+C) mole % DNA	Antigenic structure
	Cr. neoformans type A	Q-10	49.0–49.8	1,2,3,7
	Cr. albidus subsp. *albidus*	Q-10	51.5–53.7	1,2,3,7
	Cr. neoformans type B	Q-10		1,2,4,5
Cryptococcus group	*Cr. neoformans* type C			1,4,6
	Cr. albidus subsp. *aerius*	Q-10	54.9	1,4,6
	Cr. neoformans type D			1,2,3,8
	Cr. humicola	Q-10	58.8–63.2	1,2,3,8
	C. curvata	Q-10	59.8–60.7	1,2,3,7,8
	Cr. albidus subsp. *diffluens*	Q-10	54.1–55.1	1,2,3
	Cr. luteolus	Q-10	51.5–59.5	1,2,3
	Cr. terreus	Q-10	54.9	1,2,3
Schiz. group	*Schiz. pombe*	Q-10	42	1,3
	C. lipolytica	Q-9	49.5–50.2	1,2,4

[a] Antigens 2 and 3 being relatively common are omitted or included in antigen 1 and antigen 14 is also omitted except for in Group V species. (G+C) mole % DNA data are from Komagata and Nakase and Coenzyme Q data are from Yamada *et al.* Alphabetical letters depict thermolabile antigen. Strains used in each examination are not necessarily the same, even if they belong to the same species.

[b] From Phaff (1970) and Phaff and Ahearn (1970).

References

Ahearn, D. G., Meyer, S. A., Mitchell, G., Nicholson, M. A. and Ibrahim, A. I. (1977). *J. Clin. Microbiol.* **5**, 494–496.

Ainsworth, G. C. (1963). "Dictionary of the Fungi", 5th edn. Commonwealth Mycological Institute, Kew.

Alexopoulos, C. J. (1962). "Introductory Mycology", 2nd edn. Wiley, New York.

Almon, L. and Stovall, W. D. (1934). *J. Infect. Dis.* **55**, 12–25.

Analysis Committee of the Institute of Brewing (1973). *J. Inst. Brew., London* **79**, 134–136.

Arx, J. A. von, Rodrigues de Miranda, L., Smith, M. Th. and Yarrow, D. (1977). *In* "Studies in Mycology", Vol. 14, pp. 1–42. Centraalbureu voor Schimmelcultures, Baarn.

Askonas, B. A., Farthing, C. P. and Humphery, J. H. (1960). *Immunology* **3**, 336–351.

Ballou, C. E., Lipke, P. N. and Raschke, W. C. (1974). *J. Bacteriol.* **117**, 461–467.

Barnett, J. A. (1960). *Nature (London)* **186**, 449–451.

Barnett, J. A. (1961). *Nature (London)* **189**, 76.

Barnett J. A. and Pankhurst, R. J. (1974). "A New Key to the Yeasts", pp. 15–23, 71–125. North-Holland Publ., Amsterdam.

Benham, R. W. (1931). *J. Infect. Dis.* **49**, 183–215.

Benham, R. W. (1935). *J. Infect. Dis.* **57**, 255–274.

Biguet, J., Havez, R., Tran Van Ky, P. and Degaey, R. (1961). *Ann. Inst. Pasteur* **100**, 13–24.
Biguet, J., Tran Van Ky, P. and Andrieu, S. (1962). *Mycopathol. Mycol. Appl.* **17**, 239–254.
Biguet, J., Tran Van Ky, P., Andrieu, S. and Fruit, J. (1964). *Ann. Inst. Pasteur* **106**, 72–97.
Biguet, J., Tran Van Ky, P., Andrieu, S. and Fruit, J. (1965a). *Mycopathol. Mycol. Appl.* **26**, 241–256.
Biguet, J., Tran Van Ky, P., Andrieu, S. and Degaey, R. (1965b). *Sabouraudia* **4**, 148–159.
Björndal, H., Hellerqvist, C. G., Lindberg, B. and Svensson, S. (1970). *Angew. Chem. Int. Ed. Engl.* **9**, 610–619.
Buesching, W. J., Kurck, K. and Roberts, G. D. (1979). *J. Clin. Microbiol.* **9**, 565–569.
Campbell, I. (1968a). *In* "Europ. Brew. Conv. Proc. XI Cong., Madrid" pp. 145–154. Elsevier Publ. Co., Amsterdam.
Campbell, I. (1968b). *J. Appl. Bacteriol.* **31**, 515–524.
Campbell, I. (1970). *J. Gen. Microbiol.* **63**, 189–198.
Campbell, I. (1971). *J. Appl. Bacteriol.* **34**, 237–242.
Campbell, I. and Allan, A. M. (1964). *J. Inst. Brew., London* **70**, 316–320.
Campbell, I. and Brudynski, A. (1966). *J. Inst. Brew., London* **72**, 556–560.
Campbell, I., Robson, F. O. and Hough, J. S. (1968). *J. Inst. Brew., London* **74**, 360–364.
Cherry, W. B. (1980). *In* "Manual of Clinical Microbiology" (E. H. Lennette, A. Balows, W. J. Hausler, Jr and J. P. Truant, Eds), 3rd edn, pp. 501–508. American Society for Microbiology, Washington D.C.
Coons, A. H. and Kaplan, M. (1950). *J. Exp. Med.* **91**, 1–13.
Edwards, P. R. and Ewing, W. H. (1962). "Identification of Enterobacteriaceae", Burgess, Minneapolis, Minnesota.
Emmons, C. W. (1955). *Am. J. Hyg.* **62**, 227–232.
Emmons, C. W. (1972). *In* "Proc. II Int. Specialized Symp. on Yeasts, Tokyo", pp. 3–9. University of Tokyo Press, Tokyo.
Emmons, C. W., Binford, C. H., Utz, J. P. and Kwon-Chung (1977). "Medical Mycology", 3rd edn. Lea & Febiger, Philadelphia, Pennsylvania.
Evans, E. E. (1949). *Proc. Soc. Exp. Biol. Med.* **71**, 644–646.
Evans, E. E. (1950). *J. Immunol.* **64**, 423–430.
Evans, E. E. and Kessel, J. F. (1951). *J. Immunol.* **67**, 109–114.
Everand, W. C., Marshall, J. D., Silverstein, A. M., Johnson, F. B., Iverson, L. and Winslow, D. J. (1957). *Am. J. Pathol.* **33**, 616–617.
Fahey, J. L. (1960). *Science* **131**, 500–501.
Fahey, J. L. (1961). *J. Exp. Med.* **114**, 399–413.
Fahey, J. L. and Horbett, A. P. (1959). *J. Biol. Chem.* **234**, 2645–2651.
Fukazawa, Y. and Tsuchiya, T. (1966). *In* "Proc. II Symp. on Yeasts, Bratislava", pp. 553–557. Slovak Academy of Sciences, Bratislava.
Fukazawa, Y. and Tsuchiya, T. (1969). *Antonie van Leeuwenhoek* **35**, E 7–8.
Fukazawa, Y., Shinoda, T. and Tsuchiya, T. (1968a). *J. Bacteriol.* **95**, 754–763.
Fukazawa, Y., Elinov, N. P., Shinoda, T. and Tsuchiya, T. (1968b). *Jpn. J. Microbiol.* **12**, 283–292.
Fukazawa, Y., Shinoda, T. and Tsuchiya, T. (1969). *J. Bacteriol.* **98**, 1128–1134.
Fukazawa, Y., Shinoda, T., Yomoda, T. and Tsuchiya, T. (1970). *Infect. Immun.* **1**, 219–225.

Fukazawa, Y., Shinoda, T., Nishikawa, A., Kagaya, K. and Yonezawa, Y. (1972a). *In* "Proc. II Int. Specialized Symp. on Yeasts, Tokyo", pp. 213–218. University of Tokyo Press, Tokyo.

Fukazawa, Y., Kagaya, K., Shinoda, T., Nishikawa, A. and Shimomura, H. (1972b). *In* "Proc. II Int. Specialized Symp. on Yeasts, Tokyo", pp. 207–212. University of Tokyo Press, Tokyo.

Fukazawa, Y., Nakase, T., Shinoda, T., Kagaya, K. and Nishikawa, A. (1974). *In* "Proc. IV Int. Symp. on Yeasts, Vienna", pp. 273–274.

Fukazawa, Y., Nakase, T., Shinoda, T., Nishikawa, A., Kagaya, K. and Tsuchiya, T. (1975). *Int. J. Syst. Bacteriol.* **25**, 304–314.

Fukazawa, Y., Shinoda, T., Nishikawa, A. and Nakase, T. (1980a). *Int. J. Syst. Bacteriol.* **30**, 196–205.

Fukazawa, Y., Nishikawa, A., Suzuki, M. and Shinoda, T. (1980b). *In* "Medical Mycology" (H. J. Preusser, Ed.), Proc. Mycol. Symp. of XII Int. Congr. Microbiol., Munich, 1978, Zbl. Bakt., Suppl. 8, pp. 127–136. Fischer Verlag, Stuttgart.

Fuson, G. B., Price, C. W. and Phaff, H. J. (1979). *Int. J. Syst. Bacteriol.* **29**, 64–69.

Gentles, J. C. and La Touche, C. J. (1969). *In* "The Yeasts" (A. H. Rose and J. S. Harison, Eds), pp. 107–182. Academic Press, London and New York.

Gerencser, M. A. (1979). *In* "Methods in Microbiology" (T. Bergan and J. R. Norris, Eds), Vol. 13, pp. 287–321. Academic Press, London and New York.

Goldmann, M. (1968). "Fluorescent Antibody Methods". Academic Press, London and New York.

Gordon, M. A. (1958a). *Proc. Soc. Exp. Biol. Med.* **97**, 694–698.

Gordon, M. A. (1958b). *J. Invest. Dermatol.* **31**, 123–125.

Gordon, M. A. (1962). *In* "Fungi and Fungous Diseases" (G. Dalldorf, Ed.), pp. 207–219. C. Thomas, Springfield, Illinois.

Gordon, M. A., Elliott, J. C. and Hawkins, T. W. (1967). *Sabouraudia* **5**, 323–328.

Gorin, P. A. J. and Spencer, J. F. T. (1968). *Can. J. Chem.* **46**, 2299–2304.

Gorin, P. A. J. and Spencer, J. F. T. (1970). *In* "Advances in Applied Microbiology" (D. Perlman, Ed.), Vol. 13, pp. 25–89. Academic Press, New York and London.

Gorin, P. A. J. and Spencer, J. F. T. (1972). *Can. J. Microbiol.* **18**, 1709–1715.

Haikara, A. and Enari, F. M. (1975). "Europ. Brew. Conv. Proc.", pp. 363–375.

Hakomori, S. (1964). *J. Biochem. (Tokyo)* **55**, 205–208.

Hart, P. D., Russell, E. Jr and Remington, J. S. (1969). *J. Infect. Dis.* **120**, 169–191.

Hasenclever, H. F. and Mitchell, W. O. (1960). *J. Bacteriol.* **79**, 677–681.

Hasenclever, H. F. and Mitchell, W. O. (1961a). *J. Bacteriol.* **82**, 570–573.

Hasenclever, H. F. and Mitchell, W. O. (1961b). *J. Bacteriol.* **82**, 574–577.

Hasenclever, H. F. and Mitchell, W. O. (1961c). *J. Bacteriol.* **82**, 578–581.

Hebert, G. A., Pittman, B. and Cherry, W. B. (1967). *J. Immunol.* **98**, 1204–1212.

Heide, K. and Schwick, H. G. (1978). *In* "Handbook of Experimental Immunology" (D. M. Weir, Ed.). Lippincott, Philadelphia, Pennsylvania.

Hellerqvist, C. G., Lindberg, B. and Svensson, S. (1968). *Carbohydr. Res.* **8**, 43–55.

Hines, L. E. (1924). *J. Infect. Dis.* **34**, 529–535.

Holzschu, D. L., Pressley, H. L., Miranda, M. and Phaff, H. J. (1979). *J. Clin. Microbiol.* **10**, 202–205.

Hořejši, J. and Smetana, R. (1956). *Acta Med. Scand.* **155**, 65–70.

Hurley, R. (1967). *Rev. Med. Vet. Mycol.* **6**, 159–176.

Ikeda, R., Shinoda, T., Fukazawa, Y. and Kaufman, L. (1982). *J. Clin. Microbiol.* **16**, 22–29.

Inoue, S., Kobayashi, Y., Yoshizaki, E., Fukunaga, M., Kawabata, S. and Kamiki, T. (1973). *Jpn. J. Med. Technol.* **22**, 350 (in Japanese).

Jonsen, J., Thjøtta, T. and Rasch, S. (1953). *Acta Pathol. Microbiol. Scand.* **33**, 86–91.

Kanno, T. and Suzuki, J. (1972). *In* "Proc. II Int. Specialized Symp. on Yeasts, Tokyo", pp. 289–293. University of Tokyo Press, Tokyo.

Kato, S. (1967). *Bull. Brew. Sci.* **13**, 19–24.

Kauffmann, F. (1954). "Enterobacteraceae", 2nd edn. Munksgaard, Copenhagen.

Kauffmann, F. (1972). "Serological Diagnosis of *Salmonella* Species, Kauffmann-White-Schema", pp. 29–37. Munksgaard, Copenhagen.

Kaufman, L. (1973). *Am. Clin. Lab. Sci.* **3**, 141–146.

Kaufman, L. (1980a). *In* "Manual of Clinical Microbiology" (E. H. Lennette, A. Balows, W. J. Hausler, Jr and J. P. Truant, Eds), 3rd edn, pp. 628–646. American Society for Microbiology, Washington, D.C.

Kaufman, L. (1980b). *In* "Manual of Clinical Immunology" (N. R. Rose and H. Friedman, Eds), 2nd edn, pp. 553–572. American Society for Microbiology, Washington, D.C.

Kaufman, L. and Kaplan, W. (1963). *J. Bacteriol.* **85**, 986–991.

Kawamura, A., Jr (1977). "Fluorescent Antibody Techniques and Their Applications". University Park Press, Baltimore, Maryland.

Kemp, G. and Solotorovsky, M. (1964). *J. Immunol.* **93**, 305–314.

Kesten, M. D. and Mott, E. (1932). *J. Infect. Dis.* **50**, 459–465.

Kocourek, J. and Ballou, C. E. (1969). *J. Bacteriol.* **100**, 1175–1181.

Kreger-van Rij, N. J. W. (1969). *In* "The Yeasts" (A. H. Rose and J. S. Harison, Eds), Vol. 1, pp. 5–78. Academic Press, London and New York.

Kreger-van Rij, N. J. W. (1970). *In* "The Yeasts" (J. Lodder, Ed.), pp. 129–156, 455–554, 1224–1228. North-Holland Publ., Amsterdam.

Kuroshima, T., Kodaira, S. and Tsuchiya, T. (1976). *Jpn. J. Microbiol.* **20**, 485–492.

Kuttin, E. S. and Beemer, A. M. (1975). *In* "Proc. VI Cong. Int. Soc. Human and Animal Mycol., Tokyo", pp. 151–155. University of Tokyo Press, Tokyo.

Lamb, J. H. and Lamb, M. L. (1935). *J. Infect. Dis.* **56**, 8–20.

Lodder, J. (1970). "The Yeasts", 2nd edn, pp. 1–33, 114–120. North-Holland Publ., Amsterdam.

Lodder, J. and Kreger-van Rij, N. J. W. (1952). "The Yeasts", North-Holland Publ., Amsterdam.

Lodder, J., Slooff, W. Ch. and Kreger-van Rij, N. J. W. (1958). *In* "The Chemistry and Biology of Yeasts" (A. H. Cook, Ed.), pp. 1–66. Academic Press, New York and London.

LoSpalluto, J., Miller, W., Dorward, B. and Fink, C. W. (1962). *J. Clin. Invest.* **41**, 1415–1421.

Mackenzie, D. W. R. (1961). *Sabouraudia* **1**, 8–13.

Martin, D. S. (1942). *Am. J. Trop. Med.* **22**, 295–303.

Mitchell, W. O. and Hasenclever, H. F. (1970). *Infect. Immun.* **1**, 61–63.

Müller, H. L. (1966). *Zentralbl. Bakteriol. Parasitenkd. Infektionskr. Abt. 1* **200**, 266–273.

Müller, H. L. and Bürger, H. (1965). *Zentralbl. Bakteriol. Parasitenkd. Infektionskr. Abt. 1* **198**, 240–242.

Müller, H. L. and Hirsch, M. (1967). *Zentralbl. Bakteriol. Parasitenkd. Infektionskr. Abt. 1* **202**, 247–256.

Müller, H. L., Siegert, R. and Pachaly, D. (1964). *Zentralbl. Bakteriol. Parasitenkd. Infektionskr. Abt. 1* **193**, 117–122.

Nakase, T. (1972). *In* "Proc. IV Intern. Ferment. Symp.: Ferment. Technol. Today", pp. 785–791. Society of Fermentation Technology, Osaka.

Nakase, T. and Komagata, K. (1966). *J. Gen. Appl. Microbiol.* **12**, 347–352.

Nakase, T. and Komagata, K. (1968). *J. Gen. Appl. Microbiol.* **14**, 345–357.

Nakase, T. and Komagata, K. (1969). *J. Gen. Appl. Microbiol.* **15**, 85–95.

Nakase, T. and Komagata, K. (1970a). *J. Gen. Appl. Microbiol.* **16**, 241–250.

Nakase, T. and Komagata, K. (1970b). *J. Gen. Appl. Microbiol.* **16**, 511–521.

Nakase, T. and Komagata, K. (1971a). *J. Gen. Appl. Microbiol.* **17**, 43–50.

Nakase, T. and Komagata, K. (1971b). *J. Gen. Appl. Microbiol.* **17**, 77 84.

Nakase, T. and Komagata, K. (1971c). *J. Gen. Appl. Microbiol.* **17**, 121–130.

Nakase, T. and Komagata, K. (1971d). *J. Gen. Appl. Microbiol.* **17**, 161–166.

Nakase, T. and Komagata, K. (1971e). *J. Gen. Appl. Microbiol.* **17**, 227–238.

Nakase, T. and Komagata, K. (1971f). *J. Gen. Appl. Microbiol.* **17**, 259–279.

Nakase, T. and Komagata, K. (1971g). *J. Gen. Appl. Microbiol.* **17**, 363–369.

Nakase, T. and Komagata, K. (1972). *In* "Proc. I Int. Symp. on Yeasts, Smolenice", pp. 399–411. Slovak Academy of Sciences, Bratislava.

Nakase, T., Fukazawa, Y. and Tsuchiya, T. (1972). *J. Gen. Appl. Microbiol.* **18**, 349–363.

Narin, R. C. (1976). "Fluorescent Protein Tracing", 4th edn. Churchill Livingstone, New York.

Nishikawa, N., Kohgo, M. and Karakawa, T. (1979a). *J. Ferment. Technol.* **57**, 474–477.

Nishikawa, N., Kohgo, M. and Karakawa, T. (1979b). *J. Ferment. Technol.* **57**, 364–368.

Oakley, C. L. (1971). *In* "Methods in Microbiology" (J. R. Norris and D. W. Ribbons, Eds), Vol. 5A, pp. 173–218. Academic Press, London and New York.

Ohtsuka, Y. and Tsuchiya, T. (1972). *In* "Proc. II Int. Specialized Symp. on Yeasts, Tokyo", pp. 219–228. University of Tokyo Press, Tokyo.

Palmer, D. F., Kaufman, L., Kaplan, W. and Cavallaro, J. J. (1977). "Serodiagnosis of Mycotic Diseases", pp. 44–106, 143–153. Thomas, Springfield, Illinois.

Phaff, H. J. (1970). *In* "The Yeasts" (J. Lodder, Ed.), pp. 209–225, 756–766, 1146–1160. North-Holland Publ., Amsterdam.

Phaff, H. J. and Ahearn, D. G. (1970). *In* "The Yeasts" (J. Lodder, Ed.), pp. 1187–1223. North-Holland Publ., Amsterdam.

Phaff, H. J. and Spencer, J. F. T. (1966). *In* "Proc. II Int. Symp. on Yeasts, Bratislava", pp. 59–65. Slovak Academy of Sciences, Bratislava.

Pospíšil, L. (1959). *Dermatologica* **118**, 65–73.

Preece, T. F. (1971). *In* "Methods in Microbiology" (J. R. Norris and D. W. Ribbons, Eds), Vol. 4, pp. 599–607. Academic Press, London and New York.

Price, C. W., Fuson, C. B. and Phaff, H. J. (1978). *Microbiol. Rev.* **42**, 161–193.

Raschke, W. C. and Ballou, C. E. (1972). *Biochemistry* **11**, 3807–3816.

Richards, M. and Cowland, T. W. (1967). *J. Inst. Brew., London* **73**, 552–558.

Rieth, H. (1958). *Arch. Klin. Exp. Dermatol.* **207**, 413–430.

Rieth, H. (1978). Mykosen, Diagnose und Therapie, Verlags GmbH, Frankfurt.

Rieth, H.and Schönfeld, J. (1959). *Arch. Klin. Exp. Dermatol.* **208**, 343–361.

Riggs, J. L., Seiwald, R. J., Burckhalter, J. H., Downs, C. H. and Metcalf, T. G. (1958). *Am. J. Clin. Pathol.* **34**, 1081–1097.

Rogers, T. J. and Balish, E. (1980). *Microbiol. Rev.* **44**, 660–682.

Seeliger, H. P. R. (1956). *J. Bacteriol.* **72**, 127–131.

Seeliger, H. P. R. (1957). *Zentralbl. Bakteriol. Parasitenkd. Infektionskr. Abt. 1* **167**, 396–408.

Seeliger, H. P. R. (1958). "Mykologische Serodiagnostik", Johann Ambrosius Barth, Leipzig.

Seeliger, H. P. R. (1959). *Ergeb. Mikrobiol. Immunitaetsforsch. Exp. Ther.* **32**, 23–72.

Seeliger, H. P. R. (1960). *Trans. Br. Mycol. Soc.* **43**, 543–555.

Seeliger, H. P. R. (1962a). In "Fungi and Fungous Diseases" (G. Dalldorf, Ed.), pp. 158–186. Thomas, Springfield, Illinois.

Seeliger, H. P. R. (1962b). *Zentralbl. Bakteriol. Parasitenkd. Infektionskr. Abt. 1* **184**, 203–226.

Seeliger, H. P. R. (1968). In "The Fungi" (G. C. Ainsworth and A. S. Sussman, Eds), Vol. III, pp. 597–624. Academic Press, London and New York.

Seeliger, H. P. R. and Schröter, R. (1963). *Sabouraudia* **2**, 248–263.

Seeliger, H. P. R., Tomsikova, A. and Török, I. (1974). *Mykosen* **18**, 51–59, 119–134, 149–160.

Sen, K. and Komagata, K. (1979). *J. Gen. Appl. Microbiol.* **25**, 127–135.

Seneca, H., Peer, P. and Nally, R. (1963). *Nature (London)* **197**, 359–361.

Senju, O., Hosoya, J., Ishii, T. and Aoki, Y. (1978). *Jpn. J. Clin. Pathol.* **16**, 623–629 (in Japanese).

Seo, M. (1972). In "Proc. II Int. Specialized Symp. on Yeasts, Tokyo", pp. 277–281. University of Tokyo Press, Tokyo.

Shinoda, T. (1972). *Jpn. J. Bacteriol.* **27**, 27–33 (in Japanese).

Shinoda, T., Ikeda, R., Fukazawa, Y. and Sato, I. (1978). *Jpn. J. Med. Mycol.* **19**, 52 (in Japanese).

Shinoda, T., Ikeda, R., Nishikawa, A. and Fukazawa, Y. (1980). *Jpn. J. Med. Mycol.* **21**, 230–238.

Shinoda, T., Kaufman, L. and Padhye, A. A. (1981). *J. Clin. Microbiol.* **13**, 513–518.

Siluyanova, N. A. (1964). "Mycological Investigation" (P. N. Kashkin, Ed.), Vol. 2, pp. 71–76.

Silva-Hutner, M. and Cooper, B. H. (1980). In "Manual of Clinical Microbiology" (E. H. Lennette, A. Balows, W. J. Hausler, Jr and J. P. Truant, Eds), 3rd edn, pp. 562–576. American Society for Microbiology, Washington, D.C.

Skinner, C. E. and Fletcher, D. W. (1960). *Bacteriol. Rev.* **24**, 397–416.

Sober, H. A., Gutter, F. J., Wyckoff, M. M. and Peterson, E. A. (1956). *J. Am. Chem. Soc.* **78**, 756–763.

Spencer, J. F. T. and Gorin, P. A. J. (1968). *J. Bacteriol.* **96**, 180–183.

Spencer, J. F. T. and Gorin, P. A. J. (1969a). *Antonie van Leeuwenhoek* **35**, 33–44.

Spencer, J. F. T. and Gorin, P. A. J. (1969b). *Can. J. Microbiol.* **15**, 375–382.

Spencer, J. F. T. and Gorin, P. A. J. (1969c). *Antonie van Leeuwenhoek* **35**, 361–378.

Spencer, J. F. T. and Gorin, P. A. J. (1970). *Antonie van Leeuwenhoek* **36**, 135–141.

Spencer, J. F. T. and Gorin, P. A. J. (1971). *Antonie van Leeuwenhoek* **37**, 75–88.

Spencer, J. F. T., Gorin, P. A. J. and Wickerham, L. J. (1970a). *Can. J. Microbiol.* **16**, 445–448.

Spencer, J. F. T., Gorin, P. A. J. and Tulloch, A. P. (1970b). *Antonie van Leeuwenhoek* **36**, 129–133.

Stenderup, A. and Bak, A. L. (1968). *J. Gen. Microbiol.* **52**, 231–236.

Stenderup, A. and Pedersen, G. T. (1962). *Acta Pathol. Microbiol. Scand.* **54**, 462–472.

Steros, P. and Taliaferro, W. H. (1959). *J. Infect. Dis.* **105**, 105–118.

Stone, K. and Garrod, L. P. (1931). *J. Pathol. Bacteriol.* **34**, 429–436.
Storck, R. and Alexopoulos, C. J. (1970). *Bacteriol. Rev.* **34**, 126–134.
Suzuki, S., Sunayama, H. and Saito, T. (1968). *Jpn. J. Microbiol.* **12**, 19–24.
Sweet, C. E. and Kaufman, L. (1970). *Appl. Microbiol.* **19**, 830–836.
Taguchi, M., Yamazaki, M. and Tsuchiya, T. (1972). *In* "Proc. II Int. Specialized Symp. on Yeasts, Tokyo", pp. 282–288. University of Tokyo Press, Tokyo.
Taguchi, M., Tsukiji, M. and Tsuchiya, T. (1979). *Sabouraudia* **17**, 185–191.
Treffers, H. P., Moore, D. H. and Heiderberger, M. (1942). *J. Exp. Med.* **75**, 135–150.
Tsuchiya, T. (1959). *In* "Proc. VI Int. Cong. Trop. Med. Malaria, Lisbon", Vol. 4, pp. 691–695. Impressa Porpuguesa, Rua Formosa, 108–116, Porto, Portugal.
Tsuchiya, T. and Fukazawa, Y. (1966). *In* "Proc. II Symp. on Yeasts, Bratislava", pp. 545–551. Slovak Academy of Sciences, Bratislava.
Tsuchiya, T. and Imai, M. (1968). *In* "Taxonomy of Microorganisms" (T. Uemura, Ed.), pp. 147–158. University of Tokyo Press, Tokyo.
Tsuchiya, T. and Fukazawa, Y. (1969). *Antonie van Leeuwenhoek* **35**, A 17–18.
Tsuchiya, T. and Taguchi, M. (1980). *In* "Medical Mycology" (H. J. Preusser, Ed.), Proc. Mycol. Symp. of XII Int. Congr. Microbiol., Munich, 1978, Zbl. Bakt., Suppl. 8, pp. 137–146. Fischer Verlag, Stuttgart.
Tsuchiya, T., Iwahara, S., Miyasaki, F. and Fukazawa, Y. (1954). *Jpn. J. Exp. Med.* **24**, 95–104.
Tsuchiya, T., Miyasaki, F. and Fukazawa, Y. (1955a). *Jpn. J. Exp. Med.* **25**, 15–21.
Tsuchiya, T., Fukazawa, Y., Miyasaki, F. and Kawakita, S. (1955b). *Jpn. J. Exp. Med.* **25**, 75–83.
Tsuchiya, T., Kamijo, K., Fukazawa, Y., Kawakita, S. and Nishikawa, Y. (1956a). *Jpn. J. Med. Sci. Biol.* **9**, 103–112.
Tsuchiya, T., Kamijo, K., Fukazawa, Y., Miyasaki, F. and Kawakita, S. (1956b). *Yokohama Med. Bull.* **7**, 127–132.
Tsuchiya, T., Kawakita, S., Hayashi, S., Sato, I. and Takahashi, M. (1957a). *Yokohama Med. Bull.* **8**, 8–14.
Tsuchiya, T., Fukazawa, Y., Hayashi, S., Hayashi, J. and Doi, M. (1957b). *Jpn. J. Microbiol.* **1**, 125–131.
Tsuchiya, T., Fukazawa, Y., Hayashi, S., Amemiya, S. and Sano, Y. (1957c). *Jpn. J. Microbiol.* **1**, 205–212.
Tsuchiya, T., Fukazawa, Y., Amemiya, S., Yonezawa, M. and Suzuki, K. (1957d). *Yokohama Med. Bull.* **8**, 215–224.
Tsuchiya, T., Fukazawa, Y., Hayashi, J., Nishikawa, Y. and Doi, M. (1957e). *Jpn. J. Microbiol.* **1**, 339–346.
Tsuchiya, T., Fukazawa, Y. and Kawakita, S. (1958a). *J. Antibiot.* **11**, 45–47.
Tsuchiya, T., Fukazawa, Y., Sato, I., Amemiya, S. and Murata, T. (1958b). *Jpn. J. Exp. Med.* **28**, 105–114.
Tsuchiya, T., Fukazawa, Y., Sato, I., Kawakita, S., Yonezawa, M. and Yamase, Y. (1958c). *Yokohama Med. Bull.* **9**, 359–370.
Tsuchiya, T., Kawakita, S., Hayashi, J. and Kobayashi, T. (1958d). *Jpn. J. Exp. Med.* **28**, 345–352.
Tsuchiya, T., Fukazawa, Y., Sato, I., Kawakita, S. and Katsuya, J. (1958e). *Jpn. J. Exp. Med.* **28**, 413–418.
Tsuchiya, T., Fukazawa, Y. and Kawakita, S. (1959). *Mycopathol. Mycol. Appl.* **10**, 191–206.
Tsuchiya, T., Fukazawa, Y., Sano, Y., Shimura, Y. and Murata, T. (1960). *Jpn. J. Microbiol.* **4**, 61–71.

Tsuchiya, T., Fukazawa, Y. and Kawakita, S. (1961a). *In* "Studies on Candidiasis in Japan" (Osaka University, Ed.), pp. 34–46.

Tsuchiya, T., Fukazawa, Y. and Kawakita, S. (1961b). *Yokohama Med. Bull.* **12**, 184–191.

Tsuchiya, T., Fukazawa, Y. and Kawakita, S. (1961c). *Sabouraudia* **1**, 145–153.

Tsuchiya, T., Fukazawa, Y. and Yamase, Y. (1961d). *Jpn. J. Microbiol.* **5**, 417–429.

Tsuchiya, T., Mori, K. and Imai, M. (1962). *Jpn. J. Med. Mycol.* **3**, 183–184 (in Japanese).

Tsuchiya, T., Kawakita, S. and Udagawa, M. (1963a). *Sabouraudia* **2**, 209–214.

Tsuchiya, T., Kawakita, S., Imai, M., Kozakai, N., Tsuchiya, T., Joh, K. and Nakamura, M. (1963b). *Jpn. J. Med. Mycol.* **4**, 158–167 (in Japanese).

Tsuchiya, T., Kawakita, S. and Yamase, Y. (1964a). *Sabouraudia* **3**, 155–163.

Tsuchiya, T., Yamase, Y. and Udagawa, M. (1964b). *Jpn. J. Microbiol.* **8**, 21–30.

Tsuchiya, T., Fukazawa, Y. and Kawakita, S. (1965a). *Mycopathol. Mycol. Appl.* **26**, 1–15.

Tsuchiya, T., Fukazawa, Y., Kawakita, S., Imai, M. and Shinoda, T. (1965b). *Jpn. J. Microbiol.* **9**, 149–159.

Tsuchiya, T., Kawakita, S., Imai, M. and Miyagawa, K. (1966). *Jpn. J. Exp. Med.* **36**, 555–562.

Tsuchiya, T., Fukazawa, Y., Shinoda, T. and Okoshi, T. (1967a). *J. Immunol.* **98**, 1085–1092.

Tsuchiya, T., Fukazawa, Y., Shinoda, T. and Imai, M. (1967b). *Jpn. J. Exp. Med.* **37**, 285–290.

Tsuchiya, T., Fukazawa, Y. and Suzuki, K. (1969). *Jpn. J. Exp. Med.* **39**, 101–107.

Tsuchiya, T., Fukazawa, Y., Taguchi, M., Nakase, T. and Shinoda, T. (1974). *Mycopathol. Mycol. Appl.* **53**, 77–92.

Tsukiji, M. (1983). *Jpn. J. Med. Mycol.* **24**. 263–273.

Vanbreuseghem, R., de Vroey, Ch. and Takashio, M. (1978). "Practical Guide to Medical and Veterinary Mycology", 2nd edn. Masson Publ. USA, New York.

van der Walt, J. P. (1970). *In* "The Yeasts" (J. Lodder, Ed.), pp. 34–113, 157–165, 316–378, 403–407, 555–718, 772–775, 863–891. North-Holland Publ., Amsterdam.

van Uden, N. (1970). *In* "The Yeasts" (J. Lodder, Ed.), pp. 893–1087, 1235–1308. North-Holland Publ., Amsterdam.

van Uden, N., Carmo-Sousa, L. D. and Farinha, M. (1958). *J. Gen. Microbiol.* **19**, 435–445.

Vera, H. D. and Power, D. A. (1980). *In* "Manual of Clinical Microbiology" (E. H. Lennette, A. Balows, J. W. J. Hausler, Jr and J. P. Truant, Eds), 3rd edn, p. 1004. American Society for Microbiology, Washington, D.C.

Vörös-Felkai, G. and Novák, E. K. (1961). *Acta Microbiol. Acad. Sci. Hung.* **8**, 89–94.

Walters, L. S. and Thiselton, M. R. (1953). *J. Inst. Brew., London* **59**, 401–404.

Wickerham, L. J. (1951). *In* "Taxonomy of Yeasts". Technical Bull., No. 1029, pp. 1–56. U.S. Department of Agriculture, Washington, D.C.

Wickerham, L. J. (1957). *J. Am. Med Assoc.* **165**, 47–48.

Wickerham, L. J. (1970). *In* "The Yeasts" (J. Lodder, Ed.), pp. 121–127, 226–315, 448–454. North-Holland Publ., Amsterdam.

Wickerham, L. J. and Burton, K. A. (1962). *Bacteriol. Rev.* **26**, 382–397.

Wilson, D. E., Bennett, J. E. and Bailey (1968). *Proc. Soc. Exp. Med.* **127**, 820–823.

Winner, H. I. (1955). *J. Hyg.* **53**, 509–512.

Winner, H. I. and Hurley, R. (1964). *In* "*Candida albicans*". Little, Brown, Boston, Massachusetts.

Wood, B. T., Thomson, S. H. and Goldstein, G. (1965). *J. Immunol.* **95**, 225–229.

Yamada, Y. and Kondo, K. (1972a). "Proc. I Int. Symp. on Yeasts, Smolenice", pp. 363–373. Slovak Academy of Sciences.

Yamada, Y. and Kondo, K. (1972b). *In* "Proc. IV Intern. Ferment. Symp: Fermentation Technology Today", pp. 781–784. Society of Fermentation Technology, Osaka.

Yamada, Y. and Kondo, K. (1973). *J. Gen. Appl. Microbiol.* **19**, 59–77. 189–208.

Yamada, Y., Arimoto, M. and Kondo, K. (1973b). *J. Gen. Appl. Microbiol.* **19**, 353–358.

Yamada, Y., Arimoto, M. and Kondo, K. (1976a). *J. Gen. Appl. Microbiol.* **22**, 293–299.

Yamada, Y., Nojiri, M., Matsuyama, M. and Kondo, K. (1976b). *J. Gen. Appl. Microbiol.* **22**, 325–337.

Yamazaki, M., Taguchi, M. and Tsuchiya, T. (1976). *Jpn. J. Med. Mycol.* **17**, 131–135 (in Japanese).

Yarrow, D. (1970). "CBS Progress Report" (J. A. von Arx, Ed.), pp. 47–48. Inst. Royal Netherlands Acad. Arts and Sci.

Yarrow, D. (1971). "CBS Progress Report" (J. A. von Arx, Ed.), pp. 66–69. Inst. Royal Netherlands Acad. Arts and Sci.

Yarrow, D. and Meyer, S. A. (1978). *Int. J. Syst. Bacteriol.* **28**, 611–615.

Yukawa, M. and Ohta, M. (1928). *Gakugei Zasshi, Kyushu Imperial Univ.* **3**, 187–199, 200–216 (in Japanese with English summary).

Yukawa, M. and Ohta, M. (1929). *Bull. Inst. Pasteur* **27**, 542–544.

Yukawa, M., Yositome, W. and Misio, S. (1929). *Gakugei Zasshi, Kyushu Imperial Univ.* **4**, 267–281 (in Japanese with English summary).

4

Biochemical and Serological Characteristics of *Aeromonas*

M. POPOFF[1] and R. LALLIER[2]

Unité des Entérobactéries, [1]*Institut Pasteur, Paris, Cedex, France,* and
[2]*Faculté de Médecine Vétérinaire, Université de Montréal, Saint-Hyacinthe,
Québec, Canada*

I.	Introduction	127
II.	Evolution of taxonomic concepts.	128
III.	Habitat, pathogenicity and isolation	129
	A. Motile *Aeromonas* species	129
	B. *A. salmonicida*	131
IV.	Taxonomy of *Aeromonas*	132
	A. Phenotypic characteristics and tests	132
	B. DNA base composition and polynucleotide sequence relatedness	135
	C. Cell wall composition of motile *Aeromonas* species	136
V.	Serotyping of *Aeromonas*	137
	A. Antigenic characterization of motile *Aeromonas* species	137
	B. Antigenic characterization of *A. salmonicida*.	140
VI.	Phage typing of *A. salmonicida*	140
	A. Introduction.	140
	B. Characteristics of *A. salmonicida* phages	141
	C. Phage typing method	141
	Appendix I.	142
	References	142

I. Introduction

The genus *Aeromonas* (Stanier, 1943) belongs to the family Vibrionaceae (Sébald and Véron, 1963; Véron, 1966). Members of the genus *Aeromonas* are Gram-negative, rod-shaped bacteria measuring 1–4 by 0.5–1 μm. They are motile by polar flagella, generally monotrichous, or non-motile. The aeromonads grow on meat extract media, are facultatively aerobic or anaerobic and ferment glucose with or without gas production. They reduce

METHODS IN MICROBIOLOGY VOL. 16
ISBN 0-12-521516-9

nitrate to nitrite and are oxidase positive. The optimum temperature for growth is 28°C. The differentiation of the genus *Aeromonas* from the genus *Vibrio* has not yet been established with certainty. Usually, aeromonads, but not vibrios, produce arginine dehydrogenase and are resistant to 2,4-diamino-6,7-diisopropyl pteridine (vibriostatic agent 0/129).

Two clearly distinguished groups are included in the genus *Aeromonas*. Psychrophilic and non-motile aeromonads are clustered in the first group, named *A. salmonicida*. The second group is formed of mesophilic and usually motile bacteria. The taxonomy of this latter group is still confused. Recent taxonomic studies indicate that the motile aeromonads can be separated into three species, namely *A. hydrophila*, *A. caviae* and *A. sobria* (Popoff and Véron, 1976; Popoff *et al.*, 1981).

The purpose of this chapter is to report the methods presently used in identification and serological and phage typing of *Aeromonas* species. For convenience, the genus *Aeromonas* will be divided into four species: *A. hydrophila, A. caviae, A. sobria* and *A. salmonicida*.

II. Evolution of taxonomic concepts

Two stages can be distinguished in the evolution of the taxonomy of the genus *Aeromonas*.

The first period ranged from 1890 (description of *Bacillus punctatus* by Zimmerman, 1890) to 1957 (publication of the seventh edition of "Bergey's Manual of Determinative Bacteriology"). During this period, various species were included in the genus *Aeromonas* by several investigators. The merit of the seventh edition of "Bergey's Manual of Determinative Bacteriology" was to recognize only four species (*A. hydrophila, A. punctata, A. liquefaciens* and *A. salmonicida*) within the genus *Aeromonas*, and to draw up a list of synonyms for these four species (Snieszko, 1957). At that time, *A. hydrophila, A. punctata* and *A. liquefaciens* were differentiated mainly on the basis of their pathogenicity to poikilothermic vertebrates. Nevertheless, this classification did not take into consideration the opinion of Stanier (1943) who proposed that all motile aeromonads should be regarded as one species.

The second period, after 1957, is concerned with modern taxonomic works in which a large number of *Aeromonas* strains are described by an extended series of tests. This taxonomy based on tests of overall similarity offers a consistent way of defining species in quantitative terms. As a result of the studies of Eddy (1960), Eddy and Carpenter (1964), Ewing *et al.* (1961) and McCarthy (1975), all motile aeromonads were gathered into one species. This proposition confirmed the earlier view of Stanier (1943). However a controversy arose over the choice of the epithet for this single motile species. Eddy

(1960) was of the opinion that the specific epithet *liquefaciens* had priority over *hydrophila* and *punctata*. Ewing *et al.* (1961) suggested that *hydrophila* would be preferable to *liquefaciens* and *punctata*. Later, Eddy (1962) withdrew his original suggestion and proposed *punctata* in place of *liquefaciens* and *hydrophila*. McCarthy (1975) was firmly convinced of *punctata*'s legitimacy, but said that "universal acceptance of *hydrophila* will end this confusing and sterile argument". The conclusions of Schubert in the eighth edition of "Bergey's Manual of Determinative Bacteriology" (1974) differed considerably from those of the former workers. Schubert (1974) differentiated two species within motile aeromonads: *A. hydrophila* with three subspecies (*hydrophila, anaerogenes* and *proteolytica*) and *A. punctata* with two subspecies (*punctata* and *caviae*). Later, several workers (McCarthy, 1975; Popoff and Véron, 1976) clearly demonstrated that *A. hydrophila* subsp. *proteolytica* must be excluded from the genus *Aeromonas* because of its sodium requirement and (G + C) ratio. Recently numerical taxonomy studies, (G + C) mole % DNA determinations and polynucleotide sequence relationships (Popoff and Véron, 1976; Popoff *et al.*, 1981) indicated that motile aeromonads could be divided into three species: *A. hydrophila, A. caviae* and *A. sobria*, a new species. In the light of this review of different publications, one will not be surprised by the confusion surrounding the classification of motile aeromonads.

The taxonomy of *A. salmonicida* is also under discussion. Smith (1963) proposed the name *Necromonas salmonicida*. However serological, DNA homology and phage sensitivity data demonstrated the relationship between *A. salmonicida* and motile aeromonads, supporting the classification of both species within the genus *Aeromonas* (MacInnes *et al.*, 1979; Paterson *et al.*, 1980; Popoff, 1971a, 1971b). Physiological characteristics of *A. salmonicida* strains suggest the possibility of subdividing this group into at least two subspecies: typical and atypical *A. salmonicida* (McCarthy, 1977; McCarthy, 1978, Ph.D. thesis, Council of National Academic Awards, Great Britain; Trust *et al.*, 1980). Schubert (1974) divided this species into three subspecies (*salmonicida, achromogenes* and *masoucida*).

III. Habitat, pathogenicity and isolation

A. Motile *Aeromonas* species

The frequent occurrence of motile *Aeromonas* species in water has been reported by various investigators. Leclerc and Buttiaux (1962) isolated motile aeromonads from 33% of the 9036 water samples examined. They emphasized that these bacteria constituted an important source of error for the enumeration of coliforms in drinking water. Denis (1971) reported the presence of motile aeromonads in sea water and plankton. In the USA Hazen *et al.* (1978) isolated these micro-organisms from 90% of water samples collected in 30 states. They concluded that the presence of motile aeromonads in so many different systems would seem to indicate an important role for these bacteria in natural aquatic processes. Motile *Aeromonas* are also considered as normal inhabitants of the intestinal tract of fresh water fish (Chung and Kou, 1973; Horsley, 1977; Trust and Sparrow, 1974) and certain other poikilotherms (McCoy and Seidler, 1973; Shotts *et al.*, 1972). Boulanger *et al.* (1977) isolated *A. hydrophila* and *A. sobria* from the intestines of both healthy and moribund fish, whereas strains of *A. hydrophila* were recovered from the kidneys and lesions of moribund fish only. Motile *Aeromonas* species were occasionally isolated from apparently healthy people (Catsaras and Buttiaux, 1965; Lautrop, 1961). However a faecal origin could not explain the presence of the bacteria in surface water or sewage (Schubert *et al.*, 1972).

Motile *Aeromonas* species have long been recognized as the causal agent of "red-leg" disease in amphibians (Emerson and Norris, 1905; Russel, 1898; Shotts *et al.*, 1972). These bacteria were also considered to be responsible for diseases in reptiles (Marcus, 1971; Shotts *et al.*, 1972), fishes (Haley *et al.*, 1967), snails (Mead, 1969), cows (Wohlegemuth *et al.*, 1972) and man (Davis *et al.*, 1978). Variation in the virulence for fish between different *Aeromonas* strains is reported. Strains isolated from diseased fish are more virulent than water isolates (De Figueiredo and Plumb, 1977). It was also demonstrated that *A. sobria* strains are non-virulent for fish, whereas only some *A. hydrophila* strains are virulent and those virulent strains could be differentiated from the non-virulent strains by some surface characteristics (Lallier *et al.*, 1980; Mittal *et al.*, 1980). Several cases of fatal human septicaemia have been reported, but in all instances the patient was debilitated by some other disease, for example leukaemia (Davis *et al.*, 1978). Recently, motile aeromonads were reported to be pathogenic in man when wounds were exposed to polluted water (Davis *et al.*, 1978; Joseph *et al.*, 1979). It has been also suggested that some strains probably act as primary agents of acute diarrhoeal diseases (Bhat *et al.*, 1974; Chatterjee and Neogy, 1972; Sanyal *et al.*, 1975). A possible explanation of the enteropathogenic potential of motile *Aeromonas* species has arisen with the finding that some strains produced an enterotoxin (Boulanger *et al.*, 1977; Sanyal *et al.*, 1975; Wadström *et al.*, 1976).

Motile aeromonads grow well at 28°C on ordinary media (tryptone soy agar, for example). Blood agar supports an abundant growth and some strains produce a wide haemolytic zone surrounding colonies on this medium. Most *A. sobria* isolates produced only one zone of haemolysis, whereas the majority of the *A. hydrophila* produced two zones of haemolysis (Boulanger *et al.*, 1977; Olivier *et al.*, 1981). The growth of motile *Aeromonas* species may be inhibited on thiosulphate–citrate–bile salt–sucrose (TCBS) agar, a selective medium for the isolation of *Vibrio cholerae*. Two selective media (Drigalski and McConkey agar) used for Enterobacteriaceae can be used for the isolation of motile aeromonads. But these media do not permit the differentiation of colonies of *Aeromonas* from colonies of Enterobacteriaceae. Specific selective media have been proposed for the isolation of motile aeromonads from stools. Von Graevenitz and Zinterhofer (1970) have described a supplemented DNAase medium based on the production of DNAase by *Aeromonas* strains. The Rimler–Shotts agar (Shotts and Rimler, 1973) containing novobiocine, and the pril–xylose–ampicillin agar have also been devised for the isolation of motile aeromonads (Rogol *et al.*, 1979).

B. *A. salmonicida*

Aeromonas salmonicida is a parasite lacking the ability to exist as a saprophyte under natural conditions. Although this bacteria has been isolated from natural water, its existence in river water is very short-lived, lasting a few days at the most (McCraw, 1952). *A. salmonicida* is the causal agent of fish furonculosis. This disease, originally described from hatchery fish in Germany by Emmerich and Weibel (1894), is observed in most of the salmonid-producing countries in the world. Furonculosis is traditionally associated with Salmonidae (salmon and trout), but other fish families (Petromyzontidae, Cyprinidae, Serranidae and Anaplopomatidae) are susceptible to infection by *A. salmonicida* (Herman, 1968). Furunculosis appears to be a specific infection of fish.

Furunculosis is properly described as a general infection (septicaemia) in which focal lesions (so-called furoncles) may or may not occur. The term "furunculosis" is a misnomer, since the fish lesions are not analogous to furoncles as they occur in human patients: the so-called furoncles are areas of necrosis in which there is no outstanding leukocytic infiltration (McCraw, 1952). The pathogenicity of *A. salmonicida* may be due to its abundant growth in blood and tissues, and to the production of a leucocidin (Klontz *et al.*, 1966).

Cultures of *A. salmonicida* grow on ordinary media (tryptone soy agar in preference) at 28°C but not at 35°C. A relationship between virulence for fish and auto-aggregation was reported with both typical and atypical *A. salmonicida* strains (Trust *et al.*, 1980; Udey and Fryer, 1978). Generally the majority of the typical *A. salmonicida* strains only produced a brown soluble pigment on nutrient agar supplemented with 0.1% tyrosine, however a few atypical strains are pigmented (McCarthy, 1977). The growth is abundant with intense pigment production on serum agar (Williamson, 1928). On blood agar, haemolysis rapidly occurs, the colonies becoming greenish at about the seventh day. The culture is scanty on selective media for Enterobacteriaceae. No specific selective medium has been so far proposed for the isolation of *A. salmonicida*. In both diseased fish and healthy carriers, the kidney is the organ from which *A. salmonicida* is most readily isolated.

IV. Taxonomy of *Aeromonas*

A. Phenotypic characteristics and tests

1. *Phenotypic tests*

Standard procedures for enteric bacteria can be used to study biochemical characteristics of *Aeromonas* species. The recommended incubation temperature is 28°C for motile and 25°C for non-motile *Aeromonas*. Most of these techniques are detailed in several authoritative laboratory manuals (Cowan and Steel, 1965; Edwards and Ewing, 1972; Le Minor, 1972). However, three additional media and tests are described below.

Cysteine–iron–agar

The cysteine–iron–agar (Véron and Gasser, 1963) is used to detect hydrogen sulphide (H_2S) production from cysteine.

Composition of the medium: Bacto-Peptone (Difco), 10 g; KCl, 4 g; bacto-agar (Difco), 5 g; distilled water, 1000 ml. Check that pH is 7.2. Heat to dissolve the solids in the water. Distribute in 10-ml amounts into screw-cap tubes (17×170 mm). Sterilize by autoclaving at 115°C for 20 min. After cooling at 45°C, add aseptically 1 ml of the following solution sterilized by filtration: L-cysteine–HCl, 1 g; ammonium ferrous citrate, 1 g; distilled water, 1000 ml. Mix and allow to solidify in a vertical position.

Inoculate with a straight wire and incubate at 28°C.

Observe daily up to four days for blackening due to H_2S production.

Elastin–Agar

This medium is used to detect the production of elastase by *Aeromonas* species (Scharmann, 1972).

Composition of the medium: Bacto-Peptone (Difco), 10 g; $CaCl_2$, 11.6 g; Tris-HCl, 7.2 g; Elastin (Sigma), 10 g; Bacto-Agar (Difco), 20 g; distilled water, 1000 ml. Mix and dissolve the constituents by gentle heating. Adjust to pH 7.8. Sterilize at 115°C for 20 min. Distribute in Petri dishes, ensure a uniform suspension of the insoluble elastin.

Inoculate plates of elastin agar and incubate at 28 C.

Examine at intervals up to eight days for clearing of the medium (elastin hydrolysis) around the bacterial growth.

Procedure for nutritional tests

The M-70 medium (Véron, 1975) is employed to study the ability of *Aeromonas* species to use organic compounds as principal sources of carbon and energy. This procedure is readily accessible to most bacteriologists. The M-70 components must be of the highest purity.

(a) *Composition of the M-70 medium.* The trace element solution contains in 1000 ml of distilled water: H_3PO_4, 1960 mg; $FeSO_4.7H_2O$, 56 mg; $ZnSO_4.7H_2O$, 29 mg; $MnSO_4.4H_2O$, 22 mg; $CuSO_4.5H_2O$, 2.5 mg;- $Co(NO_3)_2.6H_2O$, 3 mg; H_3BO_3, 6 mg; $Na_2MnO_4.2H_2O$, 3 mg. Dissolve by stirring. Sterilize by filtration. Store at 4–6°C.

The complete M-70 medium contains in 990 ml of distilled water: $CaCl_2.2H_2O$, 15 mg; $MgSO_4.7H_2O$, 123 mg; KH_2PO_4, 680 mg; K_2HPO_4, 2610 mg; NaCl, 10 g; $(NH_4)_2SO_4$, 1 g. Dissolve and check that the pH is 7.2. Add 10 ml of the trace element solution. Sterilize by filtration. Store at room temperature. The complete M-70 medium can be dispensed in 5 ml volumes into screw-cap tubes (17 × 170 mm).

(b) *Preparation of organic compound solutions.* Prepare a 3% solution of organic compounds in distilled water. Sterilize by filtration. Store at room temperature.

(c) *Performance of tests.* The carbon source is added to the complete M-70 medium so that the final substrate concentration ranges between 0.25 and 0.30%.

To perform nutritional tests, inocula are taken from cultures on tryptone soy agar after incubation at 30°C for 18 h. The bacterial growth is removed with a platinum loop and suspended in 10 ml of distilled water. The suspension is agitated vigorously using a Vortex mixer. The suspension density is adjusted to approximately $1 × 10^8$ bacteria per millilitre (this matches the turbidity of the McFarland barium sulphate standard number 1). Then a 1:10 dilution of this primary suspension is prepared in distilled water (a calibrated 0.01 ml platinum loop can be used). The M-70 medium containing the carbon source is inoculated with two or three drops of the diluted suspension, using a Pasteur pipette.

Tubes are incubated at 28°C and screened daily for four days. An organic compound is used as a source of carbon and energy when a positive growth response is observed.

2. *Phenotypic characteristics of motile* Aeromonas *species*

The three motile *Aeromonas* species possess all the basic characteristics of the genus *Aeromonas*. These species are polar flagellated when motile. They ferment mannitol, maltose and trehalose and produce catalase, DNAase, RNAase, gelatinase and Tween 80 esterase. The following 19 substrates are used as principal carbon source: D-ribose, D-fructose, D-galactose, D-glucose, D-maltose, D-trehalose, D-gluconate, starch, caprylate, pelargonate, caprate, succinate, furmarate, DL-glycerate, L-malate, glycerol, D-mannitol, L-aspartate and L-glutamate. Most motile aeromonads produce β-galactosidase and indole, ferment mannose and glycerol and utilize mannose and sucrose as a carbon source. Lysine decarboxylase appears variable.

Characters negative for the three species are formation of capsule, pectinase, ornithine decarboxylase, tryptophan and phenylalanine deaminases, fermentation of xylose, sorbose, erythritol, adonitol, dulcitol, inositol and mucate, and production of H_2S on Kligler medium. Except for a few variants (Ross, 1962), they do not produce a brown pigment.

Physiological and biochemical characteristics of *A. hydrophila, A. caviae* and *A. sobria*, are listed in Table I. Although some aberrant strains occur, the three motile *Aeromonas* species may be readily differentiated on the basis of arabinose and salicin fermentation, esculin hydrolysis, growth in KCN medium, production of acetoin and gas from glucose.

Motile aeromonads may be confused with some marine vibrios, such as *Vibrio anguillarum, Beneckea splendida* biotype I and *Beneckea nereida*. These marine organisms are able to convert arginine or ornithine under anaerobic conditions by way of a constitutive arginine dihydrolase system (Baumann and Baumann, 1977). Like *Aeromonas*, they will dehydrogenate arginine but not decarboxylate lysine or ornithine. These vibrios cannot grow in peptone water containing less than 0.5% NaCl and they are sensitive to the vibriostatic agent 0/129 (Baumann *et al.*, 1971; Baumann and Baumann, 1977), whereas motile *Aeromonas* species grow well in peptone water without NaCl and are resistant to the vibriostatic agent 0/129.

3. *Phenotypic characteristics of* Aeromonas salmonicida

A. salmonicida possesses all the basic characteristics of the genus *Aeronomas*. These organisms are always non-motile, do not grow at 37°C, are negative for lysine decarboxylase, indole, citrate and H_2S. Most strains do not attack

TABLE I

Phenotypic characteristics of *A. hydrophila, A. caviae* and *A. sobria*[a]

Character	*A. hydrophila*	*A. caviae*	*A. sobria*
Growth on L-histidine	+(100)[b]	+(97)	−(92)
Esculin hydrolysis	+(100)	+(100)	−(81)
Growth on KCN medium	+(100)	+(97)	−(81)
Growth on L-arabinose	+(92)	+(100)	−(81)
Fermentation of salicin	+(92)	+(84)	−(81)
Growth on L-arginine	+(72)	+(86)	−(92)
Growth on salicin	+(88)	+(72)	−(77)
Elastase production	+(92)	−(100)	−(88)
Acetoin from glucose	+(92)	−(97)	±(+, 58)
Gas from glucose	+(92)	−(86)	+(88)
H$_2$S from cysteine	+(85)	−(93)	+(96)

[a] + : character usually positive; − : character usually negative; ± : non-significant frequency.
[b] The percentage of strains having the reaction is indicated in parentheses.

xylose, adonitol, rhamnose, sorbitol, sorbose, dulcitol, lactose, inositol, raffinose or cellobiose. Differentiation between typical and atypical *A. salmonicida* strains could be achieved by the following characteristics (McCarthy, 1978, PhD thesis, Council of National Academic Awards, Great Britain; Trust *et al.*, 1980): typical *A. salmonicida* strains are gelatinase positive, produce a brown pigment, catabolize arabinose, arbutin, esculin, galactose, glycerol, gluconate, mannitol, methyl-D-glucoside, salicin and trehalose, whereas atypical *A. salmonicida* strains are negative for all these tests but catabolize sucrose.

B. DNA base composition and polynucleotide sequence relatedness

1. *Motile* Aeromonas *species*

Aeromonas species are related but genetically diverse organisms (MacInnes *et al.*, 1979) which can be divided into seven DNA hybridization groups (Popoff *et al.*, 1981). *A. hydrophila* appears to contain three DNA hybridization groups. The (G + C) mole % of DNA from the first two groups of *A. hydrophila* is 61.1 ± 0.7 and 61.9 ± 0.3 respectively, whereas the (G + C) mole % of DNA preparations from the third group is 58.3 ± 0.5. *A. caviae* is divided into two hybridization groups. DNA of representative strains from the two *A. caviae* groups have (G + C) mole % values of 61.7 ± 0.9. Two hybridization groups can be differentiated within *A. sobria*. The base composition of DNA from *A. sobria* strains is 58.7 ± 0.6 (G + C) mole %.

However, within each phenotypically defined species, hybridization groups are indistinguishable by biochemical characters. If a reasonable taxonomic scheme is to be adopted by all concerned, DNA homology groupings must be correlated with phenotypic characteristics to be useful for classification and identification. Splitting motile aeromonads into seven species would be useless as long as they cannot be separated by biochemical tests. For the present, motile aeromonads would consist of three species: *A. hydrophila, A. caviae* and *A. sobria*. Each of these species contains more than one DNA hybridization group which may merit designation as additional knowledge is accumulated.

2. A. salmonicida

Polynucleotide sequence relatedness among *A. salmonicida* strains were studied by MacInnes *et al.* (1979). This species appears to be a genetically homogeneous group with high homology values. *A. salmonicida* subsp. *salmonicida* and *A. salmonicida* subsp. *masoucida* (Schubert, 1974) cannot be separated on the basis of DNA relationships. The (G + C) mole % content of DNA from both typical and atypical *A. salmonicida* is between 55.1 and 57.4 (De Ley, 1970; Hill, 1966; McCarthy, 1978, PhD Thesis, Council of National Academic Awards, Great Britain; Trust *et al.*, 1980).

C. Cell wall composition of motile *Aeromonas* species

Shaw and Hodder (1978) have studied the composition of core oligosaccharides from lipopolysaccharides of motile *Aeromonas* species. Analysis by gas-liquid chromatography (g.l.c.) classifies motile aeromonads into three distinct chemotypes. This classification is based on the presence of galactose, D-glycero-D-mannoheptose and L-glycero-D-mannoheptose in the oligosaccharides of lipopolysaccharides. Taxonomic results obtained by phenotypic and genomic studies are closely related to those of core oligosaccharides analysis (Table II).

TABLE II

Correlation between phenotypic classification, DNA hybridization grouping and core oligosaccharides composition of motile *Aeromonas* species

Phenotype classification	DNA hybridization grouping	Core oligosaccharides composition (mole %)[a]					
		RHA[b]	GAL[c]	GLU[d]	DDH[e]	LDH[f]	Type
A. hydrophila	Group 1	17.2	9.8	13.5	30.9	28.7	1
A. hydrophila	Group 2	—	—	34.1	—	65.8	2
A. caviae	Group 1	29.4	—	17.1	—	53.4	2
A. caviae	Group 3	19.6	—	23.0	—	57.4	2
A. sobria	Group 1	—	22.1	42.3	—	35.5	3
A. sobria	Group 2	31.9	26.1	30.6	—	5.3	3
A. sobria	Group 2	60.4	—	28.9	—	10.7	2

[a] From Shaw and Hodder (1978).
[b] RHA: rhamnose.
[c] GAL: galactose.
[d] GLU: glucose.
[e] DDH: D-glycero-D-mannoheptose.
[f] LDH: L-glycero-D-mannoheptose.

V. Serotyping of *Aeromonas*

A. Antigenic characterization of motile *Aeromonas* species

1. *Antigenic structure*

Although few papers appear on the antigenic structure of motile *Aeromonas* species, uncertainty still exists about their serology. On the one hand, with precipitating studies, three different workers were able to demonstrate specific precipitin lines within the *Aeromonas* strains (Bullock, 1966; Liu, 1961; McCarthy, 1975). On the other hand, antigenic heterogeneity and the possibility of serogrouping were demonstrated by agglutination studies (De Meuron and Peduzzi, 1979; Ewing *et al.*, 1961; Kjems, 1955; Miles and Miles, 1951; Page, 1962).

Motile *Aeromonas* contained thermostable O-, thermolabile K- and flagellar H-antigen. Ewing *et al.* (1961) reported the presence of 12 provisional O-antigen and nine H-antigen groups among 71 strains of *Aeromonas*. In addition to the O-antigen, De Meuron and Peduzzi (1979) reported the presence of K-antigen. In their study, the presence of the K-antigen partly inhibited the O-antigen reaction in two out of seven reference strains (Leblanc *et al.*, 1981). It was also reported on the differentiation between the thermostable O- and the thermolabile K-antigens of motile *Aeromonas* species. Using nearly 200 strains of motile aeromonads, it was observed that the presence of the K-antigen does not block O-agglutination. In addition, it was also reported that a same thermostable antigen could be associated with different thermolabile antigens and, whereas the majority of the strains possess only one O-antigen, more than one thermolabile antigen could be detected with one strain.

2. *Ecological significance of serogrouping*

With many bacterial species, the serotyping was found to be a useful tool for the epidemiologists, for example for motile *Aeromonas*. Kulp and Borden (1942) were the first to report two serological types of *Aeromonas* strains isolated from diseased frogs. De Meuron and Peduzzi (1979) found that 6 out of 11 *A. hydrophila* strains isolated from reptiles of the same origin were antigenically related. Variation in the virulence for fish was observed within the *Aeromonas* strains (De Figueiredo and Plumb, 1977; Lallier *et al.*, 1980). A direct relationship seems to exist between virulence and serogroup. It appears that all virulent strains tested belonged to one O-group, whereas non-virulent strains had other group antigens (Mittal *et al.*, 1980).

The hypothesis of human contamination by aquatic or fish *Aeromonas* strains has often been raised. The development of the serotyping system could eventually clarify this situation. Preliminary results obtained in our laboratory indicated that some human *Aeromonas* strains were antigenically similar to *Aeromonas* strains from fish.

3. *Methodology of serotyping motile* Aeromonas

(a) *Preparation of the antigens.*   The O-antigen was prepared by growing the strains on Trypticase Soy Agar (Difco) slants containing 0.1% glucose at 22°C for 18 h. The growth was resuspended in 0.85% saline, boiled for 1.5 h, washed three times in saline, then resuspended in saline and the optical density at 540 nm adjusted to 1.0.

The *Aeromonas* strains are also grown in Brain Heart Infusion Broth (BHI, Difco) at 22°C for 18 h, formalin added to a final concentration of 1% and the optical density at 540 nm adjusted to 0.3. This suspension is further used as whole cells (WC) antigen.

(b) *Preparation of immune sera.*   New Zealand inbred rabbits (weighing about 2 kg) are immunized with the O-antigens and WC antigens. Rabbits are injected intravenously with increasing volumes of antigen (0.1–2.0 ml). The immunization is done twice a week for three weeks and the animals bled seven days after the last injection. The blood is allowed to clot for 2 h at 37°C and then for 18 h at 4°C. After centrifugation at $10\,000 \times \mathbf{g}$ for 15 min, the sera are collected and frozen at -20°C until needed.

(c) *Slide agglutination.*   Bacterial strains are grown on TSA-agar plate plus 5% bovine blood (v/v) for 18 h at 22°C. One or two colonies are mixed with one drop of the anti-WC serum and the agglutination recorded after 1–2 min. On some occasions, agglutination appears after 4 min. These retarded reactions should be considered as negative, since no specific reaction is observed under these circumstances in tube agglutination when using anti-serum diluted 1/10.

(d) *Tube agglutination.*   All strains giving a positive slide reaction should be further tested in tubes for final serogroup confirmation. This will allow the elimination of the non-specific cross-reactions that occasionally occur on slide preparations. In fact, the majority of the strains reacting with multiple slide agglutination will react with only one anti-O serum when tested in tubes.

The serogrouping of motile *Aeromonas* could be based on two antigens: thermolabile, WC antigen and thermostable O-antigen. The thermostable O-antigen could be detected in tube agglutination with a whole cell preparation as antigen and an anti-O serum, or an O-antigen preparation against the anti-O or the anti-WC sera. Since the presence of the complete surface did not inhibit the O reaction, the term capsular antigen should be replaced by surface antigen. The WC or surface antigen could only be detected when using a whole cell preparation as antigen and an anti-WC serum.

For tube agglutination, two-fold serum dilutions are done in 0.05 ml of 0.85% saline in micro-titre plates. A same volume of antigen of optical density 0.6 is added. The results are recorded after incubation for 18 h at 37°C followed by an additional 6 h at 4°C.

(e) *Interpretation of results.*   Leblanc *et al.* (1981) using the above pro-cedures, studied nearly 200 motile *Aeromonas* strains. Using the thermolabile WC antigen and the anti-WC serum there are three to five times more cross-reactions than with the thermostable O-antigens. Such system cannot differentiate between O or WC agglutination without absorbed sera. Since motile *Aeromonas* species are flagellated and some possess fimbriae (Alken-son and Trust, 1980), it would be presumptive to identify the antigen responsible for the reaction with anti-WC scrum absorbed with the thermo-stable O-antigen as surface antigen. Further results will be obtained only when using serum prepared with purified flagella and fimbriae or anti-WC serum absorbed by the O-antigen and the purified flagella or fimbriae.

In conclusion, the serotyping of motile *Aeromonas* species should be based principally on their O-antigen; a strain is classified as a particular serogroup if the heated cells are agglutinated by one antiserum of this type but not by the antisera of other specificities.

B. Antigenic characterization of *A. salmonicida*

1. *Preparation of O-antisera*

To prepare an O-antigen suspension for rabbit immunization, a colony tested for smoothness is seeded on tryptone soy agar plates and incubated at 28°C for 48 h. The growth is suspended in saline and this suspension is treated by 0.5% CaCl$_2$ and 0.5% formaldehyde as described by Ando and Shimojo (1953). Finally the antigen suspension is adjusted to approximately 2×10^9 bacteria per millilitre. Rabbits are injected at four-day intervals with graded intravenous doses of 0.5, 1.0, 2.0 and 3.0 ml of the antigen suspension. The optimal time for bleeding is 10–12 days after the last injection. The sera are preserved with merthiolate (1 in 10 000 w/v). In these conditions, a titre against the homologous antigen strain will usually be 1/3200 or more (Popoff, 1969).

2. *Serological characterization of* A. salmonicida

For serological grouping of *A. salmonicida*, slide agglutination was used. The sera are adjusted to a dilution in which the homologous culture gives a distinct slide agglutination reaction within 10 s. The test is performed by the usual technique using overnight tryptone soy agar cultures. In these working conditions, *A. salmonicida* is a serologically homogeneous species (Popoff, 1969). Such a result was also reported by Karlsson (1964) and Spence *et al.* (1965).

Using tube agglutination, Paterson *et al.* (1980) showed that *A. salmonicida* could be divided into two serogroups.

VI. Phage typing of *A. salmonicida*

A. Introduction

Phage typing has been developed for *A. salmonicida*, although the method has had limited use.

B. Characteristics of *A. salmonicida* phages

Phages may be isolated from hatcheries, water and sewage, or from lysogenic isolates (Christison *et al.*, 1938; Paterson *et al.*, 1969; Popoff and Vieu, 1970). *A. salmonicida* phages can be divided into three morphological groups (Popoff, 1971a). Phages of the first group resemble T-even phages of *Escherichia coli* in having a head (100 nm), a long contractile tail (130 nm), a collar and a base plate with fibres. The second group contains phages with a polyhedric head (65 nm) and a long contractile tail (150 nm). Phages of the third group possess a polyhedric head (60 nm), a short tail (90 nm) with an anchor-like plate. The resistance of *A. salmonicida* phages to physical factors has been investigated (Popoff, 1971a). The phages usually resist at a temperature of 60–65°C for 10 min. Some phages also withstand exposure to 75°C. They are stable at pH 6.5–11.0. Serologically, *A. salmonicida* phages can be separated into ten types.

C. Phage typing method

Peptone water and tryptone soy agar are suitable media for phage propagation and typing, respectively. Propagation of phages is preferably carried out in Erlenmeyer flasks with 100 ml of peptone water. After inoculation with 1 ml of an overnight peptone water culture, flasks are incubated for 3 h at 28°C in a gyratory shaker. Phages are added to give 10^4 plaque forming units per millilitre (p.f.u. ml^{-1}) cell suspension. After overnight incubation at 28°C, lysates are centrifuged and sterilized through Millipore membranes (type HA; 0.45 μm). Phage suspensions are stored at 4°C over a chloroform layer. In these conditions, typing phages are stable for at least a year. Quantitation of p.f.u. is performed by the soft agar overlay method of Adams (1959).

The phage typing set of Popoff (1971a) is based upon studies with eight phages. Bacterial strains to be typed are grown for 4–6 h in nutrient broth at 28°C and seeded on tryptone soy agar plates. After drying at room temperature for 15 min, each phage diluted to the routine test dilution (RTD), is applied as a calibrated drop (0.025 ml) to a particular position on the plate. RTD is that dilution of the phage suspensions which gives confluent lysis (more than 50 plaques) corresponding to the zone covered by the drop applied. Plates are dried at room temperature until the surface liquid disappears and are then incubated overnight at 28°C. The interpretation of results is presented in Table III. Fourteen lysogroups are defined by this set of phages. Phage typing is well suited for epidemiological typing of *A. salmonicida*.

TABLE III

Lysotyping of *A. salmonicida*

Lysotype	Phage designation								Strains lysed (%)
	31	32	25	51	56	65	63	29	
1	CL	CL	CL	CL	CL	CL	CL	CL	10
2	CL	CL	—	CL	CL	CL	CL	CL	1
3	CL	CL	CL	CL	CL	CL	CL	—	1
4	CL	CL	CL	CL	CL	CL	—	CL	3
5	CL	CL	CL	CL	CL	CL	—	—	1
6	CL	CL	CL	CL	CL	—	—	—	59
7	CL	CL	CL	CL	—	—	—	—	3
8	CL	CL	CL	—	—	CL	CL	—	2
9	CL	CL	CL	—	—	—	—	CL	1
10	CL	CL	CL	—	—	CL	—	—	2
11	CL	CL	CL	—	—	—	—	—	13
12	CL	CL	—	—	—	—	—	—	1
13	CL	—	CL	—	—	—	—	—	1
14	—	CL	CL	—	—	—	—	—	2

Appendix I
List of reference strains

A. hydrophila : type strain ATCC 7966, isolated from tin of milk with a fishy odour.

A. caviae : type strain ATCC 15468, isolated from young guinea-pigs.

A. sobria : type strain CIP 7433, isolated from carp.

A. salmonicida : type strain NCMB 1102.

Abbreviations: ATCC, American Type Culture Collection, Rockville, Maryland; CIP, collection de l'Institut Pasteur, Paris, France; NCMB, National Collection of Marine Bacteria, Aberdeen, Scotland.

References

Adams, M. H. (1959). "Bacteriophages", pp. 29–30. Wiley (Interscience), New York.

Alkenson, H. H. and Trust, T. J. (1980). *Infect. Immun.* **27**, 938–946.

Ando, K. and Shimojo, Y. (1953). *Bull. O. M. S.* **9**, 575–577.

Baumann, P. and Baumann, L. (1977). *Ann. Rev. Microbiol.* **31**, 39–61.

Baumann, P., Baumann, L. and Mandel, M. (1971). *J. Bacteriol.* **107**, 268–294.

Bhat, P., Shantakumari, S. and Rajan, D. (1974). *Indian J. Med. Res.* **62**, 1051–1060.
Boulanger, Y., Lallier, R. and Cousineau, G. (1977). *Can. J. Microbiol.* **23**, 1161–1164.
Bullock, G. L. (1966). *Bull. Off. Int. Epiz.* **65**, 805–824.
Catsaras, M. and Buttiaux, R. (1965). *Ann. Inst. Pasteur (Lille)* **16**, 85–88.
Chatterjee, B. D. and Neogy, K. N. (1972). *Indian J. Med. Res.* **60**, 520–524.
Christison, M. H., Mackenzie, J. and Mackie, T. J. (1938). "*Bacillus salmonicida* Bacteriophage: with Particular Reference to its Occurrence in Water and the Question of its Application in Controlling *Bacillus salmonicida* Infection". Fisheries Scotland Salmon Fish., No. 5, H.M.S.O., Edinburgh.
Chung, H. and Kou, G. (1973). *J. Fish. Soc. Taïwan* **2**, 20–25.
Cowan, S. T. and Steel, K. J. (1965). "Manual for the Identification of Medical Bacteria", 1st edn. Cambridge University Press, London and New York.
Davis, W. A., Kane, J. G. and Garagusi, V. G. (1978). *Medicine* **57**, 267–277.
De Figueiredo, J. and Plumb, J. A. (1977). *Aquaculture* **ii**, 349–354.
De Ley, J. (1970). *J. Bacteriol.* **101**, 738–754.
De Meuron, P. A. and Peduzzi, R. (1979). *Zentralbl. Veterinaermed.* **B26**, 153–167.
Denis, F. (1971). Contribution à l'étude des bactéries hétérotrophes du milieu marin: inventaire de 2700 souches. Thèse de Sciences, Poitiers.
Eddy, B. P. (1960). *J. Appl. Bacteriol.* **23**, 216–249.
Eddy, B. P. (1962). *J. Appl. Bacteriol.* **25**, 137–146.
Eddy, B. P. and Carpenter, K. P. (1964). *J. Appl. Bacteriol.* **27**, 96–109.
Edwards, P. R. and Ewing, W. H. (1972). "Identification of Enterobacteriaceae", 3rd edn. Burgess, Minneapolis, Minnesota.
Emerson, H. and Norris, C. (1905). *J. Exp. Med.* **7**, 32–58.
Emmerich, R. and Weibel, C. (1894). *Arch. Hyg.* **21**, 1–21.
Ewing, W. H., Hugh, R. and Johnson, J. G. (1961). "Studies on the *Aeromonas* Group". US Department of Health, Education and Welfare, Communicable Disease Center, Atlanta, Georgia.
Haley, R., Davis, S. P. and Hyde, J. M. (1967). *Prog. Fish. Cult.* **29**, 193–195.
Hazen, T. C., Fliermans, C. B., Hirsch, R. P. and Esch, G. W. (1978). *Appl. Env. Microbiol.* **36**, 731–738.
Herman, R. L. (1968). *Trans. Am. Fish. Soc.* **97**, 221–230.
Hill, L. R. (1966). *J. Gen. Microbiol.* **44**, 419–437.
Horsley, R. W. (1977). *J. Fish Biol.* **10**, 529–553.
Joseph, L. W., Daily, O. P., Hunt, W. S., Seidler, R. S., Allen, D. A. and Colwell, R. R. (1979). *J. Clin. Microbiol.* **10**, 46–49.
Karlsson, K. A. (1964). *Zentralbl. Bakteriol. Parasitenkd. Infektionskr. Abt. 1* **194**, 73–80.
Kjems, E. (1955). *Acta Pathol. Microbiol. Scand.* **36**, 531–536.
Klontz, G. W., Yasutake, W. T. and Ross, A. J. (1966). *Am. J. Vet. Res.* **27**, 1455–1460.
Kulp, W. L. and Borden, D. G. (1942). *J. Bacteriol.* **44**, 673–685.
Lallier, R., Boulanger, Y. and Olivier, G. (1980). *Prog. Fish. Cult.* **42**, 199–200.
Lautrop, H. (1961). *Acta Pathol. Microbiol. Scand.* **51**, 299–301.
Leblanc, D., Mittal, K. R., Olivier, G. and Lallier, R. (1981). *Appl. Env. Microbiol.* **42**, 56–60.
Leclerc, H. and Buttiaux, R. (1962). *Ann. Inst. Pasteur* **103**, 97–100.
Le Minor, L. (1972). "Le Diagnostic de Laboratoire des Bacilles à Gram négatif. I-Entérobactéries", 4th edn. La Tourelle, St. Mandé.

Liu, P. V. (1961). *J. Gen. Microbiol.* **24**, 145–153.
McCarthy, D. H. (1975). "The Bacteriology and Taxonomy of *Aeromonas liquefaciens*, Technical Report Series". Fish Diseases Laboratory, Ministry of Agriculture, Weymouth, Dorset.
McCarthy, D. H. (1977). *Bull. Off. Int. Epiz.* **87**, 459–463.
McCoy, R. H. and Seidler, R. J. (1973). *Appl. Microbiol.* **25**, 534–538.
McCraw, B. M. (1952). "Furunculosis of Fish". US States Department of the Interior, Fish and Wildlife Service, Special Scientific Report: Fisheries No. 84.
MacInnes, J. I., Trust, T. J. and Crosa, J. H. (1979). *Can. J. Microbiol.* **25**, 578–586.
Marcus, L. C. (1971). *J. Am. Vet. Med. Assoc.* **159**, 1629–1631.
Mead, A. R. (1969). *Malacologia* **9**, 43.
Miles, E. M. and Miles, A. A. (1951). *J. Gen. Microbiol.* **5**, 298–306.
Mittal, K. R., Lalonde, G., Leblanc, D., Olivier, G. and Lallier, R. (1980). *Can. J. Microbiol.* **26**, 1501–1503.
Olivier, G., Lallier, R. and Larivière, S. (1981). *Can. J. Microbiol.* **27**, 330–333.
Page, L. A. (1962). *J. Bacteriol.* **84**, 772–777.
Paterson, W. D., Douglas, R. J., Grinyer, I. and MacDermott, L. A. (1969). *J. Fish. Res. Bd. Canada* **26**, 629–632.
Paterson, W. D., Douez, D. and Desautels, D. (1980). *Can. J. Microbiol.* **26**, 588–598.
Popoff, M. (1969). *Rech. Vet.* **3**, 49–57.
Popoff, M. (1971a). *Ann. Rech. Vet.* **2**, 33–45.
Popoff, M. (1971b). *Ann. Rech. Vet.* **2**, 137–139.
Popoff, M. and Véron, M. (1976). *J. Gen. Microbiol.* **94**, 11–22.
Popoff, M. and Vieu, J. F. (1970). *C.R. Acad. Sci.* **270**, 2219–2222.
Popoff, M., Coynault, C., Kiredjian, M. and Lemelin, M. (1981). *Curr. Microbiol.* **5**, 109–114.
Rogol, M., Sechter, I., Grinberg, L. and Gerichter, C. B. (1979). *J. Med. Microbiol.* **12**, 229–231.
Ross, A. J. (1962). *J. Bacteriol.* **84**, 590–591.
Russel, F. H. (1898). *J. Am. Med. Assoc.* **30**, 1442–1449.
Sanyal, S. C., Singh, S. J. and Sen, B. C. (1975). *J. Med. Microbiol.* **8**, 195–199.
Scharmann, W. (1972). *Zentralbl. Bakteriol. Parsitenkd. Infektionskr. Abt. 1* **220**, 435–442.
Schubert, R. H. W. (1974). In "Bergey's Manual of Determinative Bacteriology" (R. E. Buchanan and N. E. Gibbons, Eds), 8th edn. Williams and Wilkins, Baltimore, Maryland.
Schubert, R. H. W., Schafer, E. and Meiser, W. (1972). *Gas Wasserfach.* **113**, 132–134.
Sébald, M. and Véron, M. (1963). *Ann. Inst. Pasteur* **105**, 897–910.
Shaw, R. H. and Hodder, H. J. (1978). *Can. J. Microbiol.* **24**, 864–868.
Shotts, E. B. and Rimler, R. (1973). *Appl. Microbiol.* **26**, 550–553.
Shotts, E. B., Gaines, J. L., Martin, C. and Prestwood, A. K. (1972). *J. Am. Vet. Med. Assoc.* **161**, 603–607.
Smith, I. W. (1963). *J. Gen. Microbiol.* **33**, 263–274.
Snieszko, S. F. (1957). In "Bergey's Manual of Determinative Bacteriology" (R. S. Breed, E. G. D. Murray and N. R. Smith, Eds), 7th edn, pp. 189–193. Williams & Wilkins, Baltimore.
Spence, K. D., Fryer, I. L. and Pilcher, K. S. (1965). *Can. J. Microbiol.* **II**, 397–405.
Stanier, R. Y. (1943). *J. Bacteriol.* **46**, 213–214.
Trust, T. J. and Sparrow, R. A. H. (1974). *Can. J. Microbiol.* **20**, 1219–1228.

Trust, T. J., Khouri, A. G., Austen, R. A. and Ashburner, L. D. (1980). *FEMS Microbiol. Lett.* **9**, 39–42.

Udey, L. R. and Fryer, J. L. (1978). *Mar. Fish. Rev.* **40**, 12–17.

Véron, M. (1966). *Ann. Microbiol.* **III**, 671–709.

Véron, M. (1975). *Ann. Microbiol.* **126A**, 267–274.

Véron, M. and Gasser, F. (1963). *Ann. Inst. Pasteur* **105**, 524–534.

Von Gravenitz, A. and Zinterhofer, L. (1970). *Health Lab. Sci.* **7**, 124–127.

Wadström, T., Ljungh, A. and Wretlind, B. (1976). *Acta Pathol. Microbiol. Scand.* **84**, 112–114.

Williamson, I. J. F. (1928). "Furunculosis of the Salmonidae". Fishery Board for Scotland, Salmon Fisheries: No. 5.

Wohlegemuth, D., Pierce, F. L. and Kirkbride, C. A. (1972). *J. Am. Vet. Med. Assoc.* **160**, 1001–1002.

Zimmerman, O. E. R. (1890). *Ber. Naturw. Ges. Chemnitz* **I**, 38–39.

5

Separation of Species of the Genus *Leuconostoc* and Differentiation of the Leuconostocs from Other Lactic Acid Bacteria

E. I. GARVIE

National Institute for Research in Dairying, Shinfield, Reading, Berkshire, UK

I.	Introduction	148
II.	Isolation of leuconostocs and general media	149
	A. Isolation from natural habitats	149
	B. General method of cultivation	151
III.	Identification of species	153
	A. Preliminary identification and separation from other lactic acid bacteria	153
	B. Further identification by classical methods	154
	C. Amino acid and vitamin requirements	158
	D. Metabolism and enzyme studies	159
	E. Peptidoglycan types in cell walls	164
	F. Nucleic acid studies	165
	G. Serology and bacteriophage typing	169
IV.	Taxonomy	169
V.	Commercial importance	171
	A. Introduction	171
	B. Sugar industry	171
	C. Dextran formation	172
	D. Dairy industry	173
	E. Wine industry	173
	Appendix I	174
	References	176

METHODS IN MICROBIOLOGY VOL. 16
ISBN 0–12–521516–9

I. Introduction

The genus *Leuconostoc* comprises six species (Skerman *et al.*, 1980) but there is evidence that these should be reduced to four (Section IV). If this is accepted both *Leuconostoc dextranicum* and *L. cremoris* will be reduced to subspecies of *L. mesenteroides* (Garvie, 1983). In this chapter, the current names are used, but the evidence for the proposed changes in classification is given.

Leuconostocs are normally found living in association with vegetable matter with lactose-fermenting species occurring in milk and dairy products. Leuconostocs are not pathogenic to plants or animals. The genus is of considerable commercial importance with *L. paramesenteroides* as the only species without commercial significance. In nature, leuconostocs are occasionally the dominant flora, but usually they occur as minor components of a bacterial population and are either not recognized or are overlooked. Certain strains are now cultured for commercial reasons with the main emphasis on the particular chemical reaction for which they are important, and with less emphasis on identification and classification of strains.

All lactic acid bacteria depend on the fermentation of carbohydrates for energy, and all form lactate as a major end-product of fermentation of glucose. They can be divided into homofermentative and heterofermentative species depending on the end-products from the fermentation of glucose when they are growing in a good nutritive medium. The homofermentative species use the Embden–Myerhof (EM) glycolytic pathway (Fig. 1) converting glucose to fructose-1,6-diphosphate (EDP), which is split to glyceraldehyde phosphate. The end-product is two moles of lactic acid for each mole of glucose consumed. Fructose-1,6-diphosphate aldolase is a key enzyme in the EM pathway. Streptococci, pediococci and many lactobacilli are homofermentative. Some species possess alternative glycolytic pathways which will be used when, for some reason, the EM pathway is suppressed. The heterofermentative lactic acid bacteria comprise leuconostocs and some lactobacilli; fructose-1,6-diphosphate aldolase is absent and the EM pathway is not used. Instead glucose is converted to glucose-6-phosphate and then to 6-phosphogluconate which is decarboxylated. The resulting pentose is converted to lactic acid and ethanol and/or acetate (Blackwood and Blakley, 1960). This fermentation follows the pentose phosphate pathway initially and then the phosphoketolase pathway. Glucose-6-phosphate dehydrogenase and xyulose-5-phosphate phosphoketolase are key enzymes. The former has a cheap substrate and is easy to assay, the latter has an expensive substrate or can be assayed starting from ribose-5-phosphate, which is not always satisfactory. The production of CO_2 from glucose is easily assessed and, often, a very obvious property of heterofermentative lactic acid bacteria.

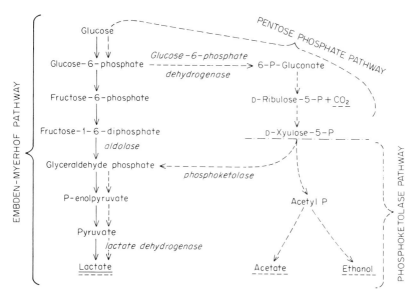

Fig. 1.   Glycolytic pathways in lactic acid bacteria. Main pathway in homofermentative lactic acid bacteria ⟶. Main pathway in heterofermentative lactic acid bacteria --→.

The differences between the homofermentative and heterofermentative bacteria are unrelated to morphology which is still used as a primary method of classification. In many ways the leuconostocs have more in common with the heterofermentative lactobacilli than with the streptococci. This will be discussed further in later sections.

II. Isolation of leuconostocs and general media

A. Isolation from natural habitats

Selective media specific for leuconostocs have not been devised and cultures are isolated from mixed populations because they grow under the conditions chosen for the selection of some other lactic acid bacteria. It is difficult to obtain an estimation of leuconostocs in mixed populations without isolating and identifying randomly picked colonies. Cavett *et al.* (1965) used a combination of 1.0% thallous acetate and 0.01% 2,3,5-triphenyltetrazolium chloride added to a peptone meat extract sucrose medium to distinguish *Streptococcus lactis* and *S. faecalis* from leuconostocs. 5.0% sucrose was included so that colonies forming dextran could be counted, these were grouped as leuconostocs. This procedure is not satisfactory as it could include dextran forming lactobacilli and would exclude non-dextran forming leuconostocs.

Leuconostocs live mainly on vegetable matter where Gram-negative organisms and aerobic spore formers may predominate and are likely to outgrow any leuconostoc present. When surface scrapings of vegetables are inoculated into a yeast glucose phosphate peptone broth (YGPB or Medium 1, Appendix I) containing 0.1% thallium acetate and 0.0005% crystal violet and incubated until turbid, the spore formers and Gram-negative rods are suppressed. Subsequent plating on agar medium of the same composition as the broth allows isolation of Gram-positive cocci. This technique does not enable any estimation of the initial population of leuconostocs to be made. Both *L. mesenteroides* and *L. paramesenteroides* will grow on acetate agar selective for lactobacilli. It is probable that *L. dextranicum* and *L. lactis* will also grow, but unlikely that *L. cremoris* would form colonies as it would be outgrown by any lactobacilli or other leuconostocs present. The pH (5.4) of the acetate agar would also act as an inhibitor of *L. cremoris*.

In a mixed population containing known species, i.e. cheese and butter starter, it is possible to monitor the numbers of each species present. Table I shows the properties which distinguish leuconostocs from starter streptococci. Leuconostocs are more fastidious and require a more complex growth medium than streptococci. The combination of the ability to use citrate and to hydrolyse arginine can be used to recognize the components of starter culture. No lactic acid bacteria use citrate as a source of energy but can break it down to form important flavour components when cultures are growing with a carbohydrate to supply energy. The breakdown of citrate results in the production of acetoin/diacetyl, i.e.

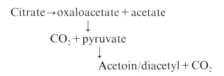

The pyruvate from all sources are pooled and the excess over that required for the regeneration of NADH is converted to acetoin. Galesloot *et al.* (1961) describe the preparation and use of a medium containing insoluble calcium citrate which can be suspended in carboxymethyl cellulose and then this suspension incorporated into agar media. The medium is opaque, but citrate-utilizing colonies form clear zones. The carbohydrate content of the medium is low, which both encourages the bacteria to use citrate and prevents acid formed by non-citrate users dissolving the calcium and causing clear zones. This medium is most useful when the components of the bacterial mixture are known, for example cheese starter cultures where any leuconostocs present

TABLE I

Some differentiating properties of cheese starter streptococci and leuconostocs

	Streptococci			Leuconostocs	
	S. lactis	S. lactis subsp. diacetylactis	S. cremoris	L. lactis	L. cremoris
Utilization of citrate	−	+	−	+[a]	+
Growth on whey agar	+	+	+	−	−
Growth on supplemented whey agar	+	+	+	+	+
Arginine hydrolysed	+	Usually +	−	−	−

[a] Not all strains use citrate, but those that do not are unlikely to be a component of starters.

will be citrate users. Not all leuconostocs use citrate and some which do may fail to grow on Galesloot's medium. The agar has limited uses with wild populations. *S. lactis* subsp. *diacetylactis* is common in cheese starter cultures and uses citrate. It normally hydrolyses arginine so that agar media with a low glucose content but containing 1.5% L-arginine will become alkaline while leuconostocs will form acid in the zone around each colony. The inclusion of both arginine and citrate is useful in separating leuconostocs from *S. lactis* subsp. *diacetylactis*.

In wine, leuconostocs and lactobacilli coexist. They can be separated from acetobacter, which will not grow under anaerobic conditions, and from spore formers which are inhibited at low pH (4.5). Yeasts will be inhibited when actidione (20 mg l^{-1}) is added to media. Leuconostocs and lactobacilli have similar growth requirements and only the isolation of colonies followed by identification will show which bacteria are present. Most workers favour a grape juice/yeast extract agar with or without added organic acids (Peynaud and Dupuy, 1964). Tomato juice can be used in place of, or in addition to, grape juice; ATB (Table I in Appendix I) is given as a medium suitable for the growth of *L. oenos*. Leuconostocs living in wine prefer a low growth temperature (20–25°C), an atmosphere containing 10% CO_2, and it can take up to seven to ten days for visible colonies to develop.

B. General method of cultivation

1. *Media*

L. oenos lives only in wine and associated environments not inhabited by other species of the genus. Different media and growth conditions are used for *L. oenos* from those for other species and, even when identification tests are done, media are adapted to suit acidophilic or non-acidophilic strains.

The composition of some general media are given in Table 1 in Appendix I. It has been found that MRS is suitable for most leuconostocs and pediococci, as well as the lactobacilli for which it was designed. MRS is not normally used for streptococci which can grow on less complex media, which only support poor growth of leuconostocs, even of the least fastidious species *L. mesenteroides*. MRS contains 2% glucose but this is not necessary for leuconostocs which cease growing at about pH 4.5, in contrast to lactobacilli and pediococci which lower the pH to 4.0 or less. The high sugar content in MRS can cause charring in autoclaving, particularly in large volumes (500 ml and over). When large volumes of any media are required it is suggested that the glucose (or other carbohydrates) is autoclaved (15 lb, 15 min) separately as an aqueous solution and added aseptically to the rest of the medium before inoculation. When Medium 2 was prepared the glucose and cysteine were dissolved in the tomato juice, Seitz filtered and added to the rest of the medium which had been autoclaved (15 lb, 15 min). In some early work a yeast glucose citrate broth (YGCB) was used but this has now been replaced by Medium 1, YGPB or MRS.

2. Growth conditions

Most leuconostocs grow at 30°C (37°C is too high) and *L. cremoris* and *L. oenos* prefer 22°C which is probably a better general temperature for the whole genus, although growth will be slower than at 30°C with many strains. *L. cremoris* and *L. oenos* grow better when 0.05% cysteine is added to the media and cysteine does not inhibit growth of other species. In broth all species will grow under atmospheric conditions but an atmosphere of $H_2 + 10\%$ CO_2 is better for colonies growing on an agar surface. At one time it was argued that leuconostocs were more aerobic than other lactic acid bacteria because they metabolized glucose by the hexose monophosphate pathway but reducing conditions have been found to improve the growth of a number of strains which grow poorly aerobically. No leuconostocs grow rapidly but many strains give good growth with overnight incubation; *L. cremoris* and isolated strains of other species may require 48 h incubation while *L. oenos* may require seven to ten days before reasonable turbidity is produced in broth.

III. Identification of species

A. Preliminary identification and separation from other lactic acid bacteria

Table II gives a few easily determined characters which should be adequate for placing any lactic acid bacteria in its correct genus. Morphology may present difficulties because both leuconostocs and streptococci can form ovoid and even rod-shaped cells, but both are usually coccoid when growing in milk or supplemented milk. Lactobacilli, on the other hand, can be short rods with pointed ends, in other words the morphology of these three genera merge and it may be difficult to be sure to which genus a particular strain belongs. Morphology combined with gas production and hydrolysis of arginine should allow for correct identification of leuconostocs, but gas production may be weak in freshly isolated strains. The technique described by Abd-el-Malek and Gibson (1948) is reliable providing a heavy inoculum of a well growing culture is used for inoculation. No leuconostocs hydrolyse

TABLE II
Differentiation of leuconostocs from other lactic acid bacteria

	Leuconos-tocs	Lactobacilli		Streptococci	Pediococci
		Hetero-fermentative	Homo-fermentative		
Morphology	Coccus-cocco-bacillus	Cocco-bacillus →rod	Cocco-bacillus →rod	Coccus→ cocco-bacillus	Coccus
Division	1 plane	1 plane	1 plane	1 plane	2 planes
Growth in litmus milk	− or slight	−	− or slight	Usually good	− or slight
Gas from glucose	+	+	−	−	−
Hydrolysis of arginine	−	− or +	−	− or +	− or +
Dextran from sucrose	− or +	− or +	−	Mostly −, a few species +	−
Type of lactic acid	D(−)	DL	D(−), DL or L(+)	L(+)	DL or L(+)
Habitat Plant	+	+	+	A few species +	+
Animal	−	+	+	+	−

arginine, whereas many species of heterofermentative lactobacilli do. The main exceptions are *Lactobacillus viridescens* and occasional strains of *Lb. confusus* and, as both these species form dextran from sucrose, their separation from leuconostocs may be difficult. The difference and similarities of these two lactobacilli to the leuconostocs are discussed further.

B. Further identification by classical methods

The basis for arguing that the heterofermentative lactic acid bacteria form a natural group has been given in Section I, and the properties of *Lb. confusus* and *Lb. viridescens* will be given along with those of leuconostocs. These two lactobacilli are biochemically closer to the non-acidophilic leuconostocs than to the other heterofermentative lactobacilli. Difficulties in identification have existed in the past but now the similarities between leuconostocs and *Lb. viridescens* and *Lb. confusus* are recognized and the characteristics of each species understood errors in identification should be few. *Lb. viridescens* was first recognized as a species by Niven and Evans in 1957 and *Lb. confusus* by Holzapfel and Kandler in 1969. Several strains of *Lb. confusus* studied by Sharpe *et al.* (1972) had been identified as *L. mesenteroides*, and it is possible that other strains have also been identified as leuconostocs.

1. *General points*

Table III gives the properties of two species of lactobacilli and those of all species of leuconostocs. *L. oenos* is the most easily recognized species because it grows in media with a low initial pH, and it is tolerant of ethanol. Some strains of *L. oenos* have a requirement for 4-0-(α-D-glucopyranosyl)-D-pantothenic acid (Amachi *et al.*, 1970). This substance is also known as tomato juice factor (TJF) (Garvie and Mabbitt, 1967). The requirement varies with different growth conditions and different strains, so that it cannot be used to identify *L. oenos*. Most strains of *L. oenos* will destroy TJF even if they do not have a high requirement for it, but no other lactic acid bacteria are known to use TJF as a form of pantothenate. The requirement for TJF can be overcome by a high level of pantothenate and also by a heavy inoculation so that media free of tomato juice or of TJF can be used for many purposes. Obviously tomato juice cannot be added to media when fermentation properties are studied, in these tests the basal broth used is simple and a high cell inoculation is used. The basal media used for other species is unsuitable for *L. oenos* but can be modified by lowering the pH to 5.2, including 0.5% agar and using 0.004% bromo-cresol green as indicator (Garvie, 1967b). *L. oenos* shows considerable variation in ability to ferment carbohydrates, which has led some workers to consider that there is more

TABLE III

Differentiation of species within the genus *Leuconostoc* and separation from *Lactobacillus confusus* and *Lactobacillus viridescens*

	Leuconostoc						Lactobacilli	
	L. mesenteroides subsp. *mesenteroides*	*L. mesenteroides* subsp. *dextranicum*	*L. mesenteroides* subsp. *cremoris*	*L. paramesenteroides*	*L. lactis*	*L. oenos*	*Lb. confusus*	*Lb. viridescens*
Hydrolysis of arginine[a]	−	−	−	−	−	−	(±)	−
Mucoid colonies on sucrose agar[a]	+	+	−	−	−	−	+	±
Lactic acid formed from glucose[a]	D(−)	D(−)	D(−)	D(−)	D(−)	D(−)	DL	DL
Growth at 45°C[a]						−	(±)	−
Growth at pH 4.8 in CMB[a]				(+)		+	NI	NI
Growth at pH 3.7 in CMB[a]				−		+	NI	NI
Growth in 10% ethanol in CMB	−	−	+	−	−	+	NI	NI
Dissimilation of citrate[a]	(+)	(+)	(−)	(+)	(+)	+	∓	+
Fermentation of:								
Arabinose[a]	+	−	−	(±)	(∓)	∓	+	−
Xylose[a]	+	(±)	−	(±)	(±)	(∓)	+	−
Fructose	(±)	(±)	+	+	+	(∓)	+	(±)
Galactose	(±)	+	−	+	(±)	(∓)	+	−
Mannose	+	(∓)	−	+	+	(±)	+	−
Cellobiose	(∓)	(±)	+	(∓)	+	−	+	−
Lactose[a]	(±)	(±)	+	(±)	+	−	∓	+
Maltose	(±)	+	−	(±)	(±)	(±)	+	−
Melibiose[a]	(±)	(±)	−	(±)	(±)	−	∓	+
Sucrose	+	+	−	+	(∓)	+	+	+
Trehalose[a]	+	+	−	+	(∓)	−	(∓)	−
Raffinose[a]	±	±	−	±	−	−	(∓)	−
Dextrin	−	−	−	(∓)	−	−	−	−
Mannitol[a]	±	(∓)	−	∓	−	NI	+	−
Amygdalin	∓	∓	−	(∓)	−	+	+	−
Aesculin	(±)	(∓)	−	+	−	(±)	+	−
Salicin	(±)	(∓)	−	−	−	−	+	−

[a] Tests suggested as the most useful in separating species.
± Variable reactions.
(±) Most strains positive, occasional strains negative.
(∓) Most strains negative, occasional strains positive.
NI, no information.

than one species of acidophilic leuconostoc. Peynaud and Domercq (1968) divided their isolates into two and Nonomura and Ohara (1967) into five species. However, despite phenotypic variation, it is probable that all acidophilic leuconostocs belong to a single genotype (Garvie and Farrow, 1980).

2. *Carbohydrate fermentations*

A basal medium for fermentation tests is given in Appendix I. It is based on that used by Garvie (1960). Sharpe *et al.* (1972) used a different basal medium which was designed for lactobacilli but is suitable for *L. mesenteroides*. The latter medium has an initial pH of 6.2 which is low for some leuconostocs, particularly *L. cremoris*. A pH of 6.7 is probably a better choice. Whittenbury (1963) used a semi-solid medium but this does not appear to be necessary. The additional information claimed from the use of such a medium is not essential to the interpretation of the fermentation patterns. The acid present in well grown cultures can change the indicator on inoculation into fermentation tests particularly when a heavy inoculation is required. Therefore, cultures should be centrifuged, the supernatant discarded and the cells taken up in sugar-free basal medium (or peptone water when the basal medium contains agar), equal to half the original volume. Up to 0.1 ml of this cell suspension can be inoculated into 5.0 ml of test medium without any change of indicator. This technique should also be used for *L. oenos*.

L. cremoris has a distinctive fermentation pattern as it can use only glucose, galactose and lactose. *Lb. viridescens* ferments fewer sugars than *Lb. confusus* so that these species resemble *L. dextranicum* and *L. mesenteroides*, respectively. *L. lactis* is adapted to live in milk and ferments lactose more readily than other species, in addition trehalose is generally not fermented. *L. paramesenteroides* has few differences from *L. mesenteroides* and was at one time considered to be a non-dextran forming variety of *L. mesenteroides*. It is easy to determine fermentation patterns but the variable results obtained with strains of the same species and the similarity of the patterns found with different species make it difficult to use fermentation patterns with confidence when identifying leuconostocs.

3. *Dextran production*

Neither *L. lactis* nor *L. paramesenteroides* form dextran, and this is a useful property in separating them from *L. mesenteroides*. Dextran production is observed on the surface of agar containing 5% sucrose. Cultures are incubated aerobically at 20–25°C for three to five days. The type of colony formed depends on the chemical structure of the dextran and McCleskey *et al.* (1947) divided strains of *L. mesenteroides* into four groups on the basis of the type of colony formed on sucrose agar. Later workers have not used this property to divide strains, and since the type of dextran formed by different cells within a culture can vary (Brooker, 1977), the type of dextran formed is not taxonomically important. The dextran formed by *Lb. confusus* and *Lb. viridescens* has not been studied and nothing is known about the variation in the type of dextran formed by different strains.

4. *Lactic acid*

The type of lactic acid formed by lactic acid bacteria has long been used to separate the different genera and species. This property was not used widely in the first half of this century as the determination of lactic acid type was tedious; it was made quick and easy with the development of enzymic techniques. L(+)-Lactate is readily determined using mammalian lactate dehydrogenase (LDH). D(−)-Lactate can be determined by a similar technique and total lactic acid can be determined chemically. Thus DL-lactate production can be studied using two enzymic methods or using total acid together with L(+)-lactate. Leuconostocs only form D(−)-lactate from glucose, whereas heterofermentative lactobacilli form DL- and streptococci L(+)-lactate. Both L(+)-LDH and D(−)-LDH are available commercially and suppliers give details of conditions under which lactate can be assayed enzymically. An alkaline pH is necessary for the conversion of lactate to pyruvate but alkaline solutions will absorb CO_2 which can interfere with the assay. Storage of reagents in a desiccator with NaOH pellets should keep solutions CO_2 free.

DL forming bacteria may not produce equal amounts of both isomers (Garvie, 1967d). Only a small amount of lactate formed by *Lb. viridescens* consists of the L(+) isomer, but *Lb. confusus* forms almost equal amounts of D- and L-lactate. The media and conditions of growth can influence the proportion of each isomer and many bacteriological media contain measurable amounts of L(+)-lactate from peptone and meat extract. A high initial content of L(+)-lactate in media can be overcome if tomato juice, which is free of lactate, is used in place of meat products (Garvie, 1967d). The medium used in the study of the production of D- and L-lactate by lactic acid bacteria may appear deficient compared with MRS, but it supports good growth of most lactic acid bacteria including *L. oenos* for which species the pH is lowered to 5.0–4.8. When traces of one isomer of lactate are detected it is advisable to make a second determination at a higher concentration because there can be slight differences in the blank value of the reagent mix. In some techniques it is suggested that the reaction goes to completion and that it is possible to calculate the amount of lactate from the increase in absorption at 340 nm. Experience with the technique has shown that it is better to include lactate standards (max 16 μg for the L isomer but 160 μg for the D isomer) in every series of determination, DL-lithium lactate can be used for both assays. All material examined has been found to contain equal amounts of each isomer. Most lactic acid bacteria reduce the pH of the medium to below 5.0 and Table IV gives an indication of the amount of lactate which will be formed when cultures are grown in dilute tomato broth (DTB, Table 1 in Appendix I). These values apply only to cultures with an initial pH of 6.7 and do not apply to *L. oenos*.

TABLE IV

Final pH, and lactate formed by leuconostocs growing in dilute tomato broth

Initial pH	Final pH	Amount of lactate (mg ml^{-1})	Dilution for testing[a]	
			L(+)-Lactate	D(−)-Lactate
6.7	5.0–4.7	2–4	1/50	1/5
	4.7–4.3	4–8	1/100	1/10
	4.3–3.9	8–16	1/200	1/20

[a] 0.2 ml of culture dilution in assay of 2.5 ml total volume.

L. oenos converts malate to lactate but it is not unique, and Alizade and Simon (1973) studied the fermentation of L- malate by *L. mesenteroides* ATCC 12291. They found that L-malate is converted to L(+)-lactate whereas glucose is converted to D(−)-lactate. The former reaction does not involve pyruvate and an LDH. It is essential, therefore, that the medium used for the determination of lactate from glucose is malate free. Other organic acids, for example citrate, which are metabolized to pyruvate can be included because pyruvate is only metabolized to lactate by an LDH.

C. Amino acid and vitamin requirements

All species of *Leuconostoc* require some preformed vitamins and amino acids. There are differences between the various species but they are not particularly useful taxonomically and any use is outweighed by the labour and expense of materials. For commercial operations, a knowledge of growth requirements can be very important both in fermentations used in making food products and in growing strains for converting sucrose to dextrans.

Garvie (1967c) described media and methods that were based on earlier studies, and gives results for several strains of every species. The two most important characters found were that *L. lactis* did not require folic acid whereas strains of other species showed a requirement, and *L. paramesenteroides* required riboflavin whereas *L. mesenteroides* did not. No purified TJF was available and tomato juice could not be incorporated in the medium. Many strains of *L. oenos* grew, and it is assumed that the conditions were correct and the pantothenate level high enough for it to be used in place of TJF.

D. Metabolism and enzyme studies

1. *General*

Modern recognition of bacterial species uses metabolic pathways and a study of key enzymes. These properties show little variation between strains of a species and are less variable than properties depending on the chemical ability of the growing cell. Obviously, the enzymes easiest to use are those of major importance to the life of the bacterium which have a readily available cheap substrate, and which can be directly assayed. Enzymes concerned with the few initial reactions, and the terminal reactions in the fermentation of glucose come into this category. A description of the pathway of glucose breakdown in leuconostocs is given in Section I. The exact method by which 6-phosphogluconate is decarboxylated and converted to ribulose-5-phosphate is uncertain (Yashima and Kitahara, 1969). Both glucose-6-phosphate dehydrogenase (G-6-PDH) and LDH can be directly assayed, are present in large amounts in cell free extracts, are stable and have relatively cheap substrates and coenzymes. The absence of an enzyme is more difficult to show than its presence but fructose-1,6-diphosphate aldolase can be used to separate heterofermentative from homofermentative lactic acid bacteria. G-6-PDH can also be used although some species of homofermentative lactobacilli possess this enzyme. LDH is present in all lactic acid bacteria and can be used to identify species of all the different genera.

2. *Lactate dehydrogenases*

Cells must be broken because the LDHs can only be assayed in cell-free extracts. Breakage can be achieved mechanically but some strains do not break readily. However, enough enzyme is usually obtained from cells of a 100 to 200-ml culture even when few cells appear disrupted. Different conditions of growth and breakage have been used by different workers but these do not affect the enzyme (Sharpe *et al.*, 1972; Hontebeyrie and Gasser, 1975). Leuconostocs have a single LDH forming D(−)-lactate whereas hetero-fermentative lactobacilli have two LDHs and form D(−)- and L(+)-lactate. All D(−)-LDHs but only some L(+)-LDHs are reversible (Table V). At one time leuconostocs were classified as streptococci but the latter genus has FDP-dependent L(+)-LDHs, enzymes with many differences from the D(−)-LDHs of leuconostocs.

TABLE V
Development of lactate dehydrogenases of some lactic acid bacteria after electrophoresis

	L(+)-LDH			D(−)-LDH	
	Lactate	Pyruvate	Pyruvate + FDP	Lactate	Pyruvate
Streptococci	−	−	+	Absent	
Leuconostocs Lactobacilli		Absent		+	+
Lb. confusus Lb. viridescens Lb. fermentum	−	+	.	+	+
Lb. reuteri Lb. brevis Lb. buchnerii	+	+	.	+	+

Enzymes can be recognized by differences in electrophoretic mobility and the technique developed for mammalian LDH was adapted for bacterial LDHs. The conditions of electrophoresis vary with different workers and details are given in the various publications (Garvie, 1969; Gasser, 1970). Where both L(+)- and D(−)-LDHs are present these can be identified using the appropriate lactate as substrate, providing the L(+)-LDH reacts with lactate. In *Lb. confusus* and *Lb. viridescens* the L(+)-LDH can only be located with pyruvate. The technique for locating the enzyme has been published (Garvie, 1969) and to achieve good results some precautions need to be taken. The amount of enzyme loaded should be carefully controlled as overloading results in a large unstained streak where the enzyme has travelled down the gel, with underloading the enzyme will be missed. These problems are less critical when using direct staining with lactate because the reaction can be stopped by immersing acrylamide gels in 5% acetic acid either after a few minutes (heavy loading) or after several hours development (underloading). With pyruvate the reaction is invisible until the gels are transferred to the phenazine methosulphate/nitroblue tetrazolium solution. The initial stage has to be timed to get enough NADH into the gel for the purple colour reaction to show but not prolonged to get a blurred area of no colour. For acrylamide gels 30 min in the pyruvate/NADH solution is recommended.

The D(−)- and L(+)-LDHs are not separately identified using pyruvate and controls with NADH but no pyruvate should be included because NADH oxidases will react if they are present. In most strains NADH oxidases are weak and cause no reaction, but in the type strain of *L. mesenteroides* NCDO 523 and in strains of *Lb. viridescens* an active NADH oxidase has been detected (Sharpe *et al.*, 1972). In the latter species the NADH oxidase can mask the following weaker L(+)-LDH.

LDHs can be assayed by following NAD reduction at 340 nm. Most crude cell extracts are suitable for use with lactate and NAD. Leuconostoc LDH can be assayed with pyruvate providing allowance is made for NADH oxidase activity, but when two LDHs are present assays using pyruvate give little information. The various techniques used for assay and electrophoresis of bacterial LDH are given in a review by Garvie (1980).

This type of study has shown that the D(−)-LDH of the non-acidophilic leuconostocs *Lb. confusus* and *Lb. viridescens* has the same electrophoretic mobility, whereas *L. oenos* has a distinct LDH (Fig. 2). The L(+)-LDH is different in the two species of lactobacilli.

Lactic acid bacteria also have NAD-independent LDHs. These have been studied in several species (Garvie, 1980). The activity is weak and they are not useful in identifying *Leuconostoc* species.

3. *Glucose-6-phosphate dehydrogenase*

Glucose-6-phosphate dehydrogenase (G-6-PDH) can be assayed in the cell-free extracts prepared for LDH. The technique for handling the two enzymes is the same and, in many strains, both enzymes can be located in the same electrophoresis gel if both lactate and glucose-6-phosphate are included in the developing solution. In leuconostocs the G-6-PDH of non-acidophilic species can use either NAD or NADP as a coenzyme but with a preference for NAD. Electrophoresis does not separate the enzyme of the different species (Fig. 2). The G-6-PDH of *Lb. confusus* and *Lb. viridescens* can also use either NAD or NADP but on electrophoresis they are separated from the G-6-PDH of leuconostocs (Garvie, 1975). The G-6-PDH of *L. oenos*, on the other hand, uses NADP as a coenzyme, is difficult to detect after electrophoresis and, in some preparations, it has not been possible to demonstrate that it is present.

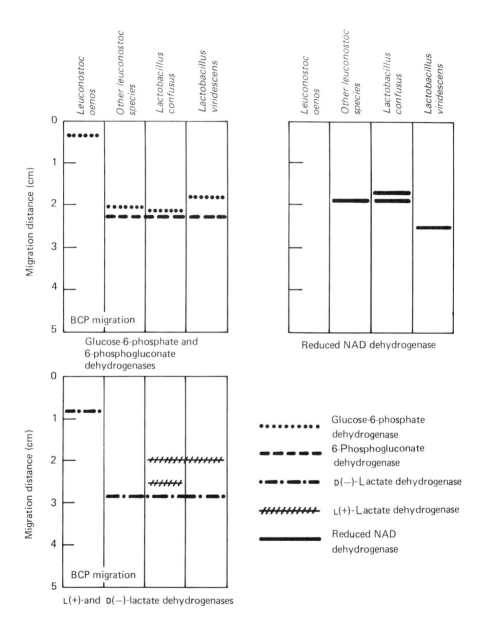

Fig. 2. The electrophoretic patterns of some dehydrogenases of the glycolytic system of leuconostocs and two species of lactobacilli.

4. *Immunological studies using dehydrogenases*

Hontebeyrie and Gasser (1973, 1975) have described methods of purifying both LDH and G-6-PDH and used the purified enzymes for inoculation into rabbits to prepare antisera active against the enzymes. New techniques using affinity chromatography have now been developed for dehydrogenases (Kelly *et al.*, 1978). Any future work should use these improved techniques.

Crude cell extracts can be tested against antisera and the techniques for this work have been described (Hontebeyrie and Gasser, 1975). Differences between the LDHs of the species of leuconostoc were found and also differences between the G-6-PDHs. The grouping of the strains examined was identical with both enzymes (Table VI). Seven groups were found: A contained *L. lactis*, F contained *L. paramesenteroides*, G contained *L. oenos*, D contained *L. dextranicum, L. cremoris* and most strains of *L. mesenteroides*. B, C and E were small groups each containing only one or two strains of *L. mesenteroides*. Many of the strains examined came from culture collections and more information, particularly using freshly isolated strains, is needed. This type of work is of great value in further separating enzymes which are electrophoretically identical. It is also useful as a basis for working out the evolutionary relationships between species.

TABLE VI

Immunological groups of D(−)-lactate dehydrogenase and glucose-6-phosphate dehydrogenases of leuconostocs

	Leuconostocs					
	L. mesenteroides					
	subsp. *mesenteroides*	subsp. *dextranicum*	subsp. *cremoris*	*L. lactis*	*L. paramesenteroides*	*L. oenos*
Lactate dehydrogenase	BCDE	D	D	A	F	G
Glucose-6-phosphate dehydrogenase	bcde	d	d	a	f	−

5. *Hybrid enzymes*

Bacteria seldom contain isoenzymes and in leuconostocs only a single band of each LDH is normally detected. Chilson *et al.* (1965) described a freeze-thaw technique for preparing hybrids using the tetrameric mammalian LDH. They showed the similarity in structure between the LDHs of several species. In leuconostocs the LDH of *L. oenos* and that of other species are well separated by electrophoresis. Using the method described for mammalian L(+)-LDHs with the LDHs of leuconostocs, it was found that *L. oenos* LDH formed a single active hybrid enzyme with the LDH of any other species of the genus and also with the LDH of *Lb. confusus* and *Lb. viridescens*, but not with the LDH of other species of lactic acid bacteria. A weak hybrid was formed with the LDH of *L. lactis* and that of some leuconostocs (Garvie, 1975).

These observations show that the structure of the LDH of leuconostocs is a dimer. It further shows that the structure of the LDH of *L. oenos* is not too different from that of other species otherwise the hybrid protein would not be an active enzyme.

More evidence from extended studies might be useful. This technique does not require any enzyme purification but enzymes with the same or very similar electrophoretic mobility clearly cannot be used together unless one protein is modified.

E. Peptidoglycan types in cell walls

Most of the work on the peptidoglycan types in bacterial cell walls has been done in Munich by various workers (Schleifer and Kandler, 1972). A wide range of bacteria have been examined and so there is good evidence that information from the work is useful in bacterial taxonomy. All species of *Leuconostoc* and *Lb. confusus* and *Lb. viridescens* have a similar type of peptidoglycan which is different from that in other species of heterofermentative lactobacilli (Table VII). Reference to the techniques used are given by

TABLE VII

Peptidoglycan types of bacterial cell walls of leuconostocs and some heterofermentative lactobacilli

Species	Peptidoglycan
L. mesenteroides	L-Lys-L-Ser-L-Ala$_2$, L-Lys-L-Ala$_2$
L. dextranicum	L-Lys-L-Ser-L-Ala$_2$
L. cremoris	L-Lys-L-Ser-L-Ala$_2$
L. lactis	L-Lys-L-Ser-L-Ala$_2$, L-Lys-L-Ala$_2$
L. paramesenteroides	L-Lys-L-Ser-L-Ala$_2$, L-Lys-L-Ala$_2$
L. oenos	L-Lys-L-Ala-L-Ser, L-Lys-L-Ser-L-Ser
Lb. viridescens	L-Lys-L-Ala-L-Ser, L-Lys-L-Ser-L-Ala$_2$
Lb. confusus	L-Lys-L-Ala$_2$, L-Lys-L-Ala
Lb. fermentum	L-Orn-D-Asp
Lb. reuteri	L-Lys-D-Asp
Lb. brevis }	L-Lys-D-Asp
Lb. buchnerii }	

Schleifer and Kandler (1972). Not all strains of leuconostocs within any species have the same murein type and there is considerable overlapping of type in different species so that the information about peptidoglycans is of greater value in showing which genera leuconostocs resemble rather than distinguishing species within the genus. *L. oenos* is not markedly different from other species which may appear to be in contrast to some of the information obtained from enzyme studies.

F. Nucleic acid studies

The study of nucleic acid in bacteria is widespread and is fundamental to any natural classification system. Indeed, the results from work with deoxyribonucleic acid (DNA) overrides conclusions from other information when there is a conflict of interpretations.

1. %(G + C) content of DNA

Good yields of high molecular weight DNA will only be obtained if cells are harvested at the correct time. This is usually the late logarithmic or early stationary phase of growth. Each strain used must be considered separately and the amount of inoculum, time of inoculum and conditions of incubation altered to suit the strain. The culture used for seeding the medium from which cells will eventually be harvested must also be considered. Different growth will be obtained if a mother culture in the early stationary phase is used as compared with one of the same optical density which has reached the stationary phase some hours before it is used. It is impossible to give precise conditions for every species for within a species different strains will require slightly different conditions.

L. mesenteroides, L. dextranicum, L. lactis and *L. paramesenteroides* do not grow rapidly and usually it is possible to obtain a suitable culture after overnight incubation (18 h) at 30°C using a 0.001% inoculum. *Lb. confusus* grows rapidly and may well be overgrown if the same conditions are used. Two techniques are possible, either a heavy inoculum followed by 3–5 h incubation or a small inoculum and the chilled inoculated media held for some hours in ice before incubation starts during the night. Slow growing species *L. cremoris* and *L. oenos* do not cause many problems. They can be left until there is reasonable opacity (between 24 and 48 h for *L. cremoris* but seven days has been used for strains of *L. oenos*). DNA extracted from *Lb. viridescens* usually has a low hypochromicity (about 0.30) and yields are low.

Leuconostocs are sensitive to lysozyme providing the enzyme has time to act. Lysozyme binds to cells washed free of salt (Metcalf and Deibel, 1973) and this is recommended, although it can make pelleting of cells on centrifuging difficult. Sometimes it is necessary to work from an initial cell slurry rather than a pellet. When Marmur (1961) developed a technique for preparing bacterial DNA he chose to work at pH 8.0 with EDTA in order to inhibit DNAase. Lysozyme has little activity at pH 8.0 and lactic acid bacteria do not have extracellular DNAase. When cells are difficult to lyse, it is better to work at pH 7.0 and it was found that 4-amino salicylate is a better additive than EDTA (Garvie, 1976). Other workers have used different conditions (Hontebeyrie and Gasser, 1977). Lysozyme is not attacked by proteolytic enzymes and, therefore, it is possible to add pronase with lysozyme and allow both enzymes to work overnight at 37°C.

Many procedures have been described for purifying DNA from contaminating protein and RNA and strains vary in the ease with which this can be done. A highly purified, high molecular weight DNA is essential for DNA/DNA hybridization work (DeLey and Tutgat, 1970). The % (G + C) content of DNA can be measured by melting temperatures even if some impurities are present, and purification is not essential when the %(G + C) content of DNA is estimated from buoyant density. Few publications give figures for protein and RNA contamination but these should be measured chemically until the technique chosen has been shown to be adequate. Generally, there should be less than 5% (w/w) protein or RNA in DNA. Some techniques involve precipitating DNA with ethanol and, if a method of this type is selected, problems may arise in dispersing the DNA for the next stage of purification. The method recommended by Kirby *et al.* (1967) is probably the most satisfactory, that is centrifuging a viscous solution adding more solute to the precipitated DNA and repeating the procedure. Complete dispersal of DNA before estimating the melting temperature is also essential. Solutions suitable for storage have an optical density at 260 nm of at least 20 and so considerable dilution is necessary. Freshly prepared DNA is sometimes difficult to disperse, whereas preparations stored for two to three months present no problems. Holding the diluted DNA at 37°C for 60 min with occasional gentle shaking will probably result in complete solution.

Escherichia coli K12 is suitable as a control DNA, but *L. mesenteroides* NCDO 768 (ATCC 12291) is probably better. The standard formula for calculating %(G + C) content of DNA from T_m is based largely on results obtained by Marmur and DeLey (DeLey, 1970). Both these laboratories found the T_m for *E. coli* was 90.6°C. Marmur also used *L. mesenteroides* ATCC 12291 and obtained a figure of 86.5°C. Other laboratories obtain figures 1°C higher so that it is necessary to use a control DNA and to adjust the calculation according to the figures obtained for the control. The reason for the discrepancy is not known.

The %(G + C) content of DNA from different species is shown in Table VIII.

TABLE VIII

The % $(G+C)$ content of the deoxyribonucleic acid of species of the genus *Leuconostoc* and of two species of lactobacilli expressed as a percentage

Species	% $(G+C)$
L. mesenteroides	38–40
L. dextranicum	38–39
L. cremoris	38–40
L. lactis	42–43
L. paramesenteroides	38–39
L. oenos	39–41
Lb. viridescens	41–42
Lb. confusus	44–45

2. DNA/DNA hybridization

Two studies have been made of the genus *Leuconostoc*. Garvie (1976) used the membrane filter method of Denhardt (1966), whereas Hontebeyrie and Gasser (1977) used an hydroxyapatite method. The results of the two studies are in agreement. The choice of technique is up to the laboratory since both methods have been well tested with DNA from many species. The results obtained from this work are summarized in Table IX and the implications are discussed in Section V.

3. RNA/DNA hybridization

Ribonucleic acid (RNA) evolves slowly and studies using RNA are valuable in showing the relationships between species and genera which are separated by DNA/DNA hybridization (Fox *et al.*, 1977). A number of genera of Gram-negative bacteria have been examined (Gillis and DeLey, 1980) but the Gram-positive bacteria, particularly the lactic acid bacteria, have not been examined. Ribosomal ribonucleic acid (rRNA) is present in large amounts in bacteria and can be extracted from lysed cells by the technique of Moore and McCarthy (1967); Garvie and Farrow (1981) have described a technique for lysing cells of Gram-positive cocci before extracting RNA. The techniques of studying RNA/DNA hybridization are still being developed. Present methods are described by DeLey and DeSmedt (1975) and by Gillespie (1968). Studies with lactic acid bacteria are required. Some preliminary work with leuconostocs has indicated that the RNA of all the non-acidophilic strains is highly related but different from that of *L. oenos* and also that of *Lb. confusus* and *Lb. viridescens*. There is only a low relationship between the two lactobacilli as judged by rRNA/DNA hybridization despite the fact that they

TABLE IX

Summary of hybridization between the deoxyribonucleic acids of strains of leuconostoc and lactobacilli

Species NCOD Number	Hybridization (%) of labelled DNA from					
	L. mesenteroides 523	L. mesenteroides 768	L. lactis 546	L. paramesenteroides 803	Lb. viridescens 403	Lb. confusus 1586
L. mesenteroides A	80–100	30–40	20–40	8–18	15–25	13–17
B	30–40	90–100	25–35	14–20	15	15
L. dextranicum	80–100	30–45	20–35	5–20	15–35	10–20
L. cremoris	70–100	45	25–35	7–8	12–21	13
L. lactis	30–50	28–60	74–100	—	20–50	18–30
L. paramesenteroides	6–20	13–21	10–15	60–100	15–20	27
L. oenos	10–15	10	0–10	5	1	1
L. viridescens	12–25	—	18–20	—	90–100	30–50
Lb. confusus	15–20	—	10–20	—	30–65	80–100

are both heterofermentative lactobacilli (Garvie, 1981).

Messenger RNA (mRNA) is not suitable for hybridization studies. It is not possible to obtain it free from large amounts of rRNA. Yields of mRNA are low and reflect only the RNA involved in the requirements of the bacteria at the time of labelling, usually in the mid logarithmic phase of growth.

G. Serology and bacteriophage typing

Separation of species and strains of non-pathogenic bacteria by serotyping is not a satisfactory technique. In the lactic acid bacteria its use is restricted to species of pathogenic streptococci and some other species also associated with animals.

Bacteriophage typing of host bacteria is restricted to those species which are readily attacked by bacteriophage. Slow growing bacteria seldom occurring in large populations are not susceptible to bacteriophage attack. Leuconostocs belong to this type of bacteria and consequently while leuconostocs have long been found in association with streptococci in cheese starter culture they are not, normally, attacked by bacteriophage. A single report (Sozzi *et al.*, 1978) described leuconostoc bacteriophage.

Leuconostocs are now cultured commercially for the production of dextran, but the dextran itself is believed to protect the bacteria from bacteriophage attack. If this is true, trouble in industry is unlikely.

Bacteriophage attacking wild populations, if they do exist, will be present in small numbers and difficult to isolate. Leuconostoc bacteriophage are, therefore, unavailable for typing strains.

IV. Taxonomy

Bacterial classification is primarily based on morphology, therefore the lactic acid bacteria fall into two families; the cocci are *Streptococcaceae* and the rods are *Lactobacillaceae*. For practical purposes this separation is not helpful. The homofermentative cocci dividing in one plane, the streptococci, occupy a different ecological niche from the cocci dividing in two planes, the pediococci, and from the heterofermentative cocci, the leuconostocs. These last two genera can be found in the same habitat as the lactobacilli (both heterofermentative and homofermentative) and in many ways a family comprising these three genera might be preferable to the present divisions.

The heterofermentative and homofermentative cocci have been accepted as belonging to different genera from the time Hucker and Pederson (1930) studied leuconostocs. The same division has not occurred in the rods— possibly because the latter occur together and have always been studied as a single group. The separation of homo- and heterofermentative organisms into different genera makes evolutionary sense since the essential pathway of sugar metabolism is different in the two groups suggesting that the organisms have been separated for a long time.

The relationship of leuconostocs to the heterofermentative lactobacilli still awaits clarification but the difficulty of separating *Lb. confusus* and *Lb. viridescens* from leuconostocs indicates that they may all belong to a single genus. The other species of heterofermentative lactobacilli have a different cell wall peptidoglycan, different LDHs and G-6-PDHs from those of *Lb. confusus* and *Lb. viridescens*, so there are arguments against forming a single genus to include all the heterofermentative lactic acid bacteria.

The first clear separation of leuconostocs into species resulted from the work of Hucker and Pederson (1930) who showed that *L. mesenteroides* fermented pentoses and sucrose. *L. dextranicum* fermented sucrose but not pentose and *L. cremoris* (*citrovorum* in 1930) did not ferment either pentose or sucrose. This separation was not wholly satisfactory and the reason is now evident because all these species have high homology using DNA/DNA hybridization and form a single geno-species. *L. lactis* and *L. paramesenteroides* were recognized later (Garvie, 1960, 1967c) but may well have been isolated long before they were recognized as separate species, and classified as either *L. mesenteroides* or *L. dextranicum*.

Confusion with named species also occurred with *L. oenos*, and isolations from wine prior to 1967 used the names of non-acidophilic species, probably incorrectly. Lactobacilli found in wine are also found in other habitats and there was no reason to think that the leuconostocs found in wine and non-acid environments were different species. Leuconostocs of wine have so far not been isolated from other environments.

Classification of bacteria from natural habitats cannot rely on elaborate techniques. In leuconostocs it is difficult to select a few tests which will be satisfactory and an attempt has been made to classify the genus using multivariate analysis. When information of the type normally used in numerical taxonomic studies was analysed a clear separation into groups was not achieved, the majority of strains of *L. mesenteroides*, *L. dextranicum*, *L. paramesenteroides* and *L. confusus* came in a single cluster. *L. lactis* and *Lb. viridescens* were in separate clusters, whereas *L. cremoris* and *L. oenos* were separated from other species but not from each other. In Table III the tests considered of most use in separating the different species are marked by a footnote.

A possible future grouping of the heterofermentative lactic acid bacteria is given below, by priority the name of the genus would be *Leuconostoc* leaving *Lactobacillus* for the homofermentative rods.

Subgenus 1. *Leuconostoc oenos*

Subgenus 2. *Leuconostoc mesenteroides* subsp. *mesenteroides* ⎱
                                    subsp. *dextranicum*   ⎬   (Garvie, 1983)
                                    subsp. *cremoris*   ⎰
            Leuconostoc paramesenteroides
            Leuconostoc lactis

Subgenus 3. *Lactobacillus confusus*
            Lactobacillus viridescens

Subgenus 4. *Lactobacillus fermentum*
            Lactobacillus reuteri
            Lactobacillus brevis
            Lactobacillus buchnerii

Classification is a developing science and as knowledge increases it may well change to accommodate the new facts.

V. Commercial importance

A. Introduction

The commercial importance of leuconostocs can be divided into two categories: natural and manufactured. In the first category (natural) the bacteria grow and the changes resulting from this growth are important; in the sugar industry spoilage occurs but in the dairy and wine industries the fermentations are beneficial. It is necessary to be able to identify the bacteria responsible; in some instances, particularly in the sugar and wine industries, the bacteria are natural contaminants, whereas in the dairy industry leuconostocs are kept as cultures, but can also occur naturally in milk and cheese. The second category (manufactured) concerns the production of dextran by *L. mesenteroides*. In this work it is necessary to be able to identify and keep stable the strain(s) used, and to ensure that they are free of contamination from wild populations.

The different species of leuconostoc are each adapted to a particular habitat and there is little overlapping, except in the dairy industry and on herbage. A knowledge of the growth characteristics of the various species will help to eliminate the sort of mistakes in identification that have occurred in the past.

B. Sugar industry

L. mesenteroides grows actively on sucrose and forms large amounts of dextran, whereas *L. dextranicum* is a less active organism and forms only small amounts of dextran. Both species can cause spoilage in sugar cane and sugar beet after harvest. The sucrose content of the crop is reduced and the dextran interferes with extraction of sucrose (Sidebotham, 1974). The bacteria are active at atmospheric temperature particularly that of the cane fields. *Lb. confusus* also forms dextran actively from sucrose and has been isolated from sugar cane (Sharpe *et al.*, 1972). The problem of spoilage has been largely overcome by matching harvesting to processing.

Dextran production by *L. mesenteroides* can be beneficial and was reported to control ergot when the bacterium was growing in the honey dew of rye (Mantle, 1965). However, this appears to be an isolated occurrence.

C. Dextran formation

Dextrans (*D. glucans*) have many commercial uses and have been extensively studied as can be judged from the bibliography compiled by Jeanes (1978). She lists 2455 papers with an additional 938 dealing with patents. Most of the workers have studied *L. mesenteroides* dextrans, all of which are formed by a D-glucose chain with substantial (1→6) linked a-D-glucopyranosyl residues, but branching of the side chain occurs (Sidebotham, 1974). Different strains form different dextrans and within a population cells may not all behave alike (Brooker, 1977). Much of the early work classifying the types of dextrans formed by *L. mesenteroides* was done in the Northern Regional Research Laboratories at Peoria (Jeanes *et al.*, 1954) and the culture collection at Peoria holds a wide variety of strains.

Dextrans are proving to be useful in research, industry and medicine. Modified dextrans have become widely used as gels used in filtration work to separate compounds of different molecular weights. They are used for purification and for determining molecular weights. The former application covers desalting and removal of small molecular weight material from larger molecules and can be used on any scale from a few millilitres to an industrial process. Derivatives of dextrans have been developed for a multiplicity of special purposes. In the medical field these include use as adjuvants, to aid interferon induction and to increase infectivity of viruses. Dextrans can be used in perfusion studies and also as carriers of pharmacologically active substances as insulin and vitamin B_{12} and for enzyme stabilization.

From this brief indication of the uses of dextran it is clear that *L. mesenteroides* has become an essential tool in many modern processes and without the ability of the bacteria to convert sucrose to dextran progress in biochemistry and medicine would probably have been slower.

D. Dairy industry

Leuconostocs occur with streptococci in milk and dairy products and in this field the two genera have not always been separated as morphologically they are indistinguishable. Lactose-fermenting species, *L. lactis* and *L. cremoris*, are the most important but other non-acidophilic species are sometimes found as minor components of milk and other dairy products. In the early part of this century leuconostocs were classified as streptococci (Hammer, 1920) but at about the same time Orla-Jensen (1919) recognized both betacocci (leuconostocs) and streptococci.

Both *L. lactis* and *L. cremoris* are included in cheese and butter starters although many starters consist entirely of streptococci. Leuconostocs are slow growing and not important in the conversion of lactose to lactic acid, but in forming flavour components in the acid ferment (Drinan *et al.*, 1976). Leuconostocs grow in milk only in association with streptococci and the flavour components are found after the pH has been lowered by the streptococci. These flavour components result partly from the heterofermentation of carbohydrate but mostly from the breakdown of citrate. Leuconostocs which cannot use citrate have no value in starters. All strains of *L. cremoris* examined can use citrate, and it is of interest to note that this species has only been isolated from dairy products. *L. mesenteroides* uses lactose very slowly and *L. cremoris*, now believed to be a variety of *L. mesenteroides*, appears to have adapted to live in milk. The ability of leuconostocs to use citrate is important to the dairy industry and few of the strains examined (Garvie, 1967a) used citrate, whereas earlier reports (Hucker and Pederson, 1930) suggested that many strains of *L. mesenteroides* were able to do so. The difference in these findings is striking but can be explained by the use of recently isolated strains in the older work as against old laboratory strains in the more recent work. Clearly, information about the stability of the ability to use citrate is important to the dairy industry.

E. Wine industry

When grape juice is made into wine the malic acid in the juice is converted to L(+)-lactate (malo-lactic fermentation) with a consequent rise in pH. This fermentation is essential in making wine, and a variety of bacteria have the necessary enzymes including both *L. oenos* and *L. mesenteroides*. When the malo-lactic fermentation was first recognized as important it was not realized that leuconostocs were a species peculiar to wine (Kunkee, 1967). *L. mesenteroides* is not important in wine fermentation as it will not grow at the low pH of grape juice nor in the ethanol formed by yeast. Peynaud and

Domercq (1968) report that *L. oenos* is the most important organism involved in malo-lactic fermentation under vinification conditions.

L. oenos is clearly important to the wine industry but much of the fermentation is left to the flora which develops naturally. *L. oenos* is a species which contains strains with different fermentation patterns, different vitamin and amino acid requirements and different growth rates in laboratory culture. All these factors could influence the growth of naturally occurring bacteria, and affect the quality of the wine. The separation of *L. oenos* into different subspecies (or possibly different species) is important if the production of wine is to be understood. More work, particularly on the metabolic pathways, the enzyme proteins and nucleic acids of *L. oenos* is needed.

To what extent the end-products (apart from lactate) formed by *L. oenos* also influence wine is not clear but if *L. cremoris* is a flavour producer in butter *L. oenos* may also be a flavour producer in wine.

Appendix I
Media used in the identification of species of the genus *Leuconostoc*

1. *Percent (w/v) of ingredient of general media*

	Medium identification code						
Ingredient	YGPB	MRS	Medium 1	Medium 2	ATB	CMB	DTB
Glucose	1.0	2.0	1.0	1.0	1.0	1.0	1.0
Peptone	1.0	1.0	1.0	1.0	1.0	1.0	0.75
Meat extract	0.8	0.8	—	—	—	—	—
Yeast extract	0.3	0.5	0.5	0.5	0.5	0.5	0.25
NaCl	0.5	—	—	—	—	—	—
KH_2PO_4	0.25	—	0.5	0.5	—	0.25	0.25
K_2HPO_4	0.25	0.2	—	—	—	—	—
$MgSO_4.7H_2O$	0.2	0.2	0.2	0.2	0.2	0.2	0.2
$MnSO_4.4H_2O$	0.005	0.005	0.005	0.005	0.005	0.005	0.005
Ammonium citrate	—	0.2	0.5	—	—	—	0.1
Citric acid	—	—	—	0.5	—	0.25	—
DL—Malic acid	—	—	—	—	—	0.25	—
Sodium acetate	—	0.5	0.25	0.25	—	—	0.25
Tween 80	—	0.1	0.1	0.1	—	0.1	0.1
Tomato juice	—	—	—	10.0	25.0	—	10.0
Usual pH	6.8	6.2	6.5	4.8	4.8	4.8	6.5

continued on p. 175

1.5% agar is added when solid media are required.
0.05% cysteine hydrochloride is added to all media when required.
Sterilization is normally at 15 lb for 15 min.
Special techniques used with different media are mentioned in the text.
YGPB: yeast glucose phosphate broth (Garvie, 1976).
ATB: acid tomato broth (Garvie and Mabbitt, 1967).
CMB: citrate-malate broth (Garvie and Mabbitt, 1967).
DTB: dilute tomato broth (Garvie, 1967d).
MRS: (DeMan *et al.*, 1960).
Mediums 1 and 2: (Garvie, 1969).

2. *Milk agar for testing for gas production*
   a. Litmus milk   800 ml
      Glucose   5.5 g
      Oxoid yeast extract   2 g
   b. Oxoid tomato juice   100 ml
   c. Nutrient agar   200 ml   (Oxoid Nutrient Broth No. 2 with
                                       1.5% agar)

To prepare, mix a and b, adjust to pH 6.8, warm to 45°C and add c melted. Dispense in 10-ml amounts and autoclave 10 lb for 10 min. Water cool. For use, melt, cool to 45 C and inoculate with 0.25 ml of culture, solidify in cold water and layer on 4.0 ml of nutrient agar.

3. *Sucrose agar for dextran production*
   Oxoid tryptone   1.0%
   Yeast extract   0.5%
   K_2HPO_4 0.5%
   Di-ammonium citrate   0.5%
   Sucrose   5.0%
   Agar   1.5%
   pH 7.0   Autoclave 15 lb for 15 min

4. *"Sugar" basal broth*
   Oxoid peptone   1.0%
   Yeast extract   0.25%
   Tween 80   0.01%
   Bromocresol purple   0.1%
     (1.6% sol in ethanol)
   pH 6.8. Tube in 5.0-ml amounts and autoclave 15 lb for 15 min
   See p. 154 for modification for use with *L. oenos*

Carbohydrate substances are prepared as 2% (w/v) solutions (1.0% for aesculin and inulin). Autoclave 10 lb for 10 min and add 0.5 ml to 5.0 ml of basal medium.

5. *Acetate agar*
   BBL Trypticase   1.0%
   Arabinose   0.5%
   Glucose   1.0%
   Sucrose   0.5%
   Yeast extract   0.5%

KH$_2$PO$_4$	0.6%
Di-ammonium hydrogen citrate	0.2%
Tween 80	0.1% (v/v)
Salt solution 1	0.25% (v/v)
Salt solution 2	0.25% (v/v)
Agar	1.5%
Salt solution 1.	MgSO$_4$.7H$_2$O

Salt solution 1.   MgSO$_4$.7H$_2$O 57.5 g
                    MnSO$_4$.4H$_2$O 12.0 g } in 250 ml of H$_2$O

Salt solution 2.   FeSO$_4$.7H$_2$O 3.4 g
                    HCl (conc)   5.0 ml } in 250 ml of H$_2$O

Buffer solution   Sodium acetate 3H$_2$O   25 g
pH 5.4   10% glacial acetic acid   40 ml } + 160 ml of H$_2$O
                    Use at 200 ml l^{-1} litre medium

To make

A Dissolve agar in water (15 g to 500 ml of H$_2$O) by steaming.
B Dissolve other ingredients for 1 litre medium in 300 ml of H$_2$O.
C Mix these two solutions and steam for 10 min (= 800 ml of H$_2$O).
  Add 200 ml of buffer solution to C while hot.
  Dispense in sterile containers. Do not autoclave.

6. *Whey serum agar with calcium citrate*
   i. Whey is made with rennet from milk.
   ii. Serum is obtained from filtering a cheese starter culture.
   iii. Whey and serum are mixed (1:1) neutralized to pH 7.3 with Ca(OH)$_2$ suspension. Steamed for 30 min, filtered and the pH readjusted to 7.3 with NaOH. MnSO$_4$ (probably 0.005% w/v, the paper is unclear) and 1.5% agar are added, the medium cleared with albumin and sterilized (15 min 10 lb).
   vi. Calcium citrate suspension—1.5% carboxymethylcellulose (viscosity 60–120 cp at 20°C 1% solution) is dissolved in water at 45–50°C. 10 g of finely powdered calcium citrate are suspended in 100 ml of carboxymethylcellulose solution. This suspension is held at 45°C for 1.5–2 h. The coarse particles which have sedimented are discarded. The fine suspension is sterilized (10 min 15 lb). (The concentration of calcium citrate can be checked by OD at 750 nm. 1/100 dilution should be 0.8–0.9 using a 1-cm light path.) 1 ml of calcium citrate suspension is added to 15 ml of whey serum agar for use. (If the calcium citrate has precipitated during storage it should be gently mixed before use.)

References

Abd-el-Malek, Y. and Gibson, T. (1948). *J. Dairy Res.* **15**, 233–248.
Alizade, M. A. and Simon, H. (1973). *Hoppe-Seylers Z. Physiol. Chem.* **354**, 163–168.
Amachi, T., Imamoto, S., Yoshizumi, H. and Senoh, S. (1970). *Tetrahedron Lett.* **56**, 4871–4874.
Blackwood, A. C. and Blakley, E. R. (1960). *J. Bacteriol.* **79**, 411–416.

Brooker, B. E. (1977). *J. Bacteriol.* **131**, 288–292.
Cavett, J. J., Dring, G. J. and Knight, A. W. (1965). *J. Appl. Bacteriol.* **28**, 241–251.
Chilson, P. D., Castello, L. A. and Kaplan, N. O. (1965). *Biochemistry N.Y.* **4**, 271–281.
DeLey, J. (1970). *J. Bacteriol.* **101**, 738–754.
DeLey, J. and DeSmedt, J. (1975). *Antonie van Leewenhoek J. Microbiol. Serol.* **41**, 287–307.
DeLey, J. and Tutgat, R. (1970). *Antonie van Leewenhoek J. Microbiol. Serol.* **36**, 461–474.
DeMan, J. C., Rogosa, M. and Sharpe, M. E. (1960). *J. Appl. Bacteriol.* **23**, 130–135.
Denhardt, D. T. (1966). *Biochem. Biophys. Res. Commun.* **23**, 641–646.
Drinan, D. F., Tobin, S. and Cogan, T. M. (1976). *Appl. Env. Microbiol.* **31**, 481–486.
Fox, G. E., Pechman, K. R. and Woese, C. R. (1977). *Int. J. Syst. Bacteriol.* **27**, 44–57.
Galesloot, Th. E., Hassing, F. and Stadhouders, J. (1961). *Neth. Milk Dairy J.* **15**, 127–150.
Garvie, E. I. (1960). *J. Dairy Res.* **27**, 283–292.
Garvie, E. I. (1967a). *J. Dairy Res.* **34**, 39–45.
Garvie, E. I. (1967b). *J. Gen. Microbiol.* **48**, 431–438.
Garvie, E. I. (1967c). *J. Gen. Microbiol.* **48**, 439–447.
Garvie, E. I. (1967d). *J. Dairy Res.* **34**, 31–38.
Garvie, E. I. (1969). *J. Gen. Microbiol.* **58**, 85–94.
Garvie, E. I. (1975). *In* "Lactic Acid Bacteria in Beverages and Food" (J. G. Carr, C. V. Cutting and G. C. Whiting, Eds), pp. 339–349. Academic Press, London and New York.
Garvie, E. I. (1976). *Int. J. Syst. Bacteriol.* **26**, 116–122.
Garvie, E. I. (1980). *Microbiol. Rev.* **44**, 106–139.
Garvie, E. I. (1981). *J. Gen. Microbiol.* **127**, 209–212.
Garvie, E. I. (1983). *Int. J. Syst. Bacteriol.* **33**, 118–119.
Garvie, E. I. and Farrow, J. A. E. (1980). *Am. J. Enol. Vitic.* **31**, 154–157.
Garvie, E. I. and Farrow, J. A. E. (1981). *Zbl. Bakt. C.* **2**, 299–310.
Garvie, E. I. and Mabbitt, L. A. (1967). *Arch. Mikrobiol.* **55**, 398–407.
Gasser, F. (1970). *J. Gen. Microbiol.* **62**, 223–239.
Gillespie, D. (1968). *Methods in Enzymol.* **12B**, 641–668.
Gillis, M. and DeLey, J. (1980). *Int. J. Syst. Bacteriol.* **30**, 7–27.
Hammer, B. W. (1920). *Res. Bull. Iowa Agric. Exp. Stn.* **63**, 59–96c.
Holzapfel, W. and Kandler, O. (1969). *Zbl. Bakt. Abt II* **123**, 657–666.
Hontebeyrie, M. and Gasser, F. (1973). *Biochimie* **55**, 1047–1056.
Hontebeyrie, M. and Gasser, F. (1975). *Int. J. Syst. Bacteriol.* **25**, 1–6.
Hontebeyrie, M. and Gasser, F. (1977). *Int. J. Syst. Bacteriol.* **27**, 9–14.
Hucker, G. J. and Pederson, C. S. (1930). *N.Y. Agric. Exp. Stn. Tech. Bull.* **167**, 3–80.
Jeanes, A. (1978). "Dextran Bibliography", U.S.D.A. Misc. Pub. 1355.
Jeanes, A., Haynes, W. C., Wilham, C. A., Rankin, J. C., Melvin, E. H., Austin, M. J., Cluskey, J. E., Fisher, B. E. and Tsuchiya, H. M. (1954). *J. Am. Chem. Soc.* **76**, 5042–5057.
Kelly, N., Delaney, M. and O'Cara, P. (1978). *Biochem. J.* **171**, 543–547.
Kirby, K. S., Fox-Carter, E. and Guest, M. (1967). *Biochem. J.* **104**, 258–262.
Kunkee, R. E. (1967). *Adv. Appl. Microbiol.* **9**, 235–279.
McClesky, C. S., Faville, L. W. and Barnett, R. O. (1947). *J. Bacteriol.* **54**, 697–708.
Mantle, P. G. (1965). *Antonie van Leewenhoek J. Microbiol. Serol.* **31**, 414–422.

Marmur, J. (1961). *J. Mol. Biol.* **3**, 208–218.

Metcalf, R. H. and Deibel, R. H. (1973). *J. Bacteriol.* **113**, 278–286.

Moore, R. L. and McCarthy, B. J. (1967). *J. Bacteriol.* **94**, 1066–1074.

Niven, C. F. Jr and Evans, J. B. (1957). *J. Bacteriol.* **73**, 758–759.

Nonomura, H. and Ohara, Y. (1967). *Mitt. Klosterneuburg Rebe Wein* **17**, 449–466.

Orla-Jensen, S. (1919). "The Lactic Acid Bacteria", pp. 1–196. Høst, Copenhagen.

Peynaud, E. and Domercq, S. (1968). *Ann. Inst. Pasteur Lille* **19**, 159–170.

Peynaud, E. and Dupuy, P. (1964). *Bull. Off. Inter. Vin.* **37**, 908–922.

Schleifer, K. H. and Kandler, O. (1972). *Bacteriol. Rev.* **36**, 407–477.

Sharpe, M. E., Garvie, E. I. and Tilbury, R. (1972). *Appl. Microbiol.* **23**, 389–397.

Sidebotham, R. L. (1974). *Adv. Carbohydr. Chem.* **30**, 371–444.

Skerman, V. A. D., McGowan, V. and Sneath, P. H. A. (1980). *Int. J. Syst. Bacteriol.* **30**, 225–420.

Sozzi, T., Poulin, J. M. and Maret, R. (1978). *J. Appl. Bacteriol.* **44**, 159–161.

Whittenbury, R. (1963). *J. Gen. Microbiol.* **32**, 375–384.

Yashima, S. and Kitahara, K. (1969). *J. Gen. Appl. Microbiol.* **15**, 421–426.

6

Fatty Acid and Carbohydrate Cell Composition in Pediococci and Aerococci, and Identification of Related Species

T. BERGAN,[1] R. SOLBERG[1] AND O. SOLBERG[2]

[1] *Department of Microbiology, Institute of Pharmacy, University of Oslo, Oslo, Norway*
[2] *Department of Bacteriology, National Institute of Public Health, Oslo, Norway*

I.	Introduction	179
II.	Taxonomic status	180
III.	Materials and methods	182
	A. Bacterial strains	182
IV.	Results	188
	A. Fatty acids	188
	B. Carbohydrates	193
	C. Clustering	193
V.	Discussion	196
	A. Gas chromatography	196
	B. Identification of isolates	205
	References	210

I. Introduction

Gram-positive cocci constitute a significant part of environmental contaminants, for instance in pharmaceutical production, brewing and wine making. Aerococci and pediococci have, for instance, been isolated from medicine bottles (Clausen, 1964), but may be present in a number of plant raw materials used in production and aerococci constitute an important part of air contaminants. These organisms are found in brewers yeast, beer, wine and fermenting mashes such as sauerkraut, pickles and silage. These

METHODS IN MICROBIOLOGY VOL. 16
ISBN 0–12–521516–9

organisms may serve as indicators of poor hygiene, although product deterioration is a possibility. For instance *Pediococcus damnosus* destroys beer by contaminating the yeast and fermentation tanks in breweries. These microbes do not have an established role as human pathogens and thus are often referred to only as contaminants in medical microbiology.

The differentiation and taxonomy of the respective Gram-positive cocci have been chaotic. In the pharmaceutical industry one does as a routine, unfortunately, usually not identify organisms, but is more concerned with numbers. Nevertheless, for differentiation between potential pathogens, products deteriorators and microbes that primarily may serve as indicators of the level of hygiene, identification of bacterial species is preferable. For the Gram-positive organisms, the taxonomic situation until recently may explain why one limits oneself to identifying quantities rather than qualities, that is bacterial species determination.

In this chapter we present an analysis of cell fatty acids and carbohydrates of aerococci and pediococci, a review of their taxonomy and a summary of the biochemical–cultural properties suitable for bacterial identification.

II. Taxonomic status

Bacteria of the genera *Pediococcus* and *Aerococcus* are Gram-positive organisms, forming diplococci or irregular agglutinates or tetrad-like packets of cells [cell arrangement depends upon cultural conditions; tetrad formation is not sufficiently stable to be used as a taxonomic criterion (Judicial Commission, 1973)]. They are aerobic and may be micro-aerophilic, and anaerobic growth is absent or slow (Evans, 1974; Kitahara, 1974). They are homofermentative converting glucose to 90% or more D(+)-lactate or DL-lactate, often with the dextrorotatory (+) enantiomorph dominating. This resembles the situation for streptococci and has motivated their inclusion in the same family, previously *Lactobacillaceae*, and now *Streptococcaceae*. The %(G+C) content of the DNA from enterococci has been 34–38 (Kocur *et al.*, 1971), which is the lower range exhibited by aerococci [36–40%(G+C)] (Evans, 1974) and pediococci [34–44%(G+C)] (Kitahara, 1974). The view has been presented that similar physiological characteristics, particularly as related to tolerance to environmental conditions, indicate a taxonomic relationship between enterococci and aerococci/pediococci (Whittenbury, 1965). But pediococci and aerococci exhibit cell division along several planes and so do not form chains of cells. This is generally regarded sufficiently distinctive to justify separation between these taxa and the streptococci. In addition comes the finding from comparative studies that aerococci and pediococci have a fatty acid composition different from that of *Streptococcus lactis*, *S. faecalis* and five *Lactobacillus* species (Uchida and Mogi, 1972).

The number of conflicting taxonomic schemes for aerococci and pediococci, though, appears to be equal to the number of working centres engaged in this field. Thus, for instance, considerable dispute has focused on whether aerococci should be recognized as a separate genus or included among the pediococci, for example as *P. homari* (Deibel and Niven, 1960). The taxon *Aerococcus viridans* is fairly homogeneous and easily recognizable. The genus *Aerococcus* was suggested after a number of isolates from air were considered different from pediococci, which were described as more micro-aerophilic, more packet forming, producing more acid from for example glucose and tolerating more tellurite or crystal violet. They are *a*-haemolytic and exhibit distinct utilization of sugars, but parallel cultivation studies with pediococci were not made (Williams *et al.*, 1953). The proposal of the genus *Aerococcus* stated that "it remains to be shown how the aerococci differ from species of *Pediococcus*". Deibel and Niven (1960) studied aerococci and pediococci in parallel and found the same "relation to oxygen in these groups". The eighth edition of "Bergey's Manual of Determinative Bacteriology" (Kitahara, 1974; Evans, 1974) describes both these groups of bacteria as micro-aerophilic. It appears to overlap with what is known as *Gaffkya homari* or *P. homari* (the lobster pathogen) (Deibel and Niven, 1960). The genus name *Gaffkya* has been referred to the list of *nomina rejicienda* (Opinion 39, Judicial Commission, 1971). There appears to be general acceptance among the laboratories engaged in this field that *A. viridans* should be accepted, and the taxon is recognized as such in the Approved List of Bacterial Names (Skerman *et al.*, 1980). The suggestion has been made that the catalase-positive strains might form a distinct species (Clausen, 1964) *A. catalasicus*. For lack of a designated type strain, the name has found restricted use. A later paper showed that the %(G + C) was similar for strains with and without catalase production (Boháček *et al.*, 1969). Cowan and Steel (1970) explicitly consider catalase as a variable characteristic in *A. catalasicus*.

A number of disputes have arisen over which species epithet names to recognize. Distinct taxonomic hierarchies have been recognized for *Pediococcus* by Nakagawa and Kitahara (1959), by Coster and White (1964), by Sakaguchi and Mori (1969), in the eighth edition of "Bergey's Manual of Determinative Bacteriology" (Evans, 1974; Kitahara, 1974) and by Back (1978). The *P. soyae* recognized in the last paper is included in *P. halophilus* in the eighth edition of "Bergey's Manual of Determinative Bacteriology" (Kitahara, 1974). Coster and White (1964) recognized both *P. cerevisiae* and *P. damnosus* as distinct entities together with *P. parvulus*, *P. halophilus* and *A. viridans*. The dispute about whether to call the type species of the genus *Pediococcus*, *P. cerevisiae* or *P. damnosus* has been solved by the Approved

List of Bacterial Names. This has implicitly considered the epithet *cerevisiae* not validly published giving recognition to *P. damnosus* Claussen 1903. This is in accordance with the proposal of Garvie (1974) as followed by the ruling of Opinion 52 of The Judicial Commission (1976). In addition to this species, Garvie (1974) recognized the species *P. acidilactici, P. parvulus* and *P. pentosaceus*. In the eighth edition of "Bergey's Manual of Determinative Bacteriology" the *damnosus* taxon is presented under the name *P. cerevisiae*, and in addition *P. urinae-equi* is recognized. Back (1978) and Back and Stackebrandt (1978) have recognized *P. damnosus, P. acidilactici, P. parvulus* and *P. pentosaceus*, and in addition have defined a *P. pentosaceus* subsp. *intermedius* and advocated the taxa *P. inopinatus, P. dextrinicus* and *P. halophilus* for which type strains were suggested. The eighth edition of "Bergey's Manual of Determinative Bacteriology" recognized *P. cerevisiae* and not *P. damnosus*, and included *P. urinae-equi* among *Pediococcus*. The other species were *P. acidilactici, P. halophilus* and *P. pentosaceus* (Kitahara, 1974). The only species in *Aerococcus* was *A. viridans* (Evans, 1974). The Approved List of Bacterial Names has recognized as validly published and recognizable *A. viridans, P. acidilactici, P. damnosus, P. dextrinicus, P. halophilus, P. parvulus* and *P. pentosaceus*.

The arguments for and against individual species being recognized as distinct taxonomic entities based on their degree of difference are subjective. The analysis of the $\%(G+C)$ content of their DNA and DNA/DNA hybridization techniques have helped clarify the situation, although still subjective, for example in the selection of the strains studied.

We have expanded the analysis to determining cell composition of fatty acids and carbohydrate moieties in order to see to what extent the, mainly, cell wall substances provide evidence for a given taxonomic scheme.

III. Materials and methods

A. Bacterial strains

The bacterial strains studied and their designation are shown in Table I. In all, 33 strains have been examined.

1. *Cell culture*

The strains were grown on tomato agar described elsewhere (Solberg and Clausen, 1973). This is an optimal medium for these bacteria and furnishes particularly trace compounds required for pediococci growth. Pediococci have a lower pH requirement than aerococci; the former were grown at pH 5.9 and the latter at pH 7.3. Incubation took place in normal air at 25°C before harvesting by suspension of agar growth in redistilled, sterile water. The plates were harvested after stationary phase cultures had been reached. Uchida and Mogi (1972) showed that the relative amount of acid found was constant after three days, which was, therefore, taken to represent the minimum incubation time.

2. *Chemical procedure*

The procedure used for separation of fatty acids has been described (Jantzen *et al.*, 1974, 1975). Lyophilized bacterial cells were methanolysed for 20 h with 2 N hydrochloric acid in anhydrous methanol at 80°C under nitrogen. The fatty acid methyl esters, and methyl glycosides, were derivatized by trifluoroacetic anhydride in acetonitril (1:1) and selectively extracted by hexane. The mixture was kept at 80°C for 2–3 min and then at room temperature for 30 min. Extraction then proceeded with 1 ml of hexane three times and adjusted to 300 μl by evaporation in N_2. This represented the fatty acid fraction. The bottom phase after hexane extraction contained glycosides and was adjusted by evaporation and acetonitril to 150 μl.

The gas-liquid chromatography (GLC) columns used for assay of fatty acids were both non-polar and polar, respectively 10% UCW 982 on Gas Chrom Q and 10% EGA-PS on Chromosorb WAW, both 100/120 mesh and for carbohydrates SP2401 (Supelco Inc., Bellefonte, Pennsylvania, USA).

The Hewlett-Packard Gas Chromatograph model 5830A with flame ionization detector and a 2-m long glass capillary column of 2 mm inner diameter was used. Nitrogen carrier gas has a flow of 30 ml min^{-1}. The sample volume was 2–5 μl. Temperature programming was employed with an increment of 2°C/min, the UCW 982 column from 140°C to 240°C and the EGA-PS material from 140°C to 210°C. The temperature of the injector and detector blocks was +30°C above the column maximum.

The peak areas were calculated using a digital integrator (Hewlett-Packard 18850 A GC-Terminal) and the relative area of each peak calculated as a percentage of the total. Each strain was assayed two to four times.

Peak identification used standards of known contents and differential retention times in the two column materials. The standards were NHI-C, (4-7028) with C:8,0; C:10,0; C:12,0; C:14,0; C:16,0; C:18,0 and C:20,0 (the terminology is explained in Table II); GLC 10 (4-7038) with C:16,0; C:18,0; C:18,1; C:18,2 and C:18,3; GLC 90 (4-7046) with C:13,0; C:15,0; C:17,0; C:19,0 and C:21,0; 2-OH-C:12 α-hydroxydodecanoic acid (4-5398). In addition standards for monounsaturated acids with double bonds in the w-7 and w-9 position of C:16,1 and C:18,1 were used. All were supplied by Supelco Inc. (Bellefonte, Pennsylvania, USA); 3-OH-C:14 β-hydroxymethyl-

TABLE I

Strains examined for cellular composition of fatty acids and carbohydrates

Strain identification number	Species	Other designations	Source	Habitat	%(G + C) DSM	%(G + C) Boháček et al.[a]
1	Aerococcus viridans*	NCTC 8251, DSM20340, ATCC 11563, CCM1914	O. G. Clausen	Air	38.5	41.8–42.7
2	A. viridans[b]	C-1[c]	O. G. Clausen	Medicine bottle		41.8–42.5
3	A. viridans	C-2, CCM1911, NCIB9644	O. G. Clausen	Medicine bottle	37.1	41.2–42.0
4	A. viridans	C-3	O. G. Clausen	Medicine bottle		
5	A. viridans	C-4	O. G. Clausen	Medicine bottle		43.2–43.5
6	A. viridans[b]	C-5	O. G. Clausen	Medicine bottle		43.2–43.5
7	A. viridans[b,d]	C-6, CCM2452, NCIB9642	O. G. Clausen	Medicine bottle	39.4	42.7–44.0
8	A. viridans	C-7	O. G. Clausen	Medicine bottle		
9	A. viridans	C-8	O. G. Clausen	Medicine bottle		
10	A. viridans	C-9	O. G. Clausen	Medicine bottle		
11	A. viridans	C-10	O. G. Clausen	Medicine bottle		
12	A. viridans	C-11	O. G. Clausen	Medicine bottle		
13	Pediococcus urinae-equi	DSM 20341	DSM	Horse urine	39.5	
14	P. damnosus*	S-5, DSM 20331, NCDO1832	O. Solberg	Beer yeast		
15	P. damnosus	S-1	O. Solberg	Beer		
16	P. damnosus	S-2, CCM2464	O. Solberg	Beer		38.0–38.8
17	P. damnosus	S-3	O. Solberg	Beer		
18	P. damnosus	S-6	O. Solberg	Beer		
19	P. damnosus	S-7	O. Solberg	Beer		41.2–41.8
20	P. damnosus	S-8	O. Solberg	Beer		
21	P. damnosus	L-583	O. Solberg	Beer		38.5–40.7

					mol% G + C	
22	*P. damnosus*	DSM 20291, ATCC 25248, CCM2465	DSM	Beer		
23	*P. damnosus*	DSM 20292	DSM	Beer		
24	*P. damnosus*[c]	S-4	O. Solberg	Beer		37.8–39.2
25	*P. dextrinicus*	DSM 20293	DSM	Beer	39.9	
26	*P. halophilus**	DSM 20339, NCIB 9735, NCDO1635	DSM	Anchovis		
27	*P. parvulus**	DSM 20332, NCDO1634, ATCC19371, NCIB9447	DSM	Silage	41.6	
28	*P. inopinatus**[f]	DSM 20285	DSM	Beer yeast	39.2	
29	*P. acidilactici**	DSM 20333, NCDO1859. Type strain	DSM	Sake sap	39.2	
30	*P. acidilactici*	NCIB 6990, ATCC8042, CCM2425, DSM 20238, NCTC 6990	O. G. Clausen		38.0	38.0–38.5
31	*Pediococcus* sp.[d]	NCTC 8066, ATCC10791, F166	O. G. Clausen	Fermenting vegetable	38.8	38.0
32	*P. pentosaceus**	DSM 20336, NCD0990	DSM	Beer yeast	35.1	
33	*P. pentosaceus* (subsp. inter-medius)	DSM 20283	DSM	Barley	39.0	

ATCC: American Type Culture Collection, Rockville, Maryland, USA.

CCM: Czechoslovak Collection of Microorganisms, Brno, Czechoslovakia.

DSM: Deutsche Sammlung von Mikroorganismen, Munich, West Germany. %(G + C) data from DSM Catalogue, 1977.

F: Pederson, C. S. (Pederson, 1949).

NCDO: National Collection of Dairy Organisms, Reading, UK.

NCIB: National Collection of Industrial Bacteria, Aberdeen, UK.

*Type strain and species listed in Approved List of Bacterial Names (Skerman *et al.*, 1980).

[a] Boháček *et al.* (1969).

[b] Strain has a positive catalase reaction and is the holotype of the strains described as *A. catalasicus* (Clausen, 1964).

[c] Strains with letter C studied by Clausen (1964), and strains with letters L and S studied by Solberg and Clausen (1973).

[d] Intended as type strain for *A. catalasicus* (Clausen, 1964).

[e] Has also been named *P. cerevisiae* subsp. *dextrinicus* (Boháček *et al.*, 1969), and has been considered equal to *P. dextrinicus* (Back, 1978) and *P. damnosus* (Solberg and Clausen, 1973); biochemical cultural reactions are given by Solberg and Clausen (1973).

[f] Proposed as type strain of *P. inopinatus* (Back, 1978). Used as *P. parvulus* (Group IV) (Back and Stackebrandt, 1978).

[g] Suggested working type strain for *P. cerevisiae* (Sneath and Skerman, 1966).

TABLE II

Fatty acid contents of strains of the genera *Pediococcus* and *Aerococcus*[a]

Strain[b] No.	C:12	C:14	C:14,1	X:1	C:16	C:16,1	C:17cy	C:18	C:18,1	C:19cy	Others
Group I											
1	0.5	11.0	1.9		40.1	17.1		6.2	16.4		6.8
2	0.5	14.7	3.8		41.8	25.9		2.8	10.9		0.7
3	0.1	5.2	1.3		39.1	14.9		8.4	20.1		10.9
4	0.0	5.8	1.6		42.1	18.9		7.3	16.9		7.4
5	0.1	5.2	1.7		41.4	20.6		5.2	17.5		8.3
6	0.1	5.1	2.5		37.8	27.3		2.5	15.7		9.0
7	0.3	8.6	3.7		28.9	33.8		2.2	16.2		6.3
8	0.6	14.0	5.2		30.3	33.7		1.6	10.8		3.8
9	0.1	6.3	2.9		39.0	27.3		3.2	16.6		4.6
10	0.2	3.8	1.8		31.9	29.0		3.1	25.5		5.0
11	0.1	7.0	2.6		38.6	24.8		3.9	16.8		6.2
12	0.4	8.9	3.0		41.5	22.6		5.5	14.7		3.4
13	0.0	3.1	0.2		31.6	13.3		11.3	29.4		11.1
Group II											
14		2.6			58.5	12.0	3.7	4.9	6.0	6.1	6.2
15		1.4			48.9	9.0	5.6	5.8	8.5	7.7	13.1
16		4.6			44.6	7.7	5.4	2.6	11.1	2.4	21.6
17		2.1			53.3	7.2	5.2	6.4	8.5	4.1	13.2

Strain								
18		1.6	51.7	12.6	4.2	4.6	6.8	7.1	11.4
19		8.7	44.3	4.9	1.0	7.7	8.7	2.4	22.3
20		2.1	49.9	6.0	4.7	7.5	11.0	8.6	10.2
21		1.9	47.7	5.9	4.7	6.1	11.0	5.7	17.0
22		7.2	45.7	2.7	1.1	6.5	13.2	4.3	19.3
23		6.9	47.1	5.2	1.5	7.9	11.5	6.1	13.8
24		1.7	60.7	5.3	0.4	8.2	9.4	3.9	10.4
25		7.4	50.2	7.2	0.1	4.4	17.8	0.0	12.9
26		5.0	37.4	6.5	0.3	4.3	27.0	7.3	12.2
27		4.3	47.6	5.0	0.0	10.7	14.8	5.1	12.5
28		8.5	45.9	5.3	0.5	8.2	13.0	4.2	14.4
Group III									
29	0.9	0.3	46.9	10.9		8.0	28.0		5.0
30	0.8	0.2	31.1	10.8		10.1	39.3		5.7
31	0.9	0.2	40.2	10.4		8.6	33.5		6.2
32	1.3	0.3	44.3	11.3		6.6	33.3		2.9
33	0.4	0.1	38.1	11.4		8.7	38.8		2.8

[a] The fatty acids are quantitated as relative contribution, i.e. relative concentration in the samples and represent the mean of the two column responses.

[b] Strain numbers correspond to those employed in Table I.

[c] Key to fatty acid designations: the number after the colon designates the number of carbon atoms in the chain, the following numeral indicates the number of double bonds. The double bond positions are undetermined. The letter X denotes unidentified moiety labelled sequentially after the dash. C:12 = n-dodecanoic = lauric;   C:14 = n-tetradecanoic = myristic;   C:14.1 = tetradecenoic;   C:16 = n-hexadecanoic = palmitic;   C:16.1 = hexadecenoic; C:17cy = n-heptadecanoic cyclopropane acid; C:18 = n-octadecanoic = stearic; C:18.1 = octadecenoic; C:19cy = n-nonadecanoic cyclopropane acid.

Open spaces denote lack of components.

teradecanoate (21750) obtained from Applied Science Laboratory Inc. These were TFA treated before GLC, as the bacterial preparations. Bacterial Acid Methyl Ester Mixture (4-5436) (Supelco), which i. a. contains cyclopropane acids was also used. The established linear relationship between the logarithm of the retention time and the number of carbon atoms per chain in chemically homologous series of fatty acid methyl esters was used for identification.

3. *Numerical analysis*

The fatty acid and carbohydrate data were grouped by numerical taxonomic methods combining the Youle transformation coefficient with the unweighted and the weighted–pair-group–average group clustering procedures.

IV. Results

A. Fatty acids

The cellular fatty acid contents of the strains are presented in Table II. The major components were the even-numbered moieties C:16,0; C:16,1 and C:18,1. They contributed 60–80% of the components in each of the strains. Significant quantities were also contributed by C:14,0 and C:18 in all the strains. Odd-numbered, branched or OH-bonded fatty acids were not observed.

The strains in Table II have been grouped into three groups (Groups I, II and III) according to their fatty acid patterns. Group I consisted of *A viridans* and the reference strain [type strain not proposed, but a typical strain derived from the collection upon which the species was proposed (Mees,1934)] of *P. urinae-equi*. The last mentioned strain would have been a suitable holotype strain and has been included in a number of extensive taxonomic studies, for example by Uchida and Mogi (1972), and Whittenbury (1965). Group I was characterized by the presence of C:12 (small amounts), C:14,1 and unidentified components with a peak between C:12 and C:14. A typical example of the fatty acid chromatogram of Group I is shown in Fig. 1. The aerococci and *P. urinae-equi* have an unidentified peak of 12 or 13 C-atom acids, not observed in any of the pediococci of Groups II or III. The fatty acid pattern of *A. viridans* was similar for all isolates, except for one instance where C:12 was missing. The strain of *P. urinae-equi* was also missing C:12, and was otherwise distinguished by a low level of C:14,1 and a high level of C:18 and C:18,1.

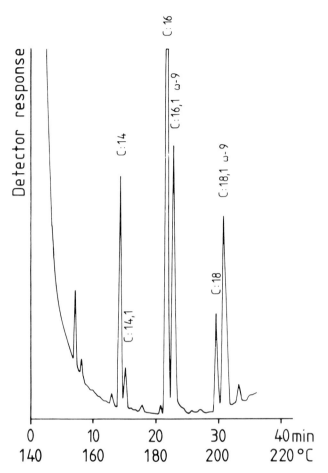

Fig. 1. Representative fatty acid profile of type strain of *Aerococcus viridans* NCTC 8251.

Distinctive of Group II (Fig. 2) were the cyclopropanes (cy) C:17cy and C:19cy, which were not detected in either of the two other groups of strains. Group II contained strains of the species *P. damnosus, P. dextrinicus, P. halophilus, P. parvulus* and *P. inopinatus.*

Group III lacks all the four fatty acid components typical of Groups I and II (C:12; C:14,1; C:17cy; C:19cy), but Group III was the only one with a peak designated X:1 appearing at position C:15 in both column chromatograms, although in rather small amounts, as shown in the examples in Fig. 3. Additional typical aspects of the fatty acid patterns are very low C:14 in

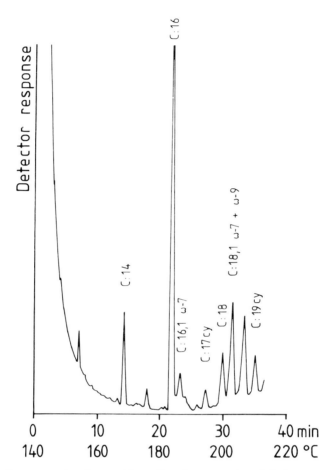

Fig. 2. Representative fatty acid profile of type strain of *Pediococcus damnosus* DSM 20291.

Group III strains, 1.3% or less, with more than 1.4% in the other two groups. Group III strains were *P. acidilactici*, *P. pentosaceus* and the strain which has served as the working type strain of the species *P. cerevisiae*. It is by biochemical–cultural properties that it does not entirely conform with the typical *P. damnosus*, but it has been designated as such in accordance with the Approved List of Bacterial Names (Skerman *et al.*, 1980).

In Group I strains, C:14 contributed more than 3.0%. The amount of C:16 was higher in Group II strains than in the others. Group I strains had an abundance of C:16,1, more than 10% in all strains except the type strain of *A. viridans*, one other *A. viridans* and the one entered as *P. urinae-equi*. The

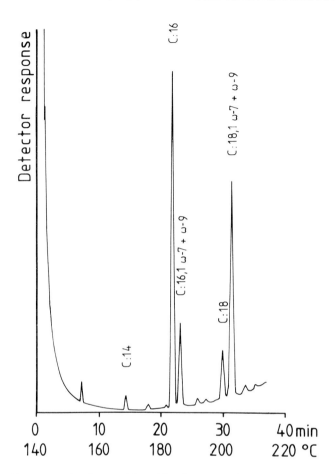

Fig. 3. Representative fatty acid profile of type strain of *Pediococcus pentosaceus* DSM 20336.

C:18,1 was particularly abundant in Group III strains, including the strain designated as the neotype strain for *P. cerevisiae*. On the same level, only slightly underneath, there were only two other strains, the one of *P. urinae-equi* and the type strain of *P. halophilus*. It is notable that *P. dextrinicus* deviates from the other Group II strains by a high amount of C:18,1 and no C:19cy. The only other deviation was in the type strain of *P. parvulus* which lacked the other cyclopropane C:17cy.

A particular finding that the ratio of saturated to unsaturated fatty acids and the ratio of C:16 to C:18 is useful as an indicator of group relationships (Table III). The ratio of C:16 to C:16,1 was a particularly good indicator. All

TABLE III

Ratio of saturated and unsaturated fatty acid contents of strains of *Aerococcus* and *Pediococcus*

Strain No.	C:16/C:16,1	C:18/C:18,1	C:16/C:18
Group I			
1	2.35	0.38	6.47
2	1.61	0.26	14.93
3	2.62	0.42	4.65
4	2.23	0.43	5.77
5	2.01	0.30	7.96
6	1.38	0.16	15.12
7	0.86	0.14	13.14
8	0.90	0.15	18.94
9	1.43	0.19	12.19
10	1.10	0.12	10.29
11	1.56	0.23	9.90
12	1.84	0.37	7.55
13	2.38	0.38	2.80
Group II			
14	4.88	0.82	11.94
15	5.43	0.68	8.43
16	5.79	0.23	17.15
17	7.40	0.75	8.33
18	4.10	0.68	11.24
19	9.04	0.89	5.75
20	8.32	0.68	6.65
21	8.08	0.55	7.82
22	16.93	0.49	7.03
23	9.06	0.69	5.96
24	11.45	0.87	7.40
25	6.97	0.25	11.41
26	5.94	0.16	8.70
27	9.52	0.72	4.45
28	8.66	0.63	5.60
Group III			
29	4.30	0.29	5.86
30	2.88	0.26	3.08
31	3.87	0.26	4.67
32	3.92	0.20	6.71
33	3.34	0.22	4.38

Group I strains had a ratio below 2.62, whereas Group II strains had values in the range 4.1–16.9 compared to 2.9–4.3 for Group III isolates. This ratio can thus be employed as an index indicating group allocation.

The strains of *P. damnosus* appeared homogeneous. The type strains of *P. dextrinicus* had no C:19cy and was thus distinctly deviant from the other strains of Group II, which had relative contents from 2.4 to 8.6% of this component. Strain No. 24 was at one time considered *P. cerevisiae* subsp. *dextrinicus* (Boháček *et al.*, 1969) but after Garvie presented her arguments favouring acceptance of the epithet *damnosus* the isolate was considered a *P. damnosus* (Solberg and Clausen, 1973). The type strain of *P. halophilus* combined a relatively lower content of C:16 than the others, although the contribution was abundant, it was more of the order exhibited by the Group I and III strains. *P. halophilus* further combined a low C:17cy with a high C:18,1. *P. parvulus* was void of C:17cy and was high in C:18. The type strains of *P. inopinatus* resembled the pattern of *P. parvulus*, which was its previous species assignation (DSM Catalogue), except that a small amount, 0.5%, of C:17cy was identified.

All Group III strains were similar regardless of species. It is notable that this also applies to the strain designated previously as the type strain of *P. cerevisiae*.

B. Carbohydrates

The glycoside composition was examined by what is essentially a qualitative technique. After hydrolysis of the cell material, the monosaccharides are present. The ensuing profile serves as finger printing the carbohydrate composition without attempted identification of the more complicated saccharides in the native cells. Fifteen major carbohydrate peaks of the hydrolysate were differentiated as presented in Table IV. The general pattern is that three isolated peaks emerged during the early phase of the run followed by two groups of peaks, D–I and J–O as presented in the Figs 4–6.

We see that the sum of the peaks D–I was lower and that the sum of peaks J–O was higher in the strains of *A. viridans*. This applied to the isolate of *P. urinea-equi* as well. Group II strains lacked the component of peak K (with the exception of *P. parvulus*). Group III strains often lacked peaks E, I, J and L and were characterized by a high contribution of peaks A, G and H. Thus, the carbohydrate composition further supported the subdivision of the strains into three categories.

C. Clustering

The results of hierarchical clustering of the strains are shown in Figs 7 and 8.

TABLE IV

Carbohydrate moieties (A–O) of strains of *Pediococcus* and *Aerococcus*

Strain No.	A	B	C	D	E	F	G	H	I	Sum of D–I	J	K	L	M	N	O	Sum of J–O	Others
Group I
1	1.6	3.2	6.5	1.4		3.5	11.5	4.1	3.3	23.8	0.9	2.7	3.5	31.1	9.4	4.2	51.8	13.1
2	2.2	3.7	3.2	2.5		5.8	9.9	3.8	2.7	24.7	1.0	3.3	4.2	29.6	9.8	5.9	53.8	12.4
3	0.2	1.7	3.5	0.3	0.9	7.3	4.9	tr	9.3	22.7		5.3	4.4	26.4	11.1	14.5	61.7	10.2
4	0.4	1.8	3.1	0.9	2.0	5.5	12.9	5.0	3.2	29.5	6.1	6.9	7.7	24.5	5.8	2.6	53.6	11.6
5	1.7	2.7	5.1	7.0	1.5	7.5	14.5	6.4	3.9	40.8	1.3	1.4	2.3	22.3	3.3	3.9	34.5	15.2
6	2.3	1.8	4.8	3.8		8.2	17.2	7.8	3.6	40.6	0.9	2.1	3.1	22.9	4.3	5.0	38.3	12.5
7	1.8	2.3	3.1	1.8	2.8	10.2	16.1	7.0	4.9	42.8	1.1	1.5	2.1	21.2	3.9	4.2	34.0	16.0
8	1.7	1.7	3.3	0.7	2.0	8.0	10.1	4.7	3.6	29.1	3.1	2.1	2.7	22.9	5.1	4.5	40.4	23.8
9	1.3	1.7	4.0	1.6		8.6	14.9	6.7	2.7	34.5	2.7	3.2	3.6	25.7	4.9	3.4	43.5	15.0
10	1.3	1.5	3.4	1.2		6.6	16.3	9.3	5.5	38.9	1.3	3.1	4.1	23.2	7.2	5.5	44.4	10.5
11	1.1	2.6	4.2	2.3	2.7	8.6	13.5	5.6	4.5	37.2	4.8	3.0	4.5	25.3	4.6	4.6	46.8	8.1
12	1.1	2.4	5.3	2.1	3.1	7.9	20.3	8.0	2.6	44.0	0.7	1.4	2.1	24.7	5.6	4.0	38.5	8.7
13	2.4	3.6	6.0	0.6	3.9	5.7	8.1	3.8	1.2	23.3	1.0	2.5	5.1	28.1	4.7	5.3	46.7	18.0

Group II

No.																		
14	2.4	2.1	2.2	3.1	0.6	22.5	12.7	6.8	7.9	53.6	0.2		1.0	20.4	0.9	1.0	23.5	16.2
15	2.4	2.8	2.4	5.5	1.2	19.8	14.4	6.4	8.4	55.7	0.2			21.2	1.3	2.0	24.7	12.0
16	1.9	2.5	3.0	5.0	tr	16.9	11.4	7.0	8.4	48.7	0.3	0.4		26.5	2.7	4.4	34.3	9.6
17	2.3	2.0	2.4	2.7	4.3	22.2	11.4	5.0	8.2	53.8	0.7			20.8	1.6	1.5	24.6	14.7
18	1.5	1.9	2.5	3.0	1.2	20.0	12.6	6.5	7.0	50.3	1.3			18.0	3.4	2.4	25.1	15.2
19	2.2	1.9	2.8	1.4	3.3	12.4	10.5	3.3	8.5	39.4	2.0		2.3	22.8	1.4	1.9	30.4	23.3
20	2.5	2.8	3.6	2.0	1.8	19.9	13.1	5.7	11.1	53.6	0.3		0.1	24.3	1.8	1.9	28.4	9.1
21	2.5	1.7	3.0	1.9	3.5	16.0	8.8	3.4	7.8	41.4	0.3		1.7	23.1	2.2	3.1	31.3	20.1
22	6.9	6.5	2.8	0.9	3.4	11.7	22.1	9.6	5.6	53.3	1.2		0.2	15.7	1.4	1.7	19.3	11.2
23	4.9	3.6	3.2	0.9	3.9	15.9	16.5	6.7	6.9	50.8	0.3			21.5	2.0	2.2	26.2	11.3
24	3.4	7.0	1.1	0.1	1.5	13.5	14.6	6.1	9.0	44.8	0.5			18.8	3.4	2.4	24.8	7.4
25	2.0	2.5	2.8	0.8	4.5	23.5	7.5		13.5	49.8	0.2		0.2	18.8	1.5	1.6	22.4	19.5
26	1.2	5.8	3.1	2.1	2.6	12.1	16.3	5.7	5.8	44.6	0.3			22.8	1.7	2.3	27.0	18.3
27	8.0	6.3	1.2	0.5	2.6	8.9	21.4	10.0	4.2	47.6	0.2	0.4		14.9	4.5	1.5	21.8	15.1
28	5.5	6.5	3.4	1.4	3.4	13.4	21.2	8.7	5.5	53.5			0.5	14.3	1.8	3.2	19.3	11.7

Group III

No.																		
29	1.5	6.8	2.8	0.6	1.6	8.8	27.8	12.2	2.5	53.5	1.3	2.1	2.1	21.1	1.4	1.9	27.8	8.2
30	6.9	2.7	2.5	1.7		3.0	32.0	16.5		53.2		0.8	1.9	17.7	1.3	2.9	24.8	9.9
31	6.7	1.6	2.1	1.6	1.2	3.0	28.4	15.5		48.5				19.3	2.7	2.6	26.5	14.6
32	10.3	4.3	2.4	1.1	0.6	7.5	27.0	12.1		51.8			0.1	19.0	1.4	1.8	22.3	8.9
33	8.3	4.8	2.2	0.9	2.7	2.7	36.9	16.8	2.9	57.9		0.1		13.5	0.9	1.3	15.8	11.0

Open spaces denote lack of components.

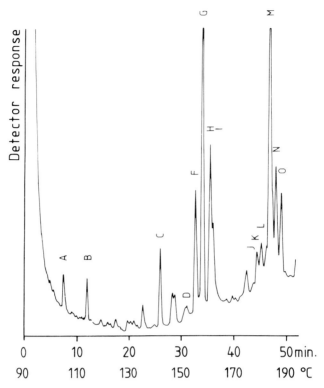

Fig. 4. Representative carbohydrate profile of type strain of *Aerococcus viridans* NCTC 8251.

V. Discussion

A. Gas chromatography

The major finding from g.l.c. determination of the fatty acid and carbohydrate composition of the aerococci and pediococci is that the bacteria are divided into three subgroups. The fatty acid composition by itself unequivocally defines Groups I, II and III both by (as shown in the key of Table V) relative quantities of selected fatty acids present and the ratios of the saturated to unsaturated moieties of C:16 and C:18. The carbohydrate constitution supports this subgrouping. All groups contained the saturated fatty acids C:14, C:16 and C:18 and the monounsaturated C:16,1 and C:18,1. Group I of *A. viridans* and *P. urinae-equi* is the only group with C:12 (small amounts) and C:14,1. No cyclopropane acid was detected, however, in Group I strains, which compensates by the presence of an unidentified peak

TABLE V

Principal distribution of fatty acids in strains of *Pediococcus* and *Aerococcus*

Group No.	C:12	C:14	C:14,1	X:1	C:16	C:16.1	C:17	C:17cy	C:18	C:18.1	C:19cy
I	(+)	+	+		+	+			+	+	
II		+			+	+		+	+	+	+
III		+		+	+	+			+	+	

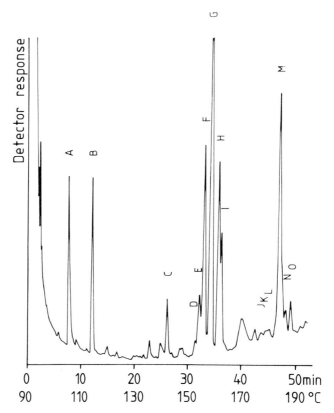

Fig. 5.   Representative carbohydrate profile of type strain of *Pediococcus damnosus* DSM 20291.

corresponding to 12 or 13 C atoms, the latter was not observed in any strain in *Pediococcus* Groups II or III. Group II with the species *P. damnosus*, *P. dextrinicus*, *P. halophilus*, *P. inopinatus* and *P. parvulus* was particularly characterized by the presence of the cyclopropane acids C:17cy and C:19cy in distinct amounts. The relatively high contribution of unidentified fatty acids in Group II was due to unidentified components of 19 or more C atoms. Group III with *P. acidilactici*, *P. pentosaceus*, *P. pentosaceus* subsp. *intermedius* and the strain designated here as *Pediococcus* sp. was characterized by the presence of small amounts of an unidentified peak similar to C:15, but otherwise only by the acids shared by both the other entities. The carbohydrate composition supports the definition of Groups I, II and III by lower relative amount of the complex of peaks designated D–I and relative high contribution of the peaks J–O. The ratio of the sum of D–I to the sum of

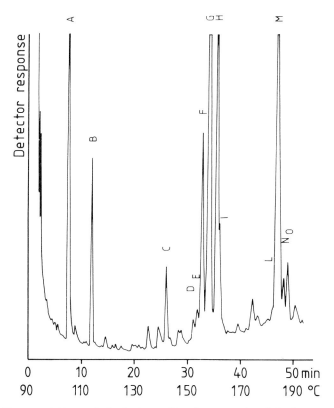

Fig. 6.   Representative carbohydrate profile of type strain of *Pediococcus pentosaceus* DSM 20336.

J–O was below 1.0 in all Group I strains, whereas the isolates of Groups II and III had ratios above 1.0. The K peak was absent in nearly all Group II strains. Group II also had higher F component than the other two groups. The Group III strains had a high contribution of the peak G component, but none or low amounts of the peak I component.

The fatty acid composition of the present strains is, with few exceptions, qualitatively in accordance with the observations of Uchida and Mogi (1972) on *A. viridans, P. damnosus* (under the name *P. cerevisiae*), *P. acidilactici, P. halophilus* (including strains received as *P. soyae*) and *P. pentosaceus*. They noted traces of C:12 in *P. halophilus* and large contributions of C:17 and C:19 cyclopropane acids in their *P. cerevisiae*. In the four strains of the latter species they found 38, 53.3, 57.3 and 78.9% of the cyclopropane acids in *P. cerevisiae*. Clearly, either the selection of their strains does not correspond to

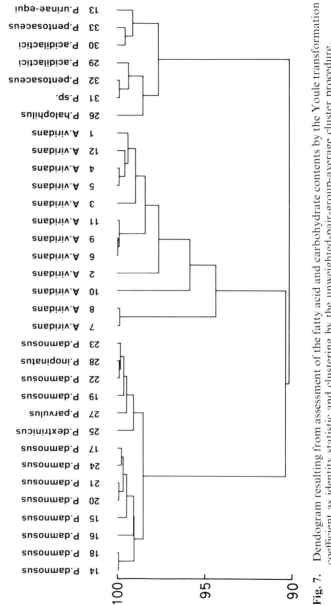

Fig. 7. Dendogram resulting from assessment of the fatty acid and carbohydrate contents by the Youle transformation coefficient as identity statistic and clustering by the unweighted-pair-group-average cluster procedure.

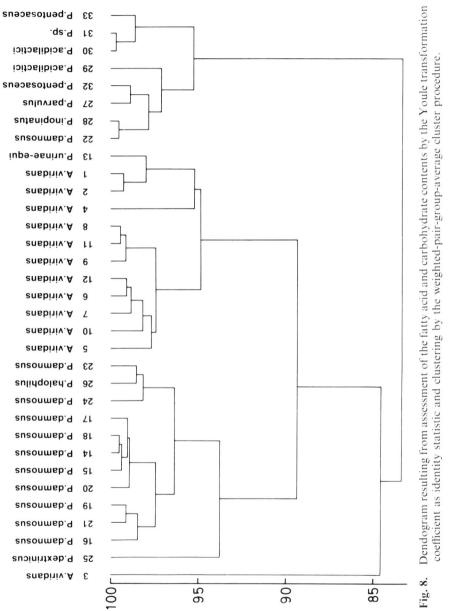

Fig. 8. Dendogram resulting from assessment of the fatty acid and carbohydrate contents by the Youle transformation coefficient as identity statistic and clustering by the weighted-pair-group-average cluster procedure.

our *P. damnosus* or the different cultivating conditions in the two studies are responsible for these marked differences. They found no C:19cy in *P. pentosaceus* or *P. acidilactici*. We could not confirm the presence of C:14,1 in our strains of *P. pentosaceus* observed as trace amounts in three, 0.1 and 0.6 in one each, and none in three strains. They reported on small amounts of C:20 and C:20,1 in some strains of Groups I and II, but without confirmation by mass spectrometry we have preferred to refrain from evaluating components above C:19.

The distribution into three subgroups according to cell composition in this study by introducing more recent species completes a similar pattern apparent from the study of Uchida and Mogi (1972) who contributed decisively to the understanding of the appropriate classification of *P. soyae* and the confusion between *P. cerevisiae, P. damnosus* and *P. pentosaceus*. It is notable that they have received strains under the name of *P. damnosus* which were classified as *P. cerevisiae*, and, as judged from their fatty acid pattern and that found in our strains, would be identical to what is presently classified as *P. damnosus*. The synonymity between *P. cerevisiae* and *P. damnosus* shown by our study and that of Uchida and Mogi (1972), however, is only partial. It is important to see that a number of the strains of Uchida and Mogi entered the study by the name of *P. cerevisiae*, were considered by them as *P. pentosaceus* and exhibited a fatty acid composition similar to that of the strain NCTC 8066 in our study. In fact, our NCTC 8066 was labelled F166, NISL 7105 by Uchida and Mogi and referred to the species *P. pentosaceus*. To avoid taxonomically illicit use of names, we have here labelled this strain *Pediococcus* sp. It is an important strain because of its previous designation as the working type strain of *P. cerevisiae* because it had particularly suitable bacteriological properties (Skerman and Sneath, 1966) presumably shown by traditional biochemical–cultural tests. Four of the eight strains of their *P. pentosaceus* formerly designated *P. cerevisiae* corresponded to our Group III.

Consequently, the relationship between strains describable as *P. cerevisiae*, as the species has become known, and Group III species may require further elucidation. Back (1978) has carried out a rather extensive study on pediococci. This excluded *A. viridans* and, consequently, the taxonomic problems associated with *P. urinae-equi* or *P. homari*. The study included 830 strains and comprised both biochemical–cultural properties and DNA hybridization. It was commented that a relative biochemical similarity exists between *P. acidilactici* and *P. pentosaceus* in spite of a DNA homology of 15–21% between the species. Because such a large number of strains was studied, it would seem unlikely that the strain NCTC 8066 represents a distinct taxonomic entity, but the findings with this taxonomically important strain were not reported specifically. Thus it is unclear whether the strain or the taxonomic problem raised by it was evaluated, although this is likely when the completeness of the study is considered (Back, 1978a,b; Back and Stackebrandt, 1978).

It may be mentioned that the original strain of *P. cerevisiae* was isolated from beer, which is the source of the great majority of *P. damnosus*, whereas *P. pentosaceus, P. pentosaceus* subsp. *intermedius* and *P. acidilactici*, which all belong to our Group III, are rare in beer, although a few such instances have been reported. Back (1978b) among the 228 strains of species which would be referred to our Group III isolated 10 from beer, beer yeast or empty beer bottles, and the rest from plant sources, and of the 512 strains of *P. damnosus* isolated 497 from beer or beer yeast and the remainder from wine. The species *P. acidilactici, P. pentosaceus* and *P. pentosaceus* ssp. *intermedius* are indistinguishable in respect to their fatty acid or carbohydrate finger print profiles. This is reflected by DNA/DNA hybridization between these species ranging from 15% to 21% of the homologous values and respective intra-species values of 83–100% (Back and Stackebrandt, 1978). In fact, these species also presented hybridization percentages which were higher than against any of the Group II species, the values against these other strains ranged from 0 to 9%. Correspondingly, Group II species showed hybridization percentages against other strains of the same species, other species within the group and other strains from Group III. No comparative value against Group I strains was obtained, but from the overall properties of these strains, it would appear that the separation in Groups I–III suggested by fatty acid composition and carbohydrate finger printing profiles parallels the grouping evidenced by DNA hybridization. The completeness and extent of parallelism in results of chemical composition of the cells and genetic data were greater than was perceivable *a priori*. This is gratifying in that the data support the value of the presently used analytical methods for taxonomic purposes.

This orderly organization of aerococci and pediococci, in fact, is also reflected by other types of data. Back (1978a,b) has demonstrated the molecular weight of lactate dehydrogenase (LDH) by polyacrylamide electrophoresis. This separates the species in a fashion identical to our Groups II and III, only with the qualification that *P. inopinatus* is closer to Group III than the other Group II species. The aerococci have one type of peptidoglycan polypeptide chains without inter-peptide bonds. The D-alanine carboxyl group is bound directly to the amino group of the adjacent L-lysine. All pediococci have inter-peptide bonds between alanine and lysine of the type L-Lys-L-Asp, with D-Asp type inter-peptide bonds between positions 3 and 4 of two peptide bridges (Kandler, 1970; DSM Catalogue). Serologically, aerococci are distinct and do not cross-react with any of the pediococci (Clausen *et al.*, 1981; Coster and White, 1964; Günther and White, 1961). Coster and White (1964) have few strains of the then recently proposed (Nakagawa and Kitahara (1959) species *P. acidilactici*, but noted this to be distinct from either the pediococci of Group II or *A. viridans* (and strains later classified as such, *Gaffkya homari*) and noted that *P. ureae-equi* and *P. homari*, which would be Group I strains, were serologically distinct from Group II or III species. Clausen *et al.* (1981) in their extensive serological study demonstrated considerable antigenic heterogeneity within Group I strains. Whereas others have used only agglutination, they in addition included carbohydrate extracts *ad modum* Lancefield to identify precipitations (ring test, Ouchterlony agarose diffusion). They studied many of the same Group I, II and III strains as we have (all with source O. G. Clausen as shown in Table I). Their results show that (a) precipitation occurs extensively between strains within and between Groups II and III, that (b) agglutination distinguishes Group III from Group II by slower and less marked reactions of *P. acidilactici* with sera prepared against *P. damnosus* and that (c) Group I is distinct from the two other groups both in precipitation tests and agglutination. Points (b) and (c) are supported by the agglutination results in an earlier study (Clausen *et al.*, 1975).

It is notable that *P. urinae-equi* was inseparable from *A. viridans* in either fatty acid or carbohydrate composition. Similar conclusions have been made from biochemical–cultural studies (Whittenbury, 1965; Sakaguchi and Mori, 1969). Uchida and Mogi (1972) studied one (and the same) strain of *P. urinae-equi* and identified the same components as we did, except for a small C:12 peak, and with a capillary column they reported on eight unidentifiable peaks also identified in two strains of *A. viridans*. *A. viridans* and *P. urinae-equi* both fail to produce acid from pentose, but do so from mannite, hydrolyse hippurate, have low salt tolerance, limited acidification ability (above pH 4.8) and lack asparaginic acid in the peptidoglycan cross-linking polypeptides (DSM Catalogue; Kandler, 1970), in distinction from all pediococci. Whittenbury (1965) suggested after determinative biochemical studies that *A. viridans* and *P. urinae-equi* would be referrable to the same species, which is consistent with the above criteria. However, he also included *P. halophilus* in the taxon; this is opposed to the classification of the salt tolerant species in our Group II. *P. halophilus* has a somewhat lower content of C:16 and higher content of C:18,1 than in other Group II strains. It was pointed out by Back (1978) that the strain of *P. halophilus* studied by Whittenbury did not attack pentoses and therefore was aberrant. The possibility of classifying *P. urinae-equi* as *A. viridans* requires support from DNA hybridization, which yet remains to be carried out, before an unequivocal evaluation can be made. However, a considerable body of data including biochemical–cultural reactions, cellular composition of fatty acids, carbohydrates and even murein composition and %(G+C) all support equality between *P. urinae-equi* and *A. viridans*. It would therefore seem unlikely that DNA hybridization data would support anything other than that *P. urinae-equi* is a later synonym of *A. viridans*.

The fatty acid composition and carbohydrate finger print profile showed homogeneity of each of the species *A. viridans* and *P. damnosus*. All isolates were similar to the type strains of the respective species. The type strains, consequently, appear to be a typical representative of their respective species by the criteria of fatty acid and carbohydrate composition.

In view of the distinct differences between the other strains of Group II, it is notable that the type strain of *P. dextrinicus*, which was earlier a subspecies of *P. cerevisiae* (i.e. *P. damnosus* biotypes) (Back, 1978b), completely lacks the cyclopropane C:19cy and has only traces of C:17cy. In this respect *P. dextrinicus* was comparable with the Group III strains, but in distinction to these, it did not have the X:1 and the relative contribution of the other acids also resemble Group II and not Group III. The carbohydrate peaks F, G, H (the only Group II strain with none present) and I were all different in relative contribution compared to the other Group II strains. The species was thus distinct in both fatty acid and carbohydrate composition.

The *P. halophilus* type strain was less different from *P. damnosus*, although low C:16 and C:17cy and high C:18,1 appeared. The carbohydrate profile was similar for the two species.

The *C. parvulus* type strains lacks C:17cy otherwise found in all Group II strains, and the carbohydrate peaks A, F, H and M were distinct. This species had a carbohydrate profile not unlike that of *P. inopinatus*. The fatty acid composition of the latter species was typical of Group II.

B. Identification of isolates

With the many types of techniques used to analyse aerococci and pediococci it appears that the core of the taxonomic system has now been laid down for these Gram-positive bacteria. From the composite data of our study on cell composition of fatty acids and carbohydrates and a number of thorough and extensive previous studies, particularly those of Back (1978a,b) and Back and Stackebrandt (1978), the separation into groups and distribution of species within them would be as follows.

Group I
  Aerococcus viridans

Group II
  Pediococcus damnosus
  P. dextrinicus   *P. inopinatus*
  P. halophilus   *P. parvulus*

TABLE VI

Biochemical–cultural reactions of aerococci and pediococci[a]

	A. viridans	P. damnosus	P. acidilactici	P. dextrinicus	P. halophilus	P. inopinatus	P. parvulus	P. pentosaceus
ACID PRODUCTION								
Amygdalin		+(−)	+	+	+	+	+(−)	+
Arabinose	+ −	−	+(−)	−	+	−	−	+
Cellobiose	+	+	−	+	+		+(−)	+
Dextrin	− +	−	−	+	+ −	−(+)	+(−)	−
m-Dulcit	−	−	−	−	−	−	−	−
Fructose	+	+(−)	+	+	+	+	+	+
Galactose	+	+(−)	+	+	+	+	+(−)	+
Glucose	+	+	+	+	+	+	+	+
Glycerol	+ −	−	+ −	−	+(−)	−	−(+)	+(−)
m-Inositol		−	−	−	−	−	−	−
Lactose	+ −	−	+(−)	−	−(+)	+	−	+
Maltotriose		−(+)	−	+	+	+(−)	+(−)	+
Maltose	+	+ −	−	+	+	+(−)	+(−)	+
Mannitol	+(−)	−	−(+)	+	−	+	+	−
Mannose	+	+(−)	+	+	+	+	+	+
Melezitose		−	−	−	−(+)	−	−	−
Melibiose		−	−	−	+	−	−	+
Methylglucoside		−(+)	−	+	+	+	− +	−
Raffinose	+	−	−(+)	−	−(+)	−	+	+
Rhamnose		−	+(−)	−	−	−	−	+ −
Ribose		−	+	−	+	−	−	+

Salicin	+	+(−)	+(−)	+	+	+	+(−)	+
Saccharose	+	−(+)	−	+(−)	+(−)	−(+)	−(+)	−
Sorbitol	+(−)	−	−	−	−(+)	−	−	−
Sorbose	−	−	−	+	−	−	−	−
Starch	−	−	+	+	−	+	+	−
Trehalose	+(−)	+(−)	+−	−(+)	+	+	+	+
Xylose	+(−)	−	+(−)	−	−	−	−	+
Amygdalin	+	+−	+	+	+	+	+	+
Inulin	−	−	−	+	−	−	−	+
GROWTH AT								
45°C	+	−	+	−(+)	−	+	+	+
10% NaCl	+	−	+(−)	+	+	−	+	+
pH 8.5	+	−	+	+	+	−	+	+
PRODUCTION OF								
Slime	−	−	−	−	−	−	−	−(+)
Acetoin	+(−)	+(−)	−(+)	−	−(+)	−(+)	+	+
Aesculin hydrolysis	+	+	+	+	+	+	+	+
Arginin dihydrolase	−	−	+	+	+	−	+	+
Catalase	+−	−	+(−)	+(−)	−	−	+	+
Hippuric acid hydrolase	+	−	−	−	+(−)	−	−	−

---

[a] Adapted from Back (1978), Kitahara (1974), Evans (1974), Clausen (1964), Cowan and Steel (1970).

Gelatin hydrolase or nitratase was not present in any of the species.

Growth conditions are indicated in Table VII.

Group III
   P. acidilactici
   P. pentosaceus
   P. pentosaceus subsp. *intermedius*

The type species of the two genera are *Aerococcus viridans* and *Pediococcus damnosus*. The status of *P. homari* and *P. urinae-equi* as later synonyms of *A. viridans* is considered established at this point. Present type strain of *P. cerevisiae* NCTC 8066 is too scant to indicate how it would be distinguishable from *P. pentosaceus*. In terms of our data on fatty acid and carbohydrate finger print patterns, these entities overlap, but in this connection we would like to point out that a similar lack of difference is found for the two *bona fide* species in Group III, which are distinguishable by DNA hybridization. Thus our findings on substance composition of Group III are not to be taken as contradicting future DNA hybridization on strains classifiable in this category.

An aerobic or micro-aerophilic Gram-positive coccal isolate with cells arranged in pairs, fours or irregular clusters, with homofermentative metabolism [identifiable e.g. by the routine GLC identification of lower carbon number fatty acids as employed for anaerobic bacteria (Sutter *et al.*, 1980)] are candidates for classification in the genera *Aerococcus* and *Pediococcus*. They do not grow as chains of cells, like streptococci.

Identification can be based on the biochemical–cultural properties shown in Table VI based in the main on the extensive studies of Back (1978a,b) and collaborators (Back and Stackebrandt, 1978) on the pediococci and other studies on *A. viridans* (Kitahara, 1974; Williams *et al.*, 1956).

A few key biochemical-cultural reactions facilitate identification. Cultures growing readily with a-haemolysis under aerobic conditions at 25°C and 37°C are most likely *A. viridans*. Pediococci grow more slowly and better at lower temperatures and form extensive cell packets. Pentoses (e.g. ribose) are acidified by the species of Group III, *P. acidilactici, P. pentosaceus* and *P. pentosaceus* subsp. *intermedius* and by only *P. halophilus* in Group II. If acid is not formed from arabinose or xylose, the strain is *P. pentosaceus* subsp. *intermedius*. If one or two of the sugars are acidified, the strain is a *P. acidilactici* if growth occurs at 50°C and *P. pentosaceus* if no such growth is observed. High salt tolerance tested by a medium with 15% NaCl signifies *P. halophilus* if growth occurs. A negative ribose reaction combined by negative arginine dehydrolase signifies a Group II strain. The further subdivision of Group II can follow the dichotomic identification scheme of Fig. 9. Melizitose acidification, although slow, occurs only with strains of *P. damnosus* and *P. halophilus*. For verification of the diagnosis reached by a dichotomic key, the reaction patterns in Table VII should be consulted.

TABLE VII

Key diagnostic characteristics of isolates belonging to *Aerococcus* and *Pediococcus*[a]

	Group I		Group II					Group III	
	A. viridans[b]	*P. damnosus*	*P. dextrinicus*	*P. halophilus*	*P. parvulus*	*P. inopinatus*	*P. acidilactici*	*P. pentosaceus*	*P. pentosaceus* subsp. *interm.*
Arabinose	+	–	–	+	–	–	+(–)	+	–
Dextrin	–	–	+	+–	+(–)	–(+)	–	–	–
Lactose	+–	+–	+–	–(+)	–	+	+(–)	+	+
Maltose	+	+–	+	+	+(–)	+	–	+	+
Melezitose		+–	–	+	–	–	–	+	–
Ribose		–	–	+	–	–	+	+	+
Saccharose	+	+–	+(–)	+	–	–	–	+–	+–
Starch	–	–	+	–	–	–	–	–	–
Temperature maximum	37– 40	25– 30	43– 45	37– 40	37– 39	37– 40	50– 53	39– 45	39– 45
6% NaCl	+	–	–	+	+(–)	+	+	+	+
7% NaCl	+	–	–	+	+(–)	+–	+	+	+
8% NaCl	+	–	–	+	+–	–(+)	+	+	+
10% NaCl	–	–	–	+	–	–	+(–)	+–	+–
15% NaCl	–	–	–	+	–	–	–	–	–
pH 8.0	+	–	+	+	–	–	+	+–	+
pH 8.5	+	–	–	+	–	–	–	+–	–
Arginin	–	–	–	+	–	–	+	+	+
Sodium gluconate (gas)	–	–	+	–	–	–	–	–	–
Lactate rotation	DL	DL	L(+)	L(+)	DL	DL	DL	DL	DL

[a] Growth is carried out at 25°C for four weeks for acid production (stoppered tubes to prevent evaporation), two weeks for nitratase, gelatinase and urease, two days for catalase (with glucose in medium), eight days for acetoin production after incubation at room temperature and arginine hydrolysis for six days. Details on media and techniques are indicated by Back (1978) and by Clausen (1964). Adapted from Back (1978), Clausen (1964) and Williams *et al.* (1956).

[b] Grows on tellurite agar.

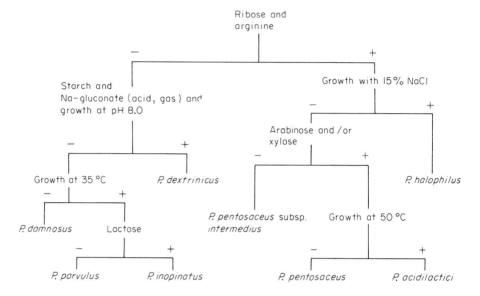

Fig. 9. Flow chart for identification of aerococci and pediococci. Result should be verified by comparison with Table VII to utilize more reactions for identification.

Serological schemes for aerococci or pediococci have not been formally worked out, although extensive individual studies have been carried out (Clausen *et al.*, 1981; Coster and White, 1964).

References

Back, W. (1978a). *Brauwissenschaft* **31**, 238–250, 312–320, 336–343 (three parts).

Back, W. (1978b). *Int. J. Syst. Bacteriol.* **28**, 523–527.

Back, W. and Stackebrandt, E. (1978). *Arch. Microbiol.* **118**, 79–85.

Bergan, T. (1972). *Acta Pathol. Microbiol. Scand.* **82B**, 64–83.

Boháček, J., Blažiček, G., Kocur, M., Solberg, O. and Clausen, O. G. (1969). *Arch. Mikrobiol.* **67**, 58–61.

Clausen, O. G. (1964). *J. Gen. Microbiol.* **35**, 1–8.

Clausen, O. G., Hegna, I. K. and Solberg, O. (1975). *J. Inst. Brew. London* **81**, 440–443.

Clausen, O. G., Hegna, I. K., Kjeldaas, L., Palmer, L. and Tjade, O. H. (1981). *J. Inst. Brew. London* **87**, 141–146.

Claussen, N. H. (1903). *C. R. Trav. Lab. Carlsberg* **6**, 64–83.

Coster, E. and White, H. (1964). *J. Gen. Microbiol.* **37**, 15–31.

Cowan, S. T. and Steel, K. J. (1970). "Manual for the Identification of Medical Bacteria", pp. 44–45, 56, 83. The University Press, Cambridge.

Deibel, R. H. and Niven, C. F. (1960). *J. Bacteriol.* **79**, 175–180.

Evans, J. B. (1974). *In* "Bergey's Manual of Determinative Bacteriology" (R. E. Buchanan and N. E. Gibbons, Eds), 8th edn, pp. 515–516. Williams & Wilkins, Baltimore, Maryland.

Garvie, E. I. (1974). *Int. J. Syst. Bacteriol.* **24**, 301–306.

Günther, H. L. and White, H. R. (1961). *J. Gen. Microbiol.* **26**, 199–205.

Günther, H. L. Coster, E. and White, H. R. (1962). *Int. Bull. Bacteriol. Nomencl. Taxon.* **12**, 189–190.

Jantzen, E., Bergan, T. and Bøvre, K. (1974). *Acta Pathol. Microbiol. Scand. Sect. B* **82**, 785–798.

Jantzen, E., Bryn, K., Bergan, T. and Bøvre, K. (1975). *Acta Pathol. Microbiol. Scand. Sect. B* **83**, 569–580.

Judicial Commission (1971). *Int. J. Syst. Bacteriol.* **21**, 104–105.

Judicial Commission (1976). *Int. J. Syst. Bacteriol.* **26**, 292.

Kandler, O. (1970). *Int. J. Syst. Bacteriol.* **20**, 491–507.

Kitahara, K. (1974). *In* "Bergey's Manual of Determinative Bacteriology" (R. E. Buchanan and N. E. Gibbons, Eds), 8th edn, pp. 513–515. Williams & Wilkins, Baltimore, Maryland.

Kocur, M., Bergan, T. and Mortensen, N. (1971). *J. Gen. Microbiol.* **69**, 167–183.

Mees. R. H. (1934). Onderzoekingen over de Biersarcina. Thesis, Delft. (cit. Whittenbury, 1965).

Nakagawa, A. and Kitahara, K. (1959). *J. Gen. Microbiol.* **5**, 95–126.

Pederson, C. S. (1949). *Bacteriol. Rev.* **13**, 225–232.

Sakaguchi, K. and Mori, H. (1969). *J. Gen. Appl. Microbiol.* **15**, 159–167.

Schultes, L. M. and Evans, J. B. (1971). *Int. J. Syst. Bacteriol.* **21**, 207–209.

Skerman, V. B. D., McGowan, V. and Sneath, P. H. A. (1980). *Int. J. Syst. Bacteriol.* **30**, 225–420.

Skerman, V. B. D. and Sneath, P. H. A. (1966). *Int. J. Syst. Bacteriol.* **16**, 1–133.

Solberg, O. and Clausen, C. O. (1973). *J. Inst. Brew. London* **79**, 227–230.

Sutter, V., Citron, D. M. and Finegold, S. (1980). "Wadsworth Anaerobic Bacteriology Manual", 3rd edn, pp. 39–114. Mosby St. Louis.

Tjeltveit, O. J. and Clausen, O. G. (1964). *J. Gen. Microbiol.* **35**, 9–12.

Uchida, K. and Mogi, K. (1972). *J. Gen. Appl. Microbiol.* **18**, 109–129.

Williams, R. E. O., Hirch, A. and Cowan, S. T. (1953). *J. Gen. Microbiol.* **8**, 475–480.

Whittenbury, R. (1965). *J. Gen. Microbiol.* **40**, 97–106.

7

Epidemiological Typing of *Klebsiella* by Bacteriocins

A. BAUERNFEIND

Max v. Pettenkofer Institut der Universität München, Munich, West Germany

I.	*Klebsiella* species and their importance in nosocomial infections	. 213
II.	Markers for subclassifying *Klebsiella* species . .	. 214
III.	Typing of *Klebsiella* by bacteriocins	. 214
IV.	Modified typing procedure	. 216
V.	Conclusions.	. 221
	References	. 222

I. *Klebsiella* species and their importance in nosocomial infections

The taxonomy and nomenclature of *Klebsiella* have been discussed by Ida Orskov (1981). In this chapter we classify the genus *Klebsiella* into four species, namely *K. pneumoniae, K. oxytoca, K. rhinoscleromatis* and *K. ozaenae* (Brenner *et al.*, 1979). Infections with *Klebsiella* are caused mainly by *K. pneumoniae* and *K. oxytoca* (in a proportion of about 2 to 1, Bauernfeind *et al.*, 1981a,b), whereas infections with *K. rhinoscleromatis* and *K. ozaenae* are less frequent and mainly localized in the nose and pharynx. Most infections with *K. pneumoniae* and *K. oxytoca* occur in hospitalized patients (Casewell *et al.*, 1977, 1978; Johnson, 1978; Cooke *et al.*, 1979; Haverkorn and Michel, 1979a,b; Montgomerie, 1979; Shinebaum *et al.*, 1978). According to the National Nosocomial Infections Study Report of the Center for Disease Control, 7.3% of nosocomial infections in the USA in 1977 (CDC, 1977) were associated with *Klebsiella* and a high percentage of these are caused by epidemic strains (Steinhauer *et al.*, 1966; Selden *et al.*, 1971; Hall, 1972; Casewell and Phillips, 1977, 1978a,b; Casewell *et al.*, 1977).

METHODS IN MICROBIOLOGY VOL. 16
ISBN 0 12 521516 9

II. Markers for subclassifying *Klebsiella* species

Nosocomial infections caused by *Klebsiella* may be analysed by capsular serotyping, but at least one additional and independent epidemiological marker should be used to prove the identity of isolates from different patients or from patients' surroundings. O-serotyping is hampered by heat stability of the capsule (Ørskov and Ørskov, 1978), and lysotyping is not useful because only 60% of strains respond and the most frequently encountered type constitutes 20% of the total (Rennie *et al.*, 1978; Slopek, 1978). Biotyping may be useful when precise standardization of substrate concentrations is observed (Barr, 1978), but in our hands the API 20E was unsatisfactory since two-thirds (of 259 non-epidemic isolates) were of one biotype.

III. Typing of *Klebsiella* by bacteriocins

The term bacteriocin is used as defined by Jacob *et al.* in 1953: bacteriocins are substances produced by bacteria (producers) in a lethal synthesis and they consist mainly of protein. After absorption to specific receptors, they can inhibit growth of other bacteria (indicators), predominantly of the same species.

Bacteriocins of *Klebsiella* were characterized in detail first by Yves Hamon and his group at the Pasteur Institute in 1963 (Hamon and Peron, 1963). Among 112 isolates investigated, 38 (34%) produced substances inhibiting growth of other strains. Most of the bacteriocinogenic strains could inhibit only a small number of other isolates of *Klebsiella*, and broad spectrum activity was exceptional. These observations were confirmed by Durlokowa *et al.* (1964), and again about one-third of the strains investigated were bacteriocinogenic. There are some differences in the names used for the bacteriocins of *Klebsiella*. Different authors call them pneumocins (Maresz-Babczyszyn *et al.*, 1964; Slopek and Maresz-Babczyszyn, 1967) as proposed by Hamon and Peron (1963), klebocins (Buffenmyer *et al.*, 1976), klebecins (Edmondson, 1979) or simply bacteriocins of *Klebsiella* (Hall, 1971). It is sometimes difficult to decide whether the name of a bacteriocin should be derived from the name of the genus or the species (according to the general rule). Objections against following this rule strictly for *Klebsiella* are the confusion of pneumocins with the bacteriocin described for *Str. pneumoniae* (Mindich, 1966) and the semantic problems arising with naming the bacteriocins of *K. rhinoscleromatis*. Use of the term bacteriocin or klebocin together with the species of origin would be a reasonable and unambiguous preliminary compromise. Bacterial isolates can be typed on a subspecies level, either by the patterns of growth inhibition exerted by their bacteriocins on a set of indicator strains (by bacteriocinogenicity) or by their sensitivity to bacteriocins synthesized by a set of producers (bacteriocinotypie).

It is possible to type isolates of *Shigella sonnei* (Abbott and Shannon, 1958; Brandis and Smarda, 1971) or *Pseudomonas aeruginosa* (Holloway, 1960; Bergan, 1975; Govan, 1978) by their bacteriogenicity, but various authors agree that, even when it is induced by mitomycin C or by ultraviolet light, synthesis of bacteriocins is not frequent enough in *Klebsiella* [between 33.6% (Hall, 1971) and 50.7% (Bauernfeind *et al.*, 1981a,b)] for isolates to be typed in this way.

Maresz-Babczyszyn *et al.* (1967) were the first to use *Klebsiella* bacteriocin for typing. In a scheme modified from Slopek and Maresz-Babczyszyn (1967), bacteriocins from eight producer strains were prepared, and 295 of 504 (58.5%) *Klebsiella* isolates were typed. In 1971, Hall analysed *Klebsiella* infections by bacteriocin typing, working out her own typing scheme by first screening 106 *Klebsiella* isolates for bacteriocinogenicity and identifying 42 (33.6%) as synthesizing bacteriocins, from which ten strong bacteriocin producers were selected to form a typing set. Although 50 distinct patterns of sensitivity were recorded in a total of 730 strains of *Klebsiella* from hospitals (mostly from patients), 37.3% belonged to only one of 16 typing patterns, even though only 7.5% of the isolates of this largest class were epidemic strains from an outbreak of *Klebsiella* infections in a single hospital. Apart from putting together about one-third of the non-epidemic strains into one single class, the value of Hall's system was impaired because as many as 21.2% of 730 isolates remained untypable. Except for weakly sensitive strains reproducibility was good. This typing scheme was useful for investigating hospital infections (e.g. epidemiological analysis of contaminated hand cream, hand washing preparations, contamination in a hospital milk kitchen and a hospital outbreak of *Klebsiella* infection).

Buffenmyer *et al.* (1976) used eight producer strains isolated by Slopek and Maresz-Babczyszyn (1967) and typed 296 cultures of *Klebsiella* from outpatients, inpatients or from the environment. In order to obtain the high titres needed for the typing method, bacteriocin production was induced by mitomycin C and typing was performed by spotting klebocin preparations onto nutrient agar plates. After a period of diffusion into the agar, exponentially proliferating cells of the strain to be typed were streaked on to the plates with cotton-tipped applicators. 67% of 296 isolates could be typed by this method but reproducibility was only 20%. This low reproducibility was due mainly to low titres of two bacteriocins from the set of eight which, however, could not be eliminated because only by these two unstable klebocins it was possible to separate 54% of the isolates into three groups; the largest group still contained 24% of all isolates.

Edmondson and Cooke (1979) screened 190 strains for bacteriogenicity.

Using mitomycin C for induction of bacteriocins, they identified 68 (35.8%) producers. A set of 15 was selected for typing, and a high proportion (96.8% of 277 non-epidemic isolates) could be typed by spotting lysates of these producers onto thoroughly standardized lawns of the strains to be typed. Typing 98 *Klebsiella* isolates twice, at an interval of one week, revealed a higher number of strong reaction differences (67%) than when freshly prepared klebecons were used. Storage of preparations at 4°C for a week reduced reproducibility further.

Deviations between consecutive typings of identical isolates were due mainly to loss of sensitivity to those four producers among the set of 15 which had low and labile bacteriocin titres. Identical problems had been encountered already by Buffenmyer *et al.* (1976); the level and stability of klebocins are evidently crucial in typing *Klebsiella* by the growth-in-broth method, whether klebocin prepared in broth is spotted onto solid media before (Buffenmyer *et al.*, 1976) or after (Edmondson and Cooke, 1979) seeding of the strains to be typed. It seems to be difficult to build up a set of producers synthesizing high titre klebocins which keep their titre for a reasonable storage time, although screening of producers from different laboratories in order to select a set of high titre, stable klebocin might improve the quality of bacteriocin typing in *Klebsiella* by the growth-in-broth method.

IV. Modified typing procedure

In the attempt to increase reproducibility, another possible approach would be the re-evaluation of the cross-streak method, the basic procedure for detecting microbial products inhibiting the growth of other micro-organisms. Bacteriocin typing by this technique has been established for *Pseudomonas aeruginosa* (Govan, 1979). In the light of the problems of working with crude preparations of bacteriocins, typing procedures for which *Klebsiella* bacteriocins would not have to be prepared should be advantageous, and so we developed a bacteriocin typing system which avoided the disadvantages both of the growth-in-broth method (preparation storage and control of bacteriocins) and the conventional cross-streak method [laborious, lower percentage of typable strains (Edmondson and Cooke, 1979)].

1. *Typing technique*

In our procedure, the bacteriocins used for typing are delivered onto and into solid medium by producer strains grown on it over a period of 16 h similar to the standard cross-streak method. The strains to be typed are then applied immediately onto the same agar plates, either manually or by a mechanical device. The producers we used were selected from 39 of 77 *Klebsiella* strains identified as bacteriocinogenic, 22 in the absence of mitomycin C and 17 after induction. In most of the bacteriocinogenic strains, the amount of klebocins was significantly augmented when mitomycin C was added to the agar medium. Concentrations of mitomycin higher than $0.25\,\mu\mathrm{g}\ \mathrm{ml}^{-1}$ did not improve typing results further (for liquid media, maximum synthesis was obtained at $1\,\mu\mathrm{g}$ mitomycin per millilitre, Edmondson and Cooke, 1979) and should not be used because among the *Klebsiella* strains to be typed there may be isolates sensitive to these concentrations of mitomycin. To exclude misinterpretation through growth inhibitions not due to klebocins, we designed a particular pattern of distribution for isolates to be typed on the surface of the agar medium as shown in Fig. 1.

Growth inhibition is only accepted as having been caused by klebocins when the control spot of the isolate outside the zone of previous producer growth is clearly discernible. For additional safety, and to allow typing of isolates slightly sensitive to mitomycin C as well, every isolate is spotted onto three different plates: one without mitomycin, one with $0.1\,\mu\mathrm{g}$ and a third containing $0.25\,\mu\mathrm{g}$ of mitomycin per millilitre of medium. In order to achieve good reproducibility of results, the following technical details should be

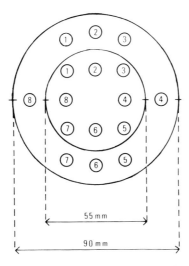

Fig. 1. Streak-and-point bacteriocin typing on solid medium. Producer strains were grown in the central area (55 mm in diameter). After their removal and killing, the isolates to be typed (Nos 1–8) were inoculated on two opposite spots both inside and outside the zone of previous producer growth.

strictly controlled: tryptic soy agar (25 ml Difco Laboratories) in 9-cm plastic Petri dishes is used. Mitomycin C (Kyowa, Japan) dissolved in methanol is incorporated into the tryptic soy agar medium at 50°C immediately before pouring. The concentration of the mitomycin C stock solution is checked regularly before use by measuring extinction at 360 nm. Producer strains are pre-grown in tryptic soy broth (Difco) at 37°C until the late logarithmic phase of growth. Densities are adjusted to 8×10^7 colony-forming units (c.f.u.) per millilitre, and 90 μl of these suspensions (7.2×10^6 c.f.u.) are inoculated by the multi-point inoculator (Denley) in a central area 55 mm in diameter (Fig. 1), using a self-made plate with a high number (85) of closely neighbouring pins. The plates are incubated for 16 h at 32°C. The bacterial lawn is removed as completely as possible by scraping with the edge of a glass microscope slide and mopping with filter paper, and any remaining viable cells are inactivated by chloroform vapours (Bauernfeind and Burrows, 1978). Indicator strains from the exponential phase of growth in tryptic soy broth are adjusted to a density of 6×10^7 c.f.u. ml^{-1}, and 2 ml of each strain are put into paired cap reservoirs (Fig. 1). The indicator strains are inoculated subsequently from these reservoirs with a multi-point inoculator so that each indicator strain is placed on two spots, one inside and one outside the zone of growth by the bacteriocinogenic strains. The plates are then incubated again for 16 h at 37°C, and the growth of the indicator strains proliferating inside and outside the zone of bacteriocins previously produced is compared and evaluated. In this procedure, neither streaking out of the producers nor spotting of the isolates to be typed are done by hand. Apart from being less laborious, mechanical inoculation allows more precise and homogeneous application of the bacteria, and this, in combination with the biological standardization to phase the growth and number of c.f.u., should produce a high degree of reproducibility.

2. *Reproducibility of typing results*

To examine reproducibility, 48 different isolates were typed independently three times by different persons within two months. Differences in typing patterns were restricted to changes in sensitivity against only one of the eight producers. Altogether, deviations in 8 of the 48 isolates (16.7%) were observed. Two were changes from $-$ to $+$, and one was a change from $+$ to $-$. The remaining six discrepancies were deviations either from weakly positive to negative or the reverse. Discrepancies involved each of the eight producers, and in no case was there more than one discrepancy in the bacteriocin pattern of a given strain. Thus the reproducibility of the results is good and compares favourably with that of biochemical identification tests. This reliability may be due to the stability of the bacteriocins synthesized by our producers, to slower inactivation rates in the agar medium as compared with those in suspensions or to the short period of time (48 h) necessary for maintaining activity.

3. *Feasibility of typing clinical isolates*

Apart from high reproducibility of results, a good typing system should allow the typing of a high percentage of clinical isolates as well as the classifying of non-epidemic isolates at least into several major groups. Typing clinical isolates of *Klebsiella* from patients in intensive care (one strain from each patient), we succeeded in classifying 241 of 259 strains (91%), the largest class including only 12% of the isolates (Fig. 2). The reliability of this method and the stability of the marker was demonstrated by typing *Klebsiella* strains consecutively isolated from patients with chronic infections. Both capsular serotyping by counter-current immunoelectrophoresis (Palfreyman, 1978) and bacteriocin typing was carried out for all the

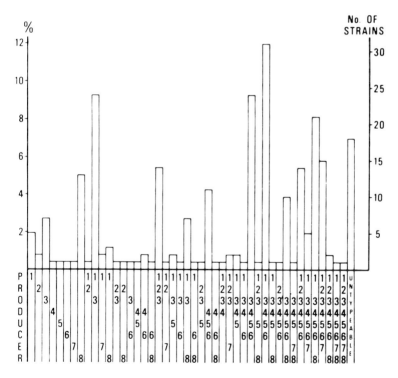

Fig. 2. Typing of *Klebsiella* strains ($n = 259$) according to bacteriocin sensitivity pattern. Percentage (left-hand scale) and corresponding number of strains inhibited by each combination of bacteriocin producers (right-hand scale). The 45 combinations actually detected are indicated beneath the abscissa.

TABLE I

Correlation of capsular serotype and bacteriocin type in *Klebsiella* isolates

Patient No.	Date of isolation of *Klebsiella*	Site of infection[a]	Capsular serotype	Bacteriocin type							
1	28 January 1980	RT	NT[b]	1	2	3	4	5	6		
	31 January 1980	RT	NT	1	2	3	4	5	6		
	7 February 1980	RT	NT	1	2	3	4	5	6		
	17 March 1980	RT	NT	1	2	3	4	5	6		
	15 April 1980	RT	NT	1	2	3	4	5	6		
2	19 December 1979	UT	31	1	2	3					
	20 December 1979	UT	31	1	2	3					
	7 January 1980	UT	51			3	4		6		
	10 January 1980	UT	51			3	4		6		
	9 February 1980	UT	51			3	4		6		
3	8 January 1980	UT	55			3					
	14 January 1980	UT	55			3					
	8 February 1980	UT	NT	1	2	3	4	5	6	7	
	5 May 1980	UT	54			3		5	6		8
4	23 January 1980	RT	7	1							8
	1 February 1980	RT	7	1							8
	6 February 1980	RT	7	1							8
	14 February 1980	RT	7	1							8

[a] RT, respiratory tract; UT, urinary tract.
[b] NT, not typable.

isolates. The results (Table I) show good coincidence of both markers, and demonstrate *in vivo* stability of bacteriocin type comparable to that of capsular serotype. Bacteriocin typing was successful even when serotyping failed. It may, therefore, be recommended as a complementary marker to the capsular serotype.

A comparison of the experience of different authors with bacteriocins from *Klebsiella* as epidemiological markers confirms their basic significance for epidemiological typing of both *K. pneumoniae* and *K. oxytoca*. Bacteriocins used by Maresz-Babczyszyn *et al.* (1967) and Edmondson and Cooke (1979) were active on both *K. rhinoscleromatis* and *K. ozaenae*. We tested only one strain of each species and observed no growth inhibition. Epidemiological typing of these species by bacteriocins needs further investigation. Sensitivity to bacteriocins constitutes a marker independent from and equivalent to those like capsular serotype or lysotype. For epidemiological typing, two different markers should be used, and bacteriocin typing may be combined with either of the others. In our experience, capsular serotype (by counter-current immunoelectrophoresis) plus bacteriocin type (by a mechanized and standardized procedure as described) are useful tools for epidemiological analysis of infections with *Klebsiella*.

4. *Characterization of the bacteriocin producers*

The producers were selected according to characteristics important for optimal typing (e.g. spectrum of inhibition, stability of patterns, resistance to mitomycin C, growth on tryptic soy agar, etc.). Additional characteristics for their identification are summarized in Table II. Some of the difficulties

TABLE II
Klebsiella producers used for the typing

No.	Species	Capsular serotype	Growth inhibited by producer No.								Sensitivity to proteases
1	*K. pneumoniae*	1		2	3	4	5	6			+
2	*K. pneumoniae*	17	1						7		+
3	*K. pneumoniae*	23	1							8	+
4	*K. pneumoniae*	24	1						7		+
5	*K. pneumoniae*	40	1	2	3	4			7		+
6	*K. pneumoniae*	7	1							8	+
7	*K. pneumoniae*	30			3	4	5	6			+
8	*K. oxytoca*	3	1								+

encountered when trying to preserve activity of klebocin preparations may be due to their colicine-like nature. In saccharose gradients our bacteriocins behave like bacteriocins of the low molecular type (Fig. 3) and do not sediment readily like the phage-tail bacteriocins of *Providencia rettgeri* (Bauernfeind, 1984a). Colicine-like bacteriocins are known to be sensitive to proteases. As expected, the klebocins were rapidly inactivated by proteases (Bauernfeind, 1984b). The high reproducibility of typing patterns reached in our system may be due to the short period of time (48 h) necessary for maintaining activity and to the higher stability of the klebocins in agar.

V. Conclusions

Nosocomial infections with *Klebsiella* may be exogenous or endogenous. Infections with bacteria from patient surroundings should be easier to

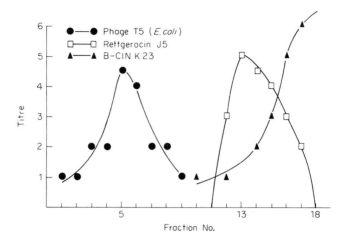

Fig. 3. Lysates of bacteriophage T5 from *Escherichia coli* and crude preparations of the phage-tail bacteriocin J5 from *Providencia rettgeri* and from klebocin were put on top of a 5–20% saccharose gradient and centrifuged for 60 min at 20 000 r/min in a SW 50.1 rotor. The centrifuge tube was punctured and fractions were collected for assay of bioactivity.

prevent by strict control of aseptic and antiseptic procedures than those infections produced by the patient's own microflora. Classifying infections by the origin of the infectious agent requires careful typing of the isolates. These data indicate an acquisition of hospital flora (colonization) by the patient and the frequency of cross-infections. Our preliminary results show a cross-infection rate with *Klebsiella* of patients in intensive care of about 35%. Markers easily available from routine diagnosis are of limited value for reliable epidemiological analysis of nosocomial infections, because the characteristics used in the identification of isolates are not useful for biotyping, and because antibiotics important for therapy are not identical with those needed for antibiotyping. Particular markers (e.g. bacteriocin typing or capsular serotyping) have to be used for precise epidemiological analysis of infections with *Klebsiella*; these methods may not be appropriate for use in routine laboratories, and typing and evaluation of the results for epidemiological analysis should therefore be performed in collaboration with reference laboratories.

References

Abbot, J. D. and Shannon, R. (1958). *J. Clin. Pathol.* **11**, 71.
Barr, J. G. (1978). *J. Med. Microbiol.* **11**, 501–511.
Bauernfeind, A. (1984a) (in preparation).

Bauernfeind, A. and Petermüller, C. (1984). *J. Clin. Microbiol.* **20**, 70–73.
Bauernfeind, A. and Burrows, J. R. (1978). *Appl. Environ. Microbiol.* **35**, 970.
Bauernfeind, A., Petermüller, C. and Burrows, J. R. (1978). *Hyg. Abt. 1 Orig. Reihe A* **240**, 271–278.
Bauernfeind, A., Petermüller, C. and Schneider, R. (1981a). *J. Clin. Microbiol.* **14**, 15–19.
Bauernfeind, A., Petermüller, C. and Jungwirth, R. (1981b). *Der Krankenhausarzt* **54**, 993–1053.
Bergan, T. (1975). *In* "Resistance of *Pseudomonas aeruginosa*" (M. R. W. Brown, Ed.), pp. 189–235. Wiley, London and New York.
Brandis, H. and Smarda, J. (1971). "Bacteriocine und Bacteriocinähnliche Substanzen". Fischer, Jena.
Brenner, D. J., Farmer, J. J. III, Hickman, F. W., Asbury, M. A. and Steigerwalt, A. G. (1979). "Taxonomie and Nomenclature Changes in Enterobacteriaceae". US Department of Health, Education and Welfare. Public Health Service, Center for Disease Control, Atlanta, Georgia.
Buffenmyer, C., Rycheck, R. R. and Yee, R. B. (1976). *J. Clin. Microbiol.* **4**, 239–244.
Casewell, M. W. and Phillips, I. (1977). *Br. Med. J.* **2**, 1315–1317.
Casewell, M. W. and Phillips, I. (1978a). *J. Hyg.* **80**, 295–297.
Casewell, M. W. and Phillips, I. (1978b). *J. Clin. Pathol.* **31**, 845–849.
Casewell, M. W., Webster, M., Dalton, M. T. and Phillips, I. (1977). *Lancet* **27**, 444–446.
CDC (1977). "National Nosocomial Infections Study Report". Center for Disease Control, Atlanta, Georgia.
Cooke, M. E., Brayson, C. J., Edmondson, A. S. and Hall, D. (1979). *J. Hyg.* **82**, 473.
Edmondson, A. S. and Cooke, E. M. (1979). *J. Hyg.* **82**, 207–223.
Govan, J. R. W. (1978). *In* "Methods in Microbiology" (T. Bergan and J. R. Norris, Eds), Vol. 10, pp. 61–91. Academic Press, London and New York.
Hall, F. A. (1971). *J. Clin. Pathol.* **24**, 712–716.
Hamon, Y. and Peron, Y. (1963). *C.R. Acad. Sci. Paris* **257**, 309–311.
Haverkorn, M. L. and Michel, M. F. (1979a). *J. Hyg.* **82**, 177.
Haverkorn, M. L. and Michel, M. F. (1979b). *J. Hyg.* **82**, 195.
Holloway, B. W. (1960). *J. Pathol. Bacteriol.* **80**, 448–450.
Jacob, F., Lwoff, A., Siminovitch, L. and Wollman, E. L. (1953). *Ann. Inst. Pasteur* **84**, 222.
Joynson, D. H. M. (1978). *J. Hyg.* **80**, 423.
Maresz-Babczyszyn, J., Przondo-Hessek, A., Lusar, Z. and Mroz-Kurpiela, E. (1964). *Arch. Immunol. Ther. Exp.* **12**, 308–318.
Maresz-Babczyszyn, J., Durlakowa, I., Mroz-Kurpiela, E., Lachowicz, Z. and Slopek, St. (1967). *Arch. Immunol. Ther. Exp.* **15**, 521.
Mindich, L. (1966). *J. Bacteriol.* **92**, 1090–1098.
Montgomerie, J. Z. (1979). *Rev. of Infect. Dis.* **1**, 736–748.
Ørskov, F. and Ørskov, I. (1978). *In* "Methods in Microbiology" (T. Bergan and J. R. Norris, Eds), Vol. 11, pp. 1–77. Academic Press, London and New York.
Ørskov, I. (1981). *In* "The Prokaryotes" (M. Starr, Ed.), Vol. 2, pp. 1160–1165. Springer-Verlag, Berlin and New York.
Palfreyman, U. (1978). *J. Hyg.* **81**, 219–225.
Reeves, P. (1972). "The Bacteriocins". Springer-Verlag, Berlin and New York.
Rennie, R. P., Nord, E. E., Sjoberg, L. and Duncan, B. R. (1978). *J. Clin. Microbiol.* **8**, 638–642.

Selden, R., Lee, S., Wen Lan Lou Wang, B. A., Bennet, J. and Eickhoff, T. C. (1971). *Ann. Int. Med.* **74**, 657–664.

Shinebaum, R. E., Cooke, M. E. and Brayson, J. C. (1978). *J. Med. Microbiol.* **12**, 201.

Slopek, S. (1978). *In* "Methods in Microbiology" (T. Bergan and J. Norris, Eds), Vol. 11, pp. 193–222. Academic Press, London and New York.

Slopek, St. and Maresz-Babczyszyn, J. (1967). *Arch. Immunol. Ther. Exp.* **15**, 525.

Steinhauer, B. W., Eickhoff, T. C., Kislak, J. W. and Finland, M. (1966). *Ann. Int. Med.* **65**, 1180.

Tagg, J. R., Dajani, A. S. and Wannamaker, L. W. (1976). *Bacteriol. Rev.* **40**, 722–756.

8

Serology of Non-gonococcal, Non-meningococcal *Neisseria* and *Branhamella* species

U. BERGER

Department of Bacteriology, Institute of Hygiene, University of Heidelberg, West Germany

I.	Introduction	225
II.	Taxonomy	226
III.	Isolation	227
IV.	Serology	229
V.	Agglutination	230
	A. Historical	230
	B. *N. flavescens*	231
	C. *N. mucosa*	232
	D. *N. sicca–N. perflava*	233
	E. *N. subflava–N. flava*	233
	F. *N. lactamica*	234
	G. *N. elongata*	234
	H. *B. catarrhalis*	235
	I. *B. caviae*	235
	J. *B. ovis*	235
VI.	Tube precipitation	235
	A. Preparation of antigens	236
	B. Results	236
VII.	Immunodiffusion	236
	A. Method	237
	B. Results	237
VIII.	Immunoelectrophoresis	239
IX.	Complement fixation reaction	240
X.	Preparation of antisera	241
XI.	Summary	243
	References	243

I. Introduction

Gram-negative aerobic diplococci other than *N. gonorrhoeae* and *N. meningitidis* have been well known since the discovery of the somewhat obscure "*Diplococcus crassus*" by Jaeger in 1895 and of "*Micrococcus catarrhalis*" by

METHODS IN MICROBIOLOGY VOL. 16
ISBN 0–12–521516–9

Frosch and Kolle in 1896. Five more species have been isolated and characterized by von Lingelsheim during the vast epidemic of meningococcal meningitis in the mining districts of Silesia and Westphalia, Germany, in 1904–1906. During a smaller outbreak of meningitis in Chicago in 1928–1930, Branham isolated a new species from the CSF of several individuals, which she called *Neisseria flavescens*. In 1953 *Neisseria caviae*, the first named species from an animal, the guinea-pig, had been isolated by Pelczar. During the years 1959–1970 eight new species partly from man, but mostly from animals have been described. These are *N. mucosa* from man (Véron *et al.*, 1959), *N. ovis* from sheep (Lindquist, 1960), *N. animalis* and *N. denitrificans* from guinea-pigs (Berger, 1960, 1962), *N. canis* from dogs (Berger, 1962), *N. cuniculi* from rabbits (Berger, 1962), *N. lactamica* from man (Hollis *et al.*, 1969) and the rod-shaped *N. elongata* from man (Bøvre and Holten, 1970).

Gram-negative aerobic diplococci other than those mentioned above, but without assignment to any named species have been isolated from cats, dogs, horses, cows, sheep, pigs, monkeys and even from geese (Pataky, 1974). Although most of the *Neisseria* species apparently are closely adapted to their specific host, human species could occasionally also be found in animals, as for example *N. mucosa* in the blowhole, mouth and throat of dolphins (Vedros *et al.*, 1973) and *N. lactamica* in the throats of chimpanzees (Kraus *et al.*, 1975). Habitats of the gram-negative aerobic diplococci are preferentially the mucosae of the throat, oropharynx, nasopharynx and nose, less frequently the genito-urinary tract and the conjunctivae.

The importance of the commensal *Neisseria* species derives from sharing their habitats with those of the pathogenic species from which they have to be differentiated and from their possible role as pathogens. As many commensals, certain strains of these species are capable of causing infections with the same localization as *N. gonorrhoeae* and *N. meningitidis*. The importance for dental caries of the species synthetizing polysaccharide from sucrose is still under discussion.

II. Taxonomy

The first attempt to classify the commensal species of gram-negative aerobic diplococci occurring in man was made by von Lingelsheim (1906). He differentiated two genera: *Micrococcus* and *Diplococcus*. The species *M. catarrhalis* and *M. cinereus* were assigned to the first genus, the species *D. pharyngis siccus, D. pharyngis flavus* I–III, *D. crassus* and *D. mucosus* (Lingelsheim, 1906) to the second genus. This classification was confirmed and slightly amended by Elser and Huntoon (1909). Later Holland (1920) assigned *M. catarrhalis* and Bergey *et al.* (1923) assigned the *Diplococcus pharyngis* group with the species designations *N. sicca, N. perflava, N. flava* and *N. subflava* to the genus *Neisseria* Trevisan (1885). All species of gram-negative, oxidase-positive diplococci discovered later (see above) as well as the rod-shaped *N. elongata* were also assigned to this genus.

After the genetic incompatibility of *N. catarrhalis, N. caviae* and *N. ovis* on the one side and of the rest of *Neisseria* spp. on the other side had been demonstrated by Catlin and Cunningham (1961), Bøvre (1965) and Kingsbury (1967), Catlin (1970) created the new genus *Branhamella* to which the three species first mentioned have been assigned and to which *N. cuniculi* probably has to be assigned as well (Bøvre and Hagen, 1981).

Both genera differ in the %(G + C) contents of the DNAs which in *Neisseria* range from 46.5 to 53.5 mole % and in *Branhamella* from 40 to 47.5 mole % (Bøvre and Hagen, 1981). That the distinction of the two genera is justified also by biochemical reasons has been confirmed by studies on carbonic anhydrase which occurs in *Neisseria*, but is lacking in *Branhamella* (Berger and Issi, 1971), on glycolytic enzymes and aspartase which are lacking in *Branhamella* as well (Holten, 1974a,b; Holten and Jyssum, 1974) and by the different fatty acid composition of the cells (Jantzen *et al.*, 1974; Lewis *et al.*, 1968; Lambert *et al.*, 1971). It remains to be clarified, whether *Branhamella* is correctly classified as a genus of the family *Neisseriaceae* or as a subgenus of the genus *Moraxella* as proposed by Henriksen and Bøvre (1968).

The differential characteristics of the 15 *Neisseria* and *Branhamella* species are shown in Table I.

III. Isolation

Apart from acting as pathogens in rare cases, the commensal *Neisseria* and *Branhamella* are always to be found as a component of autochthonous mixed bacterial population of the habitats mentioned above. They can be easily distinguished from most species of their accompanying flora by means of the oxidase reaction. Berger (1962) isolated some new species from animals by sprinkling the incubated blood agar plates with tetramethyl-*p*-phenylene diamine solution. Because of the toxicity of this oxidase reagent positive-reacting colonies have to be inoculated within one minute. The first selective media for culturing commensal *Neisseria* from the genito-urinary tract of man were used by Wax (1949, 1950) and Johnston (1951); their culture medium contained apart from ascitic fluid 20 μg of tyrothricin per millilitre. *N. lactamica* grows well on the selective medium of Thayer and Martin (1966) containing 3.75 μg of vancomycin, 7.5 μg of colistin and 12.5 units of

TABLE I

Differential characteristics of commensal *Neisseria* and *Branhamella* species

Species	Acid from						Polysaccharide from sucrose	Pigment	Haemolysis	Reduction[a] of		Gelatin	Carbonic anhydrase	Catalase
	Glucose	Maltose	Fructose	Sucrose	Lactose	Mannitol				NO_3	NO_2			
N. lactamica	+	+	−	−	+	−	−	+	−	−	N_2	−	+	+
N. mucosa subsp. *mucosa*	+	+	+	+	−	−	+	−	−	N_2	N_2	−	+	+
N. mucosa subsp. *heidelbergensis*	+	+	+	+	−	−	+	+	−	N_2	N_2	−	+	+
N. sicca	+	+	+	+	−	−	+	v	−	−	N_2	−	+	+
N. perflava	+	+	+	+	−	−	+	+	+	−	N_2	−	+	v
N. subflava subsp. *subflava*	+	+	−	−	−	−	−	+	−	−	N_2	−	+	+
N. subflava subsp. *flava*	+	+	v	+	−	−	+	+	−	−	N_2	−	+	+
N. animalis	v	−	+	+	−	+	+	−	−	−	N_2	−	+	+
N. denitrificans	+	−	−	−	−	−	−	v	−	−	N_2	−	+	+
N. flavescens	−	−	−	−	−	−	−	−	−	−	N_2	−	+	+
N. cinerea	−	−	−	−	−	−	−	v	−	−	N_2	−	+	+
N. elongata subsp. *elongata*	−	−	−	−	−	−	−	+	−	−	N_2	−	+	−
N. elongata subsp. *glycolytica*	(+)	−	−	−	−	o	o	+	−	−	+[b]	−	o	+
N. canis	−	−	−	−	−	−	−	−	−	NO_2	−	+	+	+
B. catarrhalis	−	−	−	−	−	−	−	−	−	$-NO_2\rightarrow$		−	−	+
B. caviae	−	−	−	−	−	−	−	−	+	$-NO_2\rightarrow$		−	−	+
B. ovis	−	−	−	−	−	−	−	−	−	NO_2		−	−	+
B. cuniculi	−	−	−	−	−	−	−	−	−	−		−	−	+

[a] End-products of reduction: $-NO_2\rightarrow$ = reduction of nitrate beyond nitrite without gas production; v = variable; o = not done.

[b] No gas in fermentation tube.

nystatin per millilitre (Hollis *et al.*, 1969). On the same medium not all strains of *B. catarrhalis* are inhibited (Thayer and Martin, 1966). Doern and Morse (1980) confirmed that especially pathogenic strains of this species grow well on the Thayer–Martin medium.

For isolation of *B. catarrhalis*, Berger (1961b) used heart infusion agar (Difco) containing 0.5% maltose, 0.03% water-blue and 25 μg of ristocetin per millilitre. All colourless colonies, formed by gram-negative diplococci, at that time were considered as *B. catarrhalis*, as far as they did not develop yellow pigment (*N. flavescens*). Today we know that also *N. cinerea* grows in colourless colonies on this medium. They can easily be distinguished from *B. catarrhalis* because of their inability to reduce nitrate. Similar selective media were used for the isolation of *N. flavescens* (Berger and Husmann, 1972) and *N. elongata* (Berger and Falsen, 1976). The maltose concentration was increased to 1% and lincomycin (10 μg ml^{-1}) was substituted for ristocetin which was no longer available. On this culture medium both species grew in colourless colonies. *N. flavescens* was identified due to the development of polysaccharide on 5% sucrose medium (Berger and Husman, 1972), *N. elongata* due to its rod-shaped form, reduction of nitrite and lack of catalase (Berger and Falsen, 1976). A further modification of this medium containing sucrose instead of maltose and 25 μg of lincomycin instead of 10 μg of the antibiotic per millilitre served for isolation of *N. mucosa* (Berger and Miersch, 1970). Because of its ability to form acid from sucrose, this germ grows in blue colonies. Because of its ability to denitrify nitrate it can be distinguished from *N. sicca* and *N. perflava*, which also form acid from sucrose. All these species do not require an elevated CO_2 tension during incubation, but a humid atmosphere (e.g. in a closed jar) is stimulatory.

Contrary to a common view, many strains of nearly all commensal *Neisseria* species do not grow on nutrient agar at 22°C (Wilson, 1928; Berger and Wulf, 1961). Therefore, this characteristic is of little value for differentiating them from the pathogenic species. The optimum temperature for growth is 30–37°C (Fahr and Berger, 1975).

IV. Serology

During the vast epidemic of meningitis from 1904 to 1906 von Lingelsheim tried to separate the commensal *Neisseria* species from one another and from meningococcus on the basis of their antigenic differences. A current problem at that time was to clarify the identity of the so-called Jaeger's meningococcus, which biochemically deviated from the true meningococcus Weichselbaum (acid production from lactose) but agglutinated in meningococcal antiserum and therefore was identified by Jaeger with the causative agent of epidemic meningitis (Jaeger, 1903). Lingelsheim (1906) confirmed that Jaeger's organism was agglutinated by meningococcal antiserum to a high titre, but showed at the same time that vice versa the meningococcus with an antiserum against Jaeger's coccus reacted to a very low titre or not at all. Therefore, he separated it from the meningococcus as a species of its own and called it *Diplococcus crassus*. Furthermore Lingelsheim (1906) showed that the species *Micrococcus cinereus* and *Diplococcus pharyngis flavus* I–III were serological units and shared only minor antigenic components with the meningococcus and *Micrococcus catarrhalis*. By means of serological investigations Branham later (1930) confirmed the lack of relationship between the newly described *Neisseria flavescens* and *N. meningitidis*. With serological methods (in addition to the biochemical characterization) nearly all species, described later, were determined as well, and serological methods served to differentiate serotypes (*N. elongata*) and subspecies (*N. mucosa, N. subflava*) within certain species.

Most of all the technically easiest method of agglutination was used, but there often emerged difficulties because of serum and sodium chloride agglutinability of many strains, which could be overcome by the precipitation method, especially by the immunodiffusion technique of Ouchterlony. In addition, rough strains can revert to the smooth phase and vice versa after several passages on solid media. Cary *et al.* (1958) produced suitable suspensions of spontaneously agglutinable strains by boiling heavy suspensions in 0.05 N NaOH for 10 min; after centrifugation the cell pellet was resuspended in saline and served as antigen for slide agglutination.

Because of the minor importance of this group of organisms, with very few exceptions, no sophisticated or specially adapted methods have been elaborated nor have procedures been standardized.

V. Agglutination

A. Historical

In serological investigations, which served to separate the newly described species, the tube dilution test was generally used to make possible a titre comparison of the different antigens. Lingelsheim (1906) suspended a normal loop of bacterial mass in 1 ml of every serum dilution and kept this suspension in an incubator at 37°C for 20 h. Antisera raised against *D. pharyngis flavus* I and II only agglutinated the homologous types, but not strains of the respectively other types of this species and very weakly sporadic *B. catarrhalis* strains. Antisera raised against *B. catarrhalis* with a homologous titre of 1:4000 to 1:6000 failed to agglutinate *M. cinereus* and the three types of *D. pharyngis flavus*.

Elser and Huntoon (1909) suspended 0.004 g of moist bacteria in 1 ml of sterile 0.85% saline and incubated equal parts of serum dilutions and bacterial suspensions for 2 h at 37°C and for a further 24 h at 9°C. They found that their chromogenic Groups I, II and III (later *N. perflava, N. flava* and *N. subflava*) could be distinguished from each other and that Group I contained at least two different types. (There are indeed three species producing pigment and forming acid from glucose, maltose, fructose and sucrose: *N. perflava, N. sicca* and *N. mucosa.*) Antisera raised against *N. gonorrhoeae, N. meningitidis* and *B. catarrhalis* did not agglutinate the chromogenic species. By absorption experiments, no relationship between gonococci, meningococci, *B. catarrhalis* and the chromogenic species could be established and the dissimilarity of the chromogenic Groups I, II and III could be confirmed.

Branham (1930) and most of the authors after her used bacterial suspensions obtained from solid media in buffered sodium chloride solution (pH 7.4), which was adjusted by Warner *et al.* (1952) to a density corresponding to No. 3 McFarland nephelometer standard (9×18^8 cells per millilitre). These investigators demonstrated that *N. flavescens, N. perflava, N. sicca* and *N. flava* represent serologically distinct groups, although *N. perflava* exhibited some relationship to *N. flava* as shown by cross-agglutination in *N. flava* antiserum. As demonstrated by Warner *et al.* (1952) and later investigators, strain-specific differences of the antigenic pattern could be observed within most species. Some authors preferred the slide agglutination technique because weak degrees of NaCl agglutinability are less disturbing in this than in the many hours of incubation of the tubes.

B. *N. flavescens*

Before a biochemical differentiation of this species from the other asaccharolytic *Neisseria* species was possible due to the formation of polysaccharide from sucrose, serological methods had to be employed. Thus by tube agglutination, Branham (1930) did not find any cross-reactions between *N. flavescens* and the likewise asaccharolytic *B. catarrhalis.* Wertlake and Williams (1968) identified a strain isolated from a case of septicaemia following dental surgery by means of the tube agglutination with a *N. flavescens* antiserum, where it gave a titre of 1:120; Berger and Husmann (1972) confirmed a carrier strain isolated from the nasopharynx of a soldier by means of slide agglutination with an 11-year-old *N. flavescens* antiserum. Antisera to *N. flavescens* produced by U. Berger (unpublished data) in rabbits reached an agglutinin titre to the homologous strain of 1:5120 and to the heterologous strains of 1:160 to 1:1280. The titres of two antisera produced by Warner *et al.* (1952) were within the same range.

C. *N. mucosa*

This is the species most thoroughly investigated serologically. The suspensions for agglutination were produced by Véron *et al.* (1961) by suspending the overnight growth of cultures on agar with phosphate-buffered saline (M/20, pH 7.4) and adding 0.3% formalin. They contained about 2×10^8 cells per millilitre. These suspensions generally proved to be unstable and therefore not suited for agglutination. The few stable suspensions reached an homologous titre of 1:640 to 1:5120, whereas the titres of the heterologous strains of the same species were always lower. The surface antigen of living or formalinized suspensions is thermolabile; it renders the bacteria O-inagglutinable and is destroyed above 80°C. Véron *et al.* (1961) obtained stable suspensions by heating to 120°C for 2 h. After the suspensions had been washed twice in buffered saline, they were adjusted to 2×10^8 cells per millilitre and formalinized as described above. The O-antigen is exposed in them as a granular agglutination. These suspensions can be kept at 4°C for months. For agglutination, dilution series of the antisera were produced, starting at 1:10. 0.5 ml of serum dilution was mixed with 0.5 ml of bacterial suspension, respectively, the mixture was incubated overnight at 37°C. After another 4 h at room temperature the results were read. The results suggested that two serotypes can be differentiated in *N. mucosa*. This supposition was confirmed when using absorbed antisera. The absorptions were effected with heated suspensions (2 h at 120°C) with shaking for 1 h at 20°C and repeated two or three times. The type antigens were called O:1 and O:2. Eight out of ten strains from Véron *et al.* (1961) had antigen O:1, the other two strains had antigen O:2. Apart from that a close antigenic relationship between *N. mucosa* serotype O:1 and *N. sicca* could be demonstrated which was confirmed with absorbed antisera, whereas *N. gonorrhoeae, N. meningitidis, N. perflava, N. flava, N. flavescens* and *B. catarrhalis* did not show any antigenic relations to *N. mucosa*. From these results Véron *et al.* (1961) concluded that *N. mucosa* and *N. sicca* might be two varieties of one species, the latter being distinguished by its inability to denitrify nitrate.

Berger (1971) also distinguished two subspecies in *N. mucosa*, one of them (*N. mucosa* subsp. *mucosa*) grew in whitish, larger, dry colonies, which could not be suspended homogeneously, whereas the smaller, yellow-pigmented, creamy colonies of freshly isolated strains of the other subspecies (*N. mucosa* subsp. *heidelbergensis*) were easily emulsifiable. By means of slide agglutination with suspensions of the pigmented subspecies in antisera to both a pigmented and a non-pigmented strain it could be demonstrated that both subspecies were antigenically different. These investigations were continued by the immunodiffusion technique (see below).

Vedros *et al.* (1973) isolated several pigmented strains of *N. mucosa* from dolphins (*L. obliguidens, D. bairdi*). In tube agglutination with antisera to a dolphin isolate and to a prototype *N. mucosa* strain these strains cross-reacted to a high titre, but not at all with *N. sicca* and *N. perflava*.

D. *N. sicca–N. perflava*

Using the usual tests these two species only differ in the ability to haemolyse and in the formation of pigment. As these characteristics are considered as uncertain in bacteriology, a separation of both species is sometimes difficult and not generally accepted. Attempts to separate them by giant cell formation in 5% sucrose broth (Johnson and McDonald, 1976), by growth on eosin methylene blue agar (Querido and de Araujo, 1976) and by production of an iodine-positive metabolite on trypticase soy agar (Berger and Catlin, 1975) proved to be non-reproducible. The differences in glucokinase activity (Holten, 1974a) of the electrophoretic profiles of soluble protein derived from the cells (Fox and McClain, 1974) and in the possession of plasmides (Elwell and Falkow, 1977) await confirmation with a greater number of strains.

Thus, one has to rely on serological procedures. With antisera to *N. perflava* and *N. sicca* Warner *et al.* (1952) found no cross-reactions between a lot of strains of both these species in tube agglutination, whereas by means of slide agglutination U. Berger (unpublished data) demonstrated minor cross-reactions of individual *N. sicca* and *N. perflava* strains in individual antisera to the heterologous species. With the same procedure Véron *et al.* (1961) stated that unlike *N. sicca, N. perflava* does not show an antigenic relationship to *N. mucosa*. From these results it can be concluded that *N. sicca* and *N. perflava* are two serologically distinct entities. With absorbed *N. perflava* antisera Warner *et al.* (1952) demonstrated that the six strains of this species investigated are closely related serologically, but not completely identical.

E. *N. subflava–N. flava*

These two species apparently differ only in the formation of acid from fructose by *N. flava*. By means of slide agglutination with 1:5 diluted antisera it could be demonstrated that two *N. subflava* antisera and two *N. flava* antisera agglutinated (not all) strains of both species. By means of three to five absorptions of a 1:5 diluted *N. flava* antiserum with five *N. subflava* strains, the agglutinins against all *N. subflava* and *N. flava* strains were eliminated from both antisera. From that it was concluded that both species are serologically identical and therefore should be united in the one species *N. subflava; N. flava* was lowered in rank to a subspecies (Berger and Brunhoeber, 1961). The non-absorbed *N. subflava* and *N. flava* antisera agglutinated only one of three *N. perflava* strains weakly and neither of two *N. sicca* strains (U. Berger, unpublished data). In this way the results obtained by Warner *et al.* (1952) have been essentially confirmed.

F. *N. lactamica*

In this species the usefulness of agglutination reaction is limited, because more than 70% of all strains are auto-agglutinable (Hollis *et al.*, 1969, 1970). The smooth strains are agglutinated up to a very high titre ($>$1:10 000) by homologous antisera. However, as in all *Neisseria* species, not every strain reacts in every antiserum up to the end-titre (Hollis *et al.*, 1969; U. Berger, unpublished data). Some strains have been found to cross-react in *N. meningitidis* serogroups B, C, D, X, Y and Z antisera (Mitchell *et al.*, 1965; Hollis *et al.*, 1969, 1970; Gold *et al.*, 1978; U. Berger, unpublished data), but absorptions of three *N. lactamica* antisera with *N. meningitidis* serogroups A, B, C, D, X, Y and Z strains did not result in reduction of titres to the homologous strains (Hollis *et al.*, 1969).

G. *N. elongata*

This rod-shaped, catalase-negative species, described by Bøvre and Holten in 1970, was assigned to the genus *Neisseria* because of its genetic compatibility with the "true neisseriae" and its lacking affinity to *Moraxella* and the "false neisseriae". By the work of the first investigators all strains were said not to be evenly emulsifiable in saline (Bøvre and Holten, 1970; Bøvre *et al.*, 1972). Berger and Falsen (1976) later found 5 out of 12 strains isolated by themselves to be homogeneously emulsifiable, and after several culture medium passages homogeneous suspensions could be produced from the other strains.

The same could be achieved with the type strain M2 (ATCC 25295), but not with two other strains (7823, 8554) from Holten, Oslo, which stayed rough during the whole investigation period. Using two antisera (anti-M2, anti-117) the strains could be arranged into two groups with the aid of slide agglutination, which again could be divided in two subgroups respectively (A:1, A:2, B:1, B:2) due to cross-reactions in heterologous antisera. The cross-reacting agglutinins could be eliminated by absorption. With the asaccharolytic species *N. flavescens* and *N. cinerea* no cross-reactions in the *N. elongata* antisera were observed.

Two strains (156, 180), isolated by Berger and Falsen (1976) and named *N. elongata* subsp. *intermedia*, showed remarkable similarities to *N. elongata* subsp. *glycolytica* (Henriksen and Holten, 1976) in that they were strongly catalase-positive and produced very small amounts of acid from glucose. An antiserum to one of the strains (anti-156) agglutinated not only strain 180 but also the type strain of *N. elongata* subsp. *glycolytica* (ATCC 29315) in a dilution of 1:20. On the other hand, an antiserum to strain ATCC 29315 with a homologous titre of 1:160 did not agglutinate strains 156 and 180 in a dilution of 1:10. In strain 156 antiserum (group C) both strains of *N. elongata* subgroup A:2 cross-reacted in slide agglutination.

H. *Branhamella catarrhalis*

By means of the agglutination technique some earlier investigators succeeded in differentiating *B. catarrhalis* from gono- and meningococci (Davis, 1905; Lingelsheim, 1906; Bruckner, 1908) as well as from *N. flavescens* (Branham, 1930). However, other investigators found the agglutination reaction to be unsuitable for *B. catarrhalis*, because nearly all strains spontaneously agglutinated in saline (Ghon and Pfeiffer, 1902; Elser and Huntoon, 1909; Gordon, 1921; Warner *et al.*, 1952). In a *B. catarrhalis* antiserum Warner *et al.* did not find any cross-reactions to *N. sicca, N. perflava, N. flava, N. flavescens* and *B. caviae*.

I. *Branhamella caviae*

Branhamella caviae can hardly be distinguished from *B. catarrhalis* using the conventional tests, but can be separated by agglutination (Berger, 1963). Clear serological relations could, however, be demonstrated between *B. caviae* and *N. denitrificans*, a species also isolated from the guinea-pig (Berger, 1963).

J. *Branhamella ovis*

Branhamella ovis also forms unstable suspensions according to Lindquist (1960), who first isolated this species. However, two strains, obtained by Lindquist, could be suspended homogeneously. They appeared antigenically identical in agglutination with absorbed antisera and did not cross-react with *B. caviae* (Berger, 1963).

VI. Tube precipitation

Because of the frequency of spontaneously agglutinable strains in the genera *Neisseria* and *Branhamella* Berger and colleagues preferred the tube precipitation to agglutination. Whereas preliminary experiments showed that the antigens, produced according to Lancefield and Fuller, gave unsatisfactory reactions, but the autoclaved antigens according to Rantz and Randall (1955) were used successfully.

A. Preparation of antigens

The strains were incubated in 40 ml of 0.5% glucose broth for one day at 37°C. After centrifugation the cell pellet was resuspended in 0.5 ml of saline and autoclaved for 15 min at 120°C. After another centrifugation the clear supernatant served as antigen. Two to three drops of antiserum were pipetted into small test-tubes (4 × 40 mm) and covered with the same quantity of antigen extract. The observation time was 30 min at room temperature.

B. Results

By means of the tube precipitation it could be demonstrated that *N. sicca* and *N. perflava* are heterologous species (also serologically) and that they only cross-react in exceptional circumstances (Berger and Wulf, 1961). The close serological relationship between *N. flava* and *N. subflava*, observed in agglutination tests, could also be confirmed by means of tube precipitation (Berger and Brunhoeber, 1961). Antisera to the unpigmented subspecies of *N. mucosa* only precipitated the extracts of the homologous subspecies, whereas antisera to the yellow-pigmented subspecies reacted with the extracts of both subspecies (Berger, 1971). Additionally, in both antisera there were observed weak cross-reactions with *N. sicca* and even weaker ones with *N. perflava* (U. Berger, unpublished data). Extracts of the sucrose-positive *N. animalis* were not precipitated by antisera to the equally sucrose-positive species *N. sicca, N. perflava* and *N. denitrificans* and therefore proved to be a new species (also serologically) (Berger, 1960, 1962). The differentiation of the two asaccharolytic species of man, *B. catarrhalis* and *N. cinerea*, succeeded as well with the aid of tube precipitation. Antisera to *B. catarrhalis* precipitated antigen extracts from *B. catarrhalis* only, but not from *N. cinerea*, whereas *B. catarrhalis* cross-reacted in some antisera to *N. cinerea*. Additionally, *N. cinerea* showed the serological heterogenicity, known from other *Neisseria* species: only a few of the 27 strains investigated reacted in all of the four antisera investigated, one strain reacted in none of them (Berger and Paepcke, 1962). By means of tube precipitation, the serological homogenicity of the three strains of *N. (B.) cuniculi* newly isolated from rabbits and their relationship to the subspecies *gigantea** (Berger, 1962) could be shown.

VII. Immunodiffusion

Minor differences in the antigen structure cannot be recognized using tube precipitation, and so in more recent investigations the immunodiffusion

---

*All strains of *N. cuniculi* subsp. *gigantea* are lost. The strains listed under this name in international collections are not identical with the first description.

technique (Ouchterlony) was used. With this technique the classification of *N. elongata* into three groups (A, B, C) and four subgroups (A:1, A:2, B:1, B:2), using agglutination, could be confirmed.

A. Method

Sterile microscopic slides were covered with 1% Noble agar, to which there was added 0.1% NaN$_3$. The wells punched out had a 4-mm diameter. The distance of the peripheral wells (antigens) to the central well (antiserum) also amounted to 4 mm (from edge to edge). Very heavy suspensions of living cells from overnight cultures on blood agar served as antigens. The antisera were used undiluted. Cross-reacting antibodies were absorbed by means of a suspension of the whole bacterial mass of an overnight culture on a blood agar plate (90-mm diameter) in 1 ml of undiluted antiserum, and incubated for 4 h at 50°C. The reactions were read after two days incubation at room temperature.

B. Results

With an antiserum to the type strain M2 of *N. elongata* (ATCC 25295) (Berger's subgroup A:2) strains of subgroup A:2 gave two precipitation lines, whereas strains of subgroup A:1 only gave one line, which was identical to the stronger line of subgroup A:2. By absorption with an antigen of subgroup A:1 all lines were eliminated. Strains of both A subgroups did not react with antisera to Groups B and C.

With cells of subgroup B:1 an antiserum to a strain of subgroup B:1 reacted with formation of one line, whereas cells of subgroup B:2 gave two lines, none of which was identical with the subgroup B:1 line. By absorption with subgroup B:2 cells, the two lines disappeared whereas the B:1-specific line persisted. Strains of both B subgroups did not react with antisera to Groups A and C.

With an antiserum to a strain of Group C (*N. elongata* subsp. *intermedia*) homologous strains gave two identical precipitation lines. This antiserum contained cross-reacting precipitins to *N. flavescens*, which reacted as one line apparently identical to the stronger line of Group C antigens. This identical line could not be eliminated by twofold absorption with *N. flavescens* cells, but by 1:20 dilution of the antiserum (Berger and Falsen, 1976). Group C cells did not cross-react in Group A and Group B antisera.

With an antiserum against the NADP-linked glutamate dehydrogenase from a strain of *N. meningitidis* (M6) and extracts from *N. elongata* (M2), *N. elongata* subsp. *glycolytica* (6171/75) and *N. meningitidis* (M1) Bøvre *et al.* (1977) found a reaction of partial identity between *N. meningitidis* and the two subspecies of *N. elongata* and a reaction of identity between *N. elongata* and *N. elongata* subsp. *glycolytica*. The antigen extracts were obtained by ultrasonic treatment of cells followed by centrifugation at $10\,000 \times g$ for 30 min. 1% agarose in 0.85% saline served as the gel. Incubation was at room temperature for 24 h. Some years before, Holten (1974c) had demonstrated reactions of identity between *N. meningitidis*, *N. flavescens*, *N. lactamica* and *N. cinerea* and reactions of similarity between *N. meningitidis* and all other human *Neisseria* species, using an antiserum to meningococcal NADP-dependent glutamate dehydrogenase in Ouchterlony double diffusion.

Using the same technique as above (Berger and Falsen, 1976), the existence of the two subspecies of *N. mucosa* can be confirmed (Berger, 1978). An antiserum to the subspecies *mucosa* reacted with homologous antigens by forming three precipitation lines, however, with antigens of the subspecies *heidelbergensis* there was no reaction at all. Despite that, two of the three lines could be eliminated by absorption with *N. mucosa* subsp. *heidelbergensis* cells. An antiserum to the subspecies *heidelbergensis* reacted with homologous antigens by forming a double line, with antigens of the subspecies *mucosa* by forming a single line, which could be eliminated by absorption with *N. mucosa* subsp. *mucosa* cells. *N. sicca* cross-reacted in the antisera to both subspecies of *N. mucosa* forming a strong (subsp. *mucosa* antiserum) or weak line (subsp. *heidelbergensis* antiserum). *N. perflava* only cross-reacted clearly with the subsp. *heidelbergensis* antiserum. The only technical modification compared to the experiments with *N. elongata* (Berger and Falsen, 1976) consisted in using 1.5% Noble agar and 0.1 M barbital buffer, pH 8.6, as a solvent.

Using another modification of the immunodiffusion technique, Russel *et al.* (1978) examined seven *Neisseria* species (*N. sicca*, *N. perflava*, *N. flava*, *N. subflava*, *N. flavescens*, *N. cinerea*, *N. canis*) and three *Branhamella* species (*B. catarrhalis*, *B. caviae*, *B. ovis*). A suspension of the bacterial strain to be examined was streaked in a straight line on trypticase soy agar. Following overnight incubation at 37°C, a series of wells was cut in the agar parallel to and approximately 1 cm from the edge of the bacterial growth. The bottoms of the wells were sealed with molten agar and the wells then filled with the desired antisera. The plates were kept in a sealed plastic container with wet filter paper for 48 h at room temperature.

The greatest number of precipitation lines was generally observed when an organism was tested against its homologous antiserum, but extensive cross-reactivity was also found. All *Neisseria* species reacted with the antiserum raised against *N. flava*, three species (*N. perflava*, *N. subflava*, *N. cinerea*) reacted with the *N. perflava* antiserum and three species (*N. subflava*, *N. flava*, *N. flavescens*) with the *N. subflava* antiserum. (*N. canis* apparently was a *N. subflava* strain.) Only *B. catarrhalis* reacted with the *B. catarrhalis* antiserum, *B. caviae* and *B. ovis* reacted with the *B. caviae* antiserum. (*N. sicca* apparently was a *B. caviae* strain.) *N. flava* and *N. subflava* as well as *B.*

caviae and *B. ovis* proved to be very closely related immunologically. Eliasson (1980) found that acid, but also neutral, alkaline and sonicated extracts of *B. catarrhalis* (strain 235 L) gave three to four precipitation lines in immunodiffusion against a rabbit antiserum raised to the same strain. The line closest to the antigen stained most intensely. It was called P-line, the antigenic component responsible for the precipitate P-component. The trypsin sensitivity of the P-antigen indicated that it was a protein. By absorption of rabbit antiserum to *B. catarrhalis* with a trypsinized extract of the same strain it was possible to prepare an antiserum specific for the P-antigen. When using this absorbed serum and radiolabelled protein A from *Staphylococcus aureus*, the P-antigen could be detected as a surface antigen on the intact bacteria. Out of 31 *B. catarrhalis* strains, 31 *Neisseria* species strains and 2 *Haemophilus influenzae* strains only the 31 *B. catarrhalis* strains revealed the capacity to bind anti-P antibodies abundantly, indicating that the P-antigen is characteristic of *B. catarrhalis*. This finding opens the possibility of a serological identification of *B. catarrhalis*. 69% of 90 human sera tested by means of immunodiffusion for the ability to precipitate hot acid extracts of *B. catarrhalis* formed precipitation lines which fused completely with the line formed between the P-antigen and rabbit antiserum to *B. catarrhalis*.

The hot acid extracts were prepared according to Lancefield from ten one-litre Todd-Hewitt broths in buffled flasks, in which the bacteria were incubated for 18 h at 37°C on a platform shaker, operating at 120 r/min. Cells were recovered by continuous centrifugation at $5000 \times g$, washed three times in saline and resuspended in 100 ml of saline. The pH was adjusted to 2.0 with 1 M HCl. The suspension was heated in a boiling water bath for 10 min, cooled on ice and the pH was readjusted to 7.0 with 1 M NaOH. The cell-free fluid, obtained after centrifugation for 30 min at $3000 \times g$, was stored at 20°C until used as acid extract. The immunodiffusion was performed in 0.6% (w/v) agarose in PBS. The plates were read after one and two days at $+4$°C.

VIII. Immunoelectrophoresis

By means of crossed immunoelectrophoresis of a reference antigen (Ref-ag, prepared by ultrasonic disintegration of a suspension containing *N. meningitidis* A, B and C cells) against a reference antibody (Ref-ab, produced with Ref-ag and Freud's incomplete adjuvant in rabbits) Hoff and Høiby (1978a) detected 63 precipitates, 48 of which were regularly visible. In an investigation on cross-reactivity of these 48 *N. meningitidis* antigens and antigens from 27 other bacterial species, the same authors found a close relationship between *N. meningitidis* and five commensal *Neisseria* species (*N. lactamica, N. flavescens, N. sicca, N. perflava, N. subflava*). With the exception of *N. flavescens,* which presented only 38 cross-reacting antigens, all strains cross-reacted with more than 40 meningococcal antigens. *N. subflava* showed cross-reactions with all the 48 antigens. The distant relationship of *Branhamella* to *Neisseria* was confirmed by cross-reaction of only two *B. catarrhalis* antigens with *N. meningitidis* (Hoff and Høiby, 1978b).

IX. Complement fixation reaction

By means of the CFR, Oliver (1930) demonstrated a serological relationship between *N. gonorrhoeae* and *B. catarrhalis.* Most rabbit antisera to *B. catarrhalis* yielded a positive though weaker reaction with gonococcal antigen. In infections of man due to *B. catarrhalis,* CFR with a *B. catarrhalis* antigen gave irregular and less distinct results than with rabbit immune sera. The reactions with gonococcal antigens were only rarely, and even weaker, positive.

Reyn (1943), who reviewed the earlier literature on this subject, demonstrated that five out of eight sera from rabbits, immunized with different gonococcal strains, in CFR reacted to a *B. catarrhalis* antigen with a strength of 1–6 degrees.* On the other hand, animals immunized with *B. catarrhalis* reacted with the homologous antigen to a high titre (12 degrees), but not all with gonococcal antigen. 15 out of 17 sera from man with natural gonococcal infection and positive "gono-reaction" also gave positive catarrhalis reactions (3–7 degrees). A *B. catarrhalis* antigen was able to absorb a part of the "gono-reaction" from human gono-positive sera, whereas a mixed gonococcal antigen was unable to remove any part of the catarrhalis reaction. From these experiments Reyn concluded that an unquestionable antigenic relationship exists between *N. gonorrhoeae* and *B. catarrhalis.*

In an approach to elucidate the aetiological significance of *B. catarrhalis* in acute maxillary sinusitis, Brorson *et al.* (1976) demonstrated CF antibodies against *B. catarrhalis* in the sera of 25 out of 97 patients, but in only one out of 20 control sera. The titres were low (1:7.5–1:30) and the titre changes were low. All patients with CF antibodies had also precipitating antibodies.

The antigen for complement fixation was prepared by harvesting the growth on agar plates incubated for 20 h at 37°C and washing and heating it to 56°C for 30 min. The bacteria were used in the highest concentration that did not give anti-complementary activity. The highest serum dilution giving less than 50% haemolysis was registered as the limiting value. The serum dilutions started at 1:7.5.

Using anti-glutamate dehydrogenase and anti-malate dehydrogenase rabbit sera prepared from *N. meningitidis* M6 and *N. perflava* ATCC 10555,

---

*The degree of strength is a logarithmic expression for the titre of the test (Reyn).

respectively, Holten (1974c) determined the immunological distances of the human *Neisseria* species from *N. meningitidis* by means of a quantitative micro-complement fixation test as described by Levine and Vunakis (1967). The index of dissimilarity (i.d.) for a given antigen was defined as the factor by which the serum concentration must be raised to obtain a complement fixation curve equal in height to that of the homologous system. Since the number of amino acid sequence differences seems to be proportional to the logarithm of this value, the term "immunological distance" has been introduced, defined as $100 \times \log$ i.d. The values obtained by anti-glutamate dehydrogenase permitted the classification of the strains into four groups according to their immunological distances (I.D.) from *N. meningitidis*:

Group I.	I.D. = 0–15:	*N. meningitidis, N. gonorrhoeae, N. flaves-cens, N. cinerea, N. lactamica*
Group II.	I.D. = 20–40:	*N. sicca, N. perflava, N. flava, N. subflava*
Group III.	I.D. = 70–80:	*N. mucosa*
Group IV.	I.D. = 90–105:	*N. elongata*

Using anti-malate dehydrogenase serum, all strains fell within the immunological distance of 10. No complement was fixed with either serum using extracts of the *Branhamella* species. From the taxonomical point of view the most remarkable results of this study are the close relationship of *N. flavescens, N. cinerea* and *N. lactamica* to *N. meningitidis* (and to each other), the distant relationship of the rod-shaped *N. elongata*, the different distances of the two *N. mucosa* subspecies and the similarity of *N. flava* and *N. subflava*, justifying their classification as *one* species, as proposed by Berger and Brunhoeber (1961).

X. Preparation of antisera

Most authors produced the agglutinating antisera according to the method of Warner *et al.* (1952). Living bacterial suspensions were obtained by suspending a one-day-old culture on trypticase soy agar in 0.85% buffered saline at pH 7.4. Rabbits received three intravenous injections for three days running in the first week and four injections in the second week. The dose of injection as well as the density of the suspensions were increased according to Table II.

A trial bleeding was made on the fifth day following the last injection. Antisera to *N. sicca, N. perflava, N. flava* and *N. flavescens* prepared in this way had agglutinating titres of 1:1600–1:5120 (Warner *et al.*, 1952). U. Berger (unpublished data), proceeding in this way, obtained an agglutinating *N. lactamica* antiserum with a titre of more than 1:10 000. In the case of

TABLE II
Preparation of agglutinating antisera in rabbits[a]

Week	Day	Dose	Density
1	1	0.5 ml	900×10^6 cells ml^{-1}
	2	1.0 ml	—
	3	1.0 ml	—
2	1	0.5 ml	1200×10^6 cells ml^{-1}
	2	1.0 ml	—
	3	1.5 ml	—
	4	2.0 ml	—

[a]From Warner et al. (1952).

weaker antigens it is useful to administer a booster injection in the third week and to bleed not before the fourth week (Berger and Falsen, 1976). The immunization with formalinized suspensions (0.3% of commercial 40% formaldehyde solution) apparently does not give worse results (Véron et al., 1961; Berger, 1962). The antisera can be preserved without titre loss by adding merthiolate 1:10 000.

In order to obtain an adequate titre of precipitating antibodies, it can be necessary to prolong the procedure of immunization by one week (three injections respectively during three weeks, one booster injection in the fourth week, bleeding in the fifth week) (Berger, 1978). Holten (1974c) immunized rabbits with glutamate dehydrogenase and malate dehydrogenase preparations and complete Freud's adjuvant by subcutaneous injections at multiple sites on the back, followed after four weeks by three weekly intravenous injections of the enzyme preparations. The animals were bled one week after the last injection. The preparation of the antigenic enzyme extracts must be read in the original.

As antigen for producing antisera for the complement fixation, Reyn (1942) used formalin-killed B. catarrhalis suspensions with a density of about 1000×10^6 cells per millilitre. This antigen was administered intravenously twice a week for five weeks. The first injection dose was 1 ml, all further doses were 2 ml. For the same purpose, Jessen (1934) had used suspensions of living cells of the same density, which were injected once a week for four weeks. One week after the last injection the animals were bled.

XI. Summary

In the past, serological methods such as complement fixation, agglutination in tubes and on slide, precipitation, immunodiffusion and immunoelectrophoresis were mainly used to clarify taxonomical questions within the *Neisseria–Branhamella* group. For diagnostic purposes, especially in routine laboratory work, they are less suitable, because antisera are not available commercially and their self-production is not worthwhile because of the rarity of pathogenic strains in this group of organisms.

References

Berger, U. (1960). *Z. Hyg.* **147**, 158–161.
Berger, U. (1961a). *Naturwissenschaften* **48**, 405–406.
Berger, U. (1961b). *Zbl. Bakt. I Orig.* **183**, 135–138.
Berger, U. (1962). *Zbl. Bakt. I Orig.* **148**, 445–457.
Berger, U. (1963). *Ergeb. Mikrobiol. Immunitaetsforsch. Exp. Ther.* **36**, 97–167.
Berger, U. (1971). *Z. Med. Mikrobiol. Immunol.* **156**, 154–158.
Berger, U. (1978). *Med. Microbiol. Immunol.* **165**, 169–179.
Berger, U. and Brunhoeber, H. (1961). *Z. Hyg.* **148**, 39–44.
Berger, U. and Catlin, B. W. (1975). *Zbl. Bakt. I Orig. Reihe A* **232**, 129–130.
Berger, U. and Falsen, E. (1976). *Med. Microbiol. Immunol.* **162**, 239–249.
Berger, U. and Husmann, D. (1972). *Med. Microbiol. Immunol.* **158**, 121–127.
Berger, U. and Issi, R. (1971). *Arch. Hyg.* **154**, 540–544.
Berger, U. and Miersch, M. (1970). *Z. Med. Mikrobiol.* **155**, 186–191.
Berger, U. and Paepcke, E. (1962). *Z. Hyg.* **148**, 269–281.
Berger, U. and Wulf, B. (1961). *Z. Hyg.* **147**, 257–268.
Bergey *et al.* (1923). "Bergey's Manual of Determinative Bacteriology", 1st edn. (Bergey, D. H., Harrison, F. C., Breed, R. S., Hammer, B. W. and Huntoon, F. M., Eds), Williams & Wilkins, Baltimore, Maryland.
Bøvre, K. (1965). *Acta Pathol. Microbiol. Scand.* **64**, 229–242.
Bøvre, K., Frøholm, L. O., Henriksen, S. D. and Holten, E. (1977). *Acta Pathol. Microbiol. Scand. Sect. B* **85**, 18–26.
Bøvre, K., Fuglesang, J. E. and Henriksen, S. D. (1972). *Acta Pathol. Microbiol. Scand. Sect. B* **80**, 919–922.
Bøvre, K. and Hagen, N. (1981). *In* "The Prokaryotes" (Starr, M. P., Stolp, H., Trüper, H. G., Balows, A. and Schlegel, H. G., Eds), pp. 1506–1529. Springer-Verlag, Berlin and New York.
Bøvre, K. and Holten, E. (1970). *J. Gen. Microbiol.* **60**, 67–75.
Branham, S. E. (1930). *Public Health Rep.* **45** (I), 845–849.
Brorson, J.-E., Axelsson, A. and Holm, S. E. (1976). *Scand. J. Infect. Dis.* **8**, 151–155.
Bruckner, J. (1908). *C.R. Soc. Biol.* **64**, 619–620.
Cary, S. G., Lindberg, R. G. and Faber, J. E. (1958). *J. Bacteriol.* **75**, 43–45.
Catlin, B. W. (1970). *Int. J. Syst. Bacteriol.* **20**, 155–159.
Catlin, B. W. and Cunningham, L. S. (1961). *J. Gen. Microbiol.* **26**, 303–312.
Davis, D. J. (1905). *J. Infect. Dis.* **2**, 602–619.
Doern, G. V. and Morse, S. A. (1980). *J. Clin. Microbiol.* **11**, 193–195.

Eliasson, I. (1980). *Acta Pathol. Microbiol. Scand. Sect. B* **88**, 281–286.
Elser, W. J. and Huntoon, F. M. (1909). *J. Med. Res.* **20**, 371–541.
Elwell, L. P. and Falkow, S. (1977). *In* "The Gonococcus" (R. B. Roberts, Ed.), pp. 137–154. Wiley, New York.
Fahr, A. and Berger, U. (1975). *Zbl. Bakt. I Orig. Reihe A* **230**, 551–555.
Fox, R. H. and McClain, D. E. (1974). *Int. J. Syst. Bacteriol.* **24**, 172–176.
Frosch, P. and Kolle, W. (1896). *In* "Die Mikrooganismen" (C. Flügge, Ed.), 3rd edn, pp. 154–155.
Ghon, A. and Pfeiffer, H. (1902). *Z. Klin. Med.* **44**, 262–281.
Gold, R., Goldschneider, I., Lepow, M. L., Draper, T. E. and Randolph, M. (1978). *J. Infect. Dis.* **137**, 112–121
Gordon, J. E. (1921). *J. Infect. Dis.* **29**, 462–494.
Henriksen, S. D. and Bøvre, K. (1968). *J. Gen. Microbiol.* **51**, 387–392.
Henriksen, S. D. and Holten, E. (1976). *Int. J. Syst. Bacteriol.* **26**, 478–481.
Hoff, G. E. and Høiby, N. (1978a). *Acta Pathol. Microbiol. Scand. Sect. B* **86**, 1–9.
Hoff, G. E. and Høiby, N. (1978b). *Acta Pathol. Microbiol. Scand. Sect. B* **86**, 87–92.
Holland. D. F. (1920). *J. Bacteriol.* **5**, 215–229.
Hollis, D. G., Wiggins, G. L. and Weaver, R. E. (1969). *Appl. Microbiol.* **17**, 71–77.
Hollis, D. G., Wiggins, G. L., Weaver, R. E. and Schubert, J. H. (1970). *Ann. N.Y. Acad. Sci.* **174**, 444–449.
Holten, E. (1974a). *Acta Pathol. Microbiol. Scand. Sect. B* **82**, 201–206.
Holten, E. (1974b). *Acta Pathol. Microbiol. Scand. Sect. B* **82**, 207–213.
Holten, E. (1974c). *Acta Pathol. Microbiol. Scand. Sect. B* **82**, 849–859.
Holten, E. and Jyssum, K. (1974). *Acta Pathol. Microbiol. Scand. Sect. B* **82**, 843–848.
Jaeger, H. (1895). *Z. Hyg.* **19**, 351–370.
Jaeger, H. (1903). *Zbl. Bakt. I Orig.* **33**, 23–34.
Jantzen, E., Bryn, K., Bergan, T. and Bøvre, K. (1974). *Acta Pathol. Microbiol. Scand. Sect. B* **82**, 767–779.
Jessen, J. (1934). *Zbl. Bakt. I Orig.* **133**, 75–88.
Johnson, K. G. and McDonald, I. J. (1976). *Can. J. Microbiol.* **22**, 431–434.
Johnston, J. (1951). *Am. J. Syph. Neurol.* **35**, 79–82.
Kingsbury, D. T. (1967). *J. Bacteriol.* **94**, 870–874.
Kraus, S. J., Brown, W. H. and Arko, R. J. (1975). *J. Clin. Invest.* **55**, 1349–1356.
Lambert, M. A., Hollis, D. G., Moss, C. W., Weaver, R. E. and Thomas, M. L. (1971). *Can. J. Microbiol.* **17**, 1491–1502.
Levine, L. and Vunakis, H. van (1967). *In* "Methods in Enzymology" (S. P. Colowick and N. O. Kaplan, Eds), Vol. 11, pp. 928–936. Academic Press, New York and London.
Lewis, V. J., Weaver, R. E. and Hollis, D. G. (1968). *J. Bacteriol.* **96**, 1–5.
Lindquist, K. (1960). *J. Infect. Dis.* **106**, 162–165.
Lingelsheim, W. von (1906). *Klin. Jahrb.* **15**, 373–488.
Mitchell, M. S., Rhoden, D. L. and King, E. O. (1965). *J. Bacteriol.* **90**, 560.
Oliver, J. O. (1930). *J. Hyg.* **29**, 259–272.
Pataky, M. (1974). *Acta Vet. Acad. Sci. Hung.* **24**, 355–359.
Pelczar, M. J. (1953). *J. Bacteriol.* **65**, 744.
Querido, N. B. G. and de Araujo, W. C. (1976). *Appl. Microbiol.* **31**, 612–614.
Rantz, L. A. and Randall, E. (1955). *Stanford Med. Bull.* **13**, 290–291.
Reyn, A. (1943). *Acta Pathol. Microbiol. Scand.* **20**, 257–271.
Russell, R. R. B., Johnson, K. G. and McDonald, I. J. (1978). *Can. J. Microbiol.* **24**, 189–191.

Thayer, J. D. and Martin, J. E. (1966). *Public Health Rep.* **81**, 559–562.
Vedros, N. A., Johnston, D. G. and Warren, P. I. (1973). *J. Wildlife Dis.* **9**, 241–244.
Véron, M., Thibault, P. and Second, L. (1959). *Ann. Inst. Pasteur* **97**, 497–510.
Véron, M., Thibault, P. and Second, L. (1961). *Ann. Inst. Pasteur* **100**, 166–179.
Warner, G. S., Faber, J. E. and Pelczar, M. J. (1952). *J. Infect. Dis.* **90**, 97–103.
Wax, L. (1949). *J. Vener. Dis. Inf.* **30**, 145–146.
Wax, L. (1950). *J. Vener. Dis. Inf.* **31**, 208–213.
Wertlake, P. T. and Williams, T. W. (1968). *J. Clin. Pathol.* **21**, 437–439.
Wilson, S. P. (1928). *J. Pathol. Bacteriol.* **31**, 477–492.

9

Serotyping of *Pasteurella multocida*

G. R. CARTER

Division of Pathobiology and Public Practice, Virginia–Maryland Regional College of Veterinary Medicine, Virginia Tech, Blacksburg, Virginia, USA

I.	Introduction	247
II.	Species characteristics	248
III.	Handling of cultures	248
IV.	Colony variation	250
V.	Identification of capsular substances	251
	A. Indirect haemagglutination test (IHA)	251
	B. Identification of capsular types B and E by counterimmunoelectrophoresis (CIE)	255
	C. Identification of capsular type A strains with hyaluronidase	255
	D. Identification of capsular type D cultures with acriflavine	256
VI.	Identification of somatic serogroup antigens	256
	A. Antisera	257
	B. Antigen	257
	C. Gel diffusion test	257
VII.	Designated serotypes	258
VIII.	Other typing procedures	258
	References	258

I. Introduction

In a recent communication a standard system designating serotypes of *Pasteurella multocida* was recommended (Carter and Chengappa, 1981a). It was based upon the identification of the capsular substances according to Carter's system and identification of somatic or O-antigens by the system developed by Heddleston and associates (Heddleston *et al.*, 1972a,b; Heddleston and Wessman, 1975). Thus, a serotype is identified first by a capital letter representing the capsular type followed by an Arabic number 1, 2, 3, 4, etc., representing the somatic antigen. In some instances, the somatic type may be represented by additional numbers, representing less strong reactions.

METHODS IN MICROBIOLOGY VOL. 16
ISBN 0 12 521516 9

Here procedures used for identification of the capsular and somatic antigens will be presented. The material is taken from published papers with comments based on the author's experience.

Successful serological work with this species requires considerable experience, patience, perseverance and attention to detail. Cultures may be antigenically unstable and subject to colonial variation. Rabbits in which some anti-pasteurella sera are produced frequently possess antibodies to *P. multocida* which may complicate serological tests. Difficulty is frequently experienced in obtaining antisera of sufficient antibody titre.

The procedures recommended for serotyping *P. multocida* are outlined step-by-step in Fig. 1.

II. Species characteristics

Typical cultures of *P. multocida* have the following characteristics:

Gram-negative coccobacillus
Non-haemolytic
Acid but no gas or H_2 in TSI agar
Non-motile
Indole positive*
Oxidase positive
Reduces nitrate
Acid but no gas from glucose, sucrose and mannitol**
Lactose usually negative
Maltose variable.

   *Some strains may be negative unless heart infusion broth is used.
   **Carbohydrates other than glucose may occasionally be negative.

III. Handling cultures

Cultures of *P. multocida* are best stored in the lyophilized state (preferably in the refrigerator) or in defribrinated blood in a freezer (colder the better). In the latter method organisms from blood agar cultures are suspended in small amounts of blood (0.5 ml) and placed in the freezer. Cultures can be kept temporarily on Stock Culture Agar (Difco) at room temperature. The medium is dispensed in screw-cap tubes to give a butt of about 5 cm in depth. Small vials can be used for shipment. The culture is stabbed into the butt several times and incubated for 18 h. The screw cap is sealed with tape to assure airtightness. Many cultures will remain viable at room temperature for months; however, the occasional culture will die for no obvious reason. If

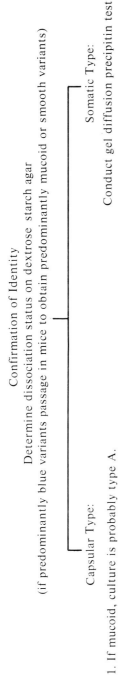

Confirmation of Identity

Determine dissociation status on dextrose starch agar

(if predominantly blue variants passage in mice to obtain predominantly mucoid or smooth variants)

Capsular Type:

1. If mucoid, culture is probably type A.
   Perform staphylococcal hyaluronidase test.

2. If not type A conduct acriflavine test for type D.

3. If not type A or D by the above tests use the IHA test.

4. If from suspected haemorrhagic septicaemia in cattle, water buffaloes, etc., perform either the IHA test or the counterimmunoelectrophoresis test.

Somatic Type:

Conduct gel diffusion precipitin test

1. On fowl cholera isolates first use pooled sera types 1, 3, 4, 5.

2. For definitive typing use all type sera.

Result

Serotype = Capsular type letter followed by somatic type number(s).

Fig. 1. Steps followed in serotyping *Pasteurella multocida.*

these or other cultures on or in media are stored in the refrigerator they will die within a week or two. Occasional cultures will undergo colonial change in stock culture agar, for example all smooth or mucoid variants may be lost leaving only blue or non-capsulated variants.

When attempting to recover organisms from lyophilized material, stock culture agar or other media, it is advisable to plate on blood agar. Any remaining lyophilized material should be transferred to a tube of broth and incubated as a back up. Small numbers of organisms will occasionally fail to grow on media without blood. Some stock cultures in stock culture agar from which organisms cannot be obtained by plating will yield growth if broth (we use Brain Heart Infusion, BHI) is poured onto the butt and the tube then incubated for several days.

IV. Colony variation

Cultures of *P. multocida* undergo considerable colonial dissociation (Elberg and Ho, 1950; Carter, 1975; Heddleston and Wessman, 1975). The variants can be readily recognized on clear agar media through which the light of a lamp is allowed to pass obliquely. Tryptose Agar (Difco), various solid media to which has been added 2–5% serum, Dextrose Starch Agar (Difco) and UPC Agar (Namioka) are satisfactory.

Smooth colonies composed of well-capsulated organisms show an iridescence in oblique light. Organisms with the largest capsules are derived from mucoid colonies; these show less iridescence and are relatively large, sticky and possess flowing margins. The capsules of organisms from the mucoid variants are composed largely of hyaluronic acid. The smooth colonies are smaller and possess organisms with appreciable capsules. Intermediate variants are seen that do not show iridescence; they possess organisms with a lesser amount of capsular material. Blue variants are the smallest and constituent organisms have little if any capsule and have a tendency to auto-agglutinate. Rough colonies are seen occasionally.

Serological procedures for the identification of capsular substances are best carried out with cultures composed of predominantly mucoid or smooth variants. Most strains recovered from disease processes possess either or both of these variants. Virulent avian cultures are usually less mucoid than bovine and porcine strains. Some cultures consisting of mostly blue and intermediate variants can be restored to a state of predominantly smooth or mucoid forms by passage in mice. Each mouse is inoculated intraperitoneally with 0.05–0.1 ml—occasionally larger amounts are required—of a 6 to 8 h broth culture. Heart's blood from dead or dying mice is plated on blood agar. A second passage is sometimes necessary. The virulence of some cultures from dogs and cats and occasionally from other animals cannot be increased by mouse passage.

V. Identification of capsular substances

Capsular types are designated A, B, D and E. They can be identified by various procedures.

A. Indirect haemagglutination test (IHA)

The IHA procedures are described in several papers, (Carter, 1955, 1961, 1972).

1. *Preparation of anti-pasteurella (capsular) sera*

Dried out or partially dehydrated plate media should not be used for preparing organisms for typing or immunization. Dehydration can be avoided by using freshly prepared media or media that have been stored in sealed plastic bags. Plates should be incubated in sealed containers or in a humidified incubator.

The selected smooth or mucoid capsular type strains are inoculated onto blood agar or dextrose starch agar so as to yield nearly confluent growth after 18 to 24 h incubation at 37°C. The bacteria are washed off with 10 ml of physiological saline (0.85%) and then separated by centrifugation. The packed cells are resuspended in sufficient 0.3% formalized saline to yield about 10^9 cells per millilitre. Suspensions are incubated overnight at 37°C to kill the bacteria. Although various immunizing schedules can be used, the schedule of injections listed in Table I has been satisfactory (*P. multocida* free rabbits are preferable). We usually inoculate at least two rabbits with each culture.

TABLE I
Injection schedule of immunization

Day	Amount	Route
1	1.0	Subcutaneous
6–7	0.5	Intravenous
10	1.0	Intravenous
14	1.5	Intravenous
18	2.0	Intravenous
22	3.0	Intravenous
26	5.0	Intravenous
30		Exsanguinate

Sera are inactivated at 56°C for 30 min and then preserved with merthiolate to give a final concentration of 1:10 000. Small amounts of antisera are stored at 4°C for regular use while larger amounts are stored in a freezer.

Organisms grown and killed (use 0.5% formalin) as described above can be emulsified with Freund's incomplete adjuvant and used to raise adequate levels of IHA antibodies to types B and E. Equal volumes of bacterial suspension (McFarland's tube 10) and Freund's incomplete adjuvant (Difco) are emulsified with a syringe. Each rabbit is inoculated subcutaneously with a total of 1.0 ml of emulsion, 0.25 ml being inoculated into each of four sites. Rabbits are trial bled at four weeks and exsanguinated if the IHA titre is satisfactory. Occasionally titres can be increased by giving a further intravenous inoculation of a saline suspension of the kind used in the schedule given above. The procedure just described has not been found adequate for preparing IHA sera for types A and D.

In the preparation of type A and D antisera it is advisable to employ several different cultures as some do not appear to be "antigenic" with regard to capsular antibodies. It seems likely that capsular haptens rather than complete antigens are involved in IHA with these types. It may be advantageous to use a type A strain that is iridescent but also does not produce capsular hyaluronic acid. Some avian cultures have this characteristic.

In practice, it is not advisable to use an anti-pasteurella serum with a titre below 1:80. Although some cross-reaction can be tolerated at a 1:20 dilution, it should not be any higher.

2. Centrifugation

Many models of clinical type centrifuges are satisfactory for the methods described below. We have found two low-priced Clay-Adams models to be satisfactory. The six place model is used for centrifuging the bacterial suspensions. The top speed (4830 r/min on AC, 5890 r/min on DC) will remove most *P. multocida* within 30 min. The four place physician's model is very convenient for washing erythrocytes. It has one speed (3400 r/min on AC only). Conical heavy duty graduated 12-ml centrifuge tubes are used for red cells and bacteria. The 15-ml size is too weak and breakage is frequent.

3. Preparation of erythrocytes

Group O human red cells are employed because they absorb the capsular antigens well and provide a carrier of known antigenic nature. For this reason, it is only necessary to absorb sera once. If blood of different species and individuals are used with the same sera a number of absorptions may be required. Other erythrocytes can be used, for example those of the chicken, if the Group O cells are not available. Outdated Group O cells can be obtained from blood banks. These can be used for at least two weeks past the expiry date. Fresh cells excluded for some reason or other from the blood bank can be obtained occasionally.

When the Group O cells are obtained small amounts (5–10 ml) are dispensed aseptically into sterile screw-cap tubes for future use. Blood is washed three times with at least ten volumes of physiological saline. Washed cells carried over from one day to another are washed once before using.

If it is difficult to obtain fresh Group O erythrocytes they can be treated with formalin (Carter and Rappay, 1962) and used for a long period. However, the IHA test is sharper and easier to read if fresh red cells are used. Sawada *et al.* (1982) have successfully used glutaraldehyde-treated sheep erythrocytes. The latter could be used for at least seven months.

4. *Absorption of sera with erythrocytes*

Those capsular typing sera that require absorption are diluted 1:10 with merthiolated (1:10 000) saline. 0.2 ml of packed washed Group O cells is added to each 2 ml of a 1:10 dilution of antiserum. After incubation at 37°C for 2 h the red cells are removed by centrifugation. This same procedure can be used for absorbing normal sera that are being tested for the presence of *P. multocida* antibodies. Occasionally sera will require additional absorption.

5. *Occurrence of capsular types*

Most cultures of *P. multocida* recovered from disease processes in farm and other animals in North and South America are type A. These frequently have a mucoid type colony, although avian cultures are usually less mucoid. Capsular type D cultures are recovered frequently from swine but much less frequently from other animals. Capsular type A strains cause almost all fowl cholera. Capsular type B and E cultures are recovered from cattle, buffaloes and occasionally other species with haemorrhagic septicaemia. The capsular type E cultures have only been recovered from cattle, water buffaloes and swine in Africa.

6. *Indirect haemagglutination procedure*

Non-mucoid cultures for capsular typing are streaked on blood agar as described in Section V.A.1. They are then washed off with 4–5 ml of saline and heated at 56°C for 30 min. The bacteria are removed by centrifugation and saved. The remainder of the procedure is performed as described below for the mucoid cultures beginning with the transfer of the supernatant and addition of red cells.

Mucoid cultures are routinely treated with hyaluronidase. The mucoid culture is washed off the plate with 3 ml of 0.1 M phosphate-buffered saline (PBS), pH 6. To the suspension is added 1 ml of PBS, pH 6, containing 15 National Formulary units (50 viscosity-reducing units) of testicular hyaluronidase. For convenience, reconstituted hyaluronidase is stored in 1-ml amounts in the frozen state at $-20°C$. The suspension is placed in a water bath at 37°C for 3–4 h. It is then heated to 56°C for 30 min, after which the bacteria are removed by centrifugation. Adequately treated organisms can be centrifuged out at $1315 \times g$ for 30 min. The supernatant is transferred to another tube and 0.1 ml of packed, washed Group O human red cells is added.

After the addition of the red cells to the supernatant (extract) and thorough mixing, the mixture is incubated at 37°C for 2 h. If it is more convenient, the treatment can be carried out overnight at 4°C. The red cells are separated by centrifugation and washed three times with 10 ml of physiological saline after which sufficient saline solution is added to yield a 0.5% suspension. The test is set up as shown in Table II.

TABLE II

Amounts of reagents and serum dilutions for indirect haemagglutination procedure

Tube[a] No.	1	2	3	4	5	6	
Saline	0.45 ml	0.25 ml	0.25 ml	0.25 ml	0.25 ml	0.25 ml	
Serum	0.05 ml						
Transfer	0.25 ml	0.25 ml	0.25 ml	0.25 ml	0.25 ml	0.25 ml	(Discard)
RBC[b] 0.5%	0.25 ml	0.25 ml	0.25 ml	0.25 ml	0.25 ml	0.25 ml	
Final dilution	1:20	1:40	1:80	1:160	1:320	1:640	

[a]Kahn tubes.
[b]RBC: red blood cells.

Twofold serial dilutions of the specific antisera are made in saline in Kahn tubes (12×75 mm). The amount of serum dilution is 0.25 ml to which is added 0.25 ml of treated erythrocytes. The final dilutions employed are 1:20 to 1:640.

The two control tubes contain (a) 0.25 ml of saline and 0.25 ml of 0.5% suspension of treated cells and (b) 0.25 ml of 1:10 dilution of the serum and 0.25 ml of 0.5% suspension of untreated erythrocytes. After it is learned that a serum does not possess agglutinins for the human cells, the second control tube is omitted.

The racks of tubes are shaken vigorously then left at room temperature for approximately 2 h, at which time a reading is made. The tubes are then left overnight at room temperature after which a second reading is made. Positive reactions are occasionally seen only at the second reading. To facilitate reading the test after 2 h, the tubes are held over a well-lighted microscope mirror. A positive reaction consists of marked agglutination, whereas a negative test shows no evidence of clumping. The following morning, the tests are read by shaking the tubes lightly to dislodge the cells. The cells are then allowed to settle and a reading is made over a mirror as described above.

The test can be carried out equally well in micro-titre plates.

B. Identification of capsular types B and E by counterimmunoelectrophoresis (CIE) (Carter and Chengappa, 1981b)

The antigen preparation for the CIE procedure consists of the same capsular extract as used for the IHA test described above. The antisera are the same as used for the IHA procedure.

The electrophoresis plates are prepared by pre-coating glass plates (10×8 cm) with 15 ml amounts of agarose/agar solution consisting of 0.5% agarose (SeaKem), 0.5% Bactoagar (Difco) and 0.015 sodium azide in 0.025 M barbitol buffer pH 8.8. Wells of 3 mm in diameter are cut with a template (Grafor Corp.). The distance, centre to centre, between wells is 7 mm. Holes may be cut by other means and no doubt other dimensions would be equally satisfactory.

A 20-μl amount of saline capsular extract is placed in the cathodal well and an equal quantity of antisera in the anodal well. The electrophoresis tank is filled (Gelman Instrument Co.) with 0.05 M barbitol buffer, pH 8.8. The antigen preparation and antisera are electrophoresed by 30 min at 150 V. The plates are then examined for precipitation lines. The presence of a clear line of precipitation is interpreted as a positive reaction.

C. Identification of capsular type A strains with hyaluronidase

The tests (Carter and Rundell, 1975) are conducted on freshly prepared dextrose starch or blood agar plates (50×12 mm). For maximal production of hyaluronic acid, it is important that the medium is not dehydrated. Each strain is inoculated transversely across the whole plate, so as to provide lines of growth approximately 3–5 mm apart. After this inoculation a hyaluronidase-producing strain of *Staphylococcus aureus* is streaked heavily at right angles to the pasteurella streak lines. The inoculated plates are incubated at 37°C and observed periodically for up to 24 h. The hyaluronidase effect when present is always evident at 24 h and frequently earlier.

The hyaluronidase effect with type A cultures is manifested as a diminution in the size of the pasteurella colonies in the region adjacent to the staphylococcal streak. With some strains the effect is apparent for up to 1 cm from the edge of the staphylococcal growth. Some type D strains show a slight reduction in colony size near the staphylococcal streak indicating that these strains may possess a small amount of peripheral hyaluronic acid. This cannot be confused with the marked effect seen with mucoid type A strains. We have found that some type A fowl cholera strains do not produce appreciable amounts of hyaluronic acid. Their capsular type can only be identified by the IHA test.

D. Identification of capsular type D cultures with acriflavine (Carter, 1983)

Only cultures consisting of predominantly smooth, capsulated variants should be examined. They are plated on blood or dextrose starch agar before inoculation into Kahn tubes (12×75 mm) containing 3 ml of Brain Heart Infusion broth (Difco). The cultures are incubated at 37°C for 18–24 h. The bacteria are removed by centrifugation, and 2.5 ml of the supernatant is removed and discarded.

0.5 ml of 1:1000 aqueous solution of acriflavine neutral (N.F.) is added to the remaining 0.5 ml of broth containing the bacteria. Because acriflavine solutions do not keep, solutions are freshly prepared each week and stored in a brown glass bottle. After mixing to resuspend the bacteria, the tube is left stationary at room temperature. If the strain being examined is a type D, a heavy flocculent precipitate becomes evident and begins to settle within 5 min. After 30 min the heavy precipitate settles, leaving a distinct clear supernatant.

Although settling may be seen with types other than D, the precipitate is fine and after shaking is diffusely resuspendable rather than coarsely flocculent as with the type D cultures. Confusion is most apt to occur with cultures consisting of predominantly blue variants. These will produce a granular precipitate with the acriflavine.

VI. Identification of somatic serogroup antigens

Heddleston et al. (1972a,b) have identified 16 different somatic serogroup types by means of their gel diffusion precipitin (GDP) test. The procedures for the GDP are given below.

A. Antisera

Antisera are prepared in 12–16-week-old male chickens. Each strain of *P. multocida* is grown for 24 h at 37°C on Dextrose Starch Agar (Difco) or other suitable media, suspended in 0.85% NaCl solution containing 0.3% formalin and adjusted to a density equivalent to $10 \times$ McFarland No. 1. The cell suspensions are emulsified in equal amounts of light mineral oil containing 3.0% Arlacel A (Difco or BBL). 1 ml of emulsified antigen is injected subcutaneously into the neck of each of three birds for each culture. The birds are exsanguinated three weeks later. The antisera are preserved using 0.01% merthiolate and 0.06% phenol. The antisera are tested against homologous and heterologous antigens. Those antisera giving the most specific reactions are preferred for further testing.

B. Antigen

Growth (18–20 h) from a heavy seeded Dextrose Starch Agar plate is suspended in 1.0 ml of 0.85% NaCl solution containing 0.3% formalin. The cells are heated in a water bath (100°C) for 1 h and sedimented by centrifugation. The supernatant is used for the antigen.

B. O. Blackburn (personal communication) found the following buffer an improvement over the 0.85% NaCl solution for the extraction of antigen.

   2.5 g of Na_2HPO_4
   0.36 g of NaH_2PO_4
   8.50 g of NaCl
   1 litre of H_2O

The buffer is autoclaved and 3 ml of formalin are added.

C. Gel diffusion tests

The agar gel consists of 0.9% Noble (Difco) agar, 8.5% NaCl* and 0.01% merthiolate in distilled water. 5 ml of agar gel are placed on a microscope slide (or in a 60×15 mm Petri dish); five wells (four outside and one in the centre) 4 mm in diameter and 5 mm from centre to centre are cut on each slide. Antisera are placed in the outer wells and antigens in centre wells. Hold overnight and observe results. A positive reaction with a particular type serum consists of clear bands of precipitation usually near the antigen (centre) well. Strong reactions with other type sera should be noted and should be recorded as, for example 3(4) or 3(4,12). The strongest reaction should be listed first.

*The high salt concentration is required for maximal precipitation of chicken sera.

VII. Designated serotypes (Carter and Chengappa, 1981a)

When the identity of the capsular substances and the somatic antigens have been determined a serotype is designated by a letter indicating the capsular identity followed by a number or numbers standing for the somatic antigen or antigens, for example B:2, A:1, A:3 (4,12), A:4 (7), etc.

Cultures representing different serotypes are available from the National Animal Disease Center, Ames, Iowa.

VIII. Other typing procedures

Namioka and associates (Namioka and Bruner, 1963; Namioka and Murata, 1961, 1964) have described an agglutination procedure for identifying O-antigens (somatic) of *P. multocida*. Because the procedures involve complicated agglutinin–absorption of sera they have not been widely used. Rimler (1978) has described a coagglutination test for the identification of type B and E strains of *P. multocida*. Carter (1972b) described a plate agglutination test for the identification of somatic types of capsular type A cultures. The cells for agglutination were pre-treated with hyaluronidase.

References

Carter, G. R. (1955). *Am. J. Vet. Res.* **16**, 481–484.
Carter, G. R. (1961). *Vet. Rec.* **73**, 1052.
Carter, G. R. (1972a). *Appl. Microbiol.* **24**, 162–163.
Carter, G. R. (1972b). *Avian Dis.* **16**, 1109–1114.
Carter, G. R. (1973). *Am. J. Vet. Res.* **34**, 293–294.
Carter, G. R. (1975). *Am. J. Vet. Res.* **18**, 210–213.
Carter, G. R. and Chengappa, M. M. (1981a). "Proc. 24th Meeting of the Amer. Assn. Vet. Lab. Diag.", pp. 37–42.
Carter, G. R. and Chengappa, M. M. (1981b). *Vet. Rec.* **108**, 145–146.
Carter, G. R. and Rappay, D. (1962). *Br. Vet. J.* **118**, 189–292.
Carter, G. R. and Rundell, S. W. (1975). *Vet. Rec.* **96**, 343.
Elberg, S. S. and Ho, C. L. (1950). *J. Comp. Path.* **60**, 41–50.
Heddleston, K. L. and Wessman, G. (1975). *J. Clin. Microbiol.* **1**, 377–383.
Heddleston, K. L., Goodson, T., Leibovitz, J. and Angstrom, C. I. (1972a). *Avian Dis.* **16**, 729–734.
Heddleston, K. L., Gallagher, J. E. and Rebers, P. A. (1972b). *Avian Dis.* **16**, 925–936.
Namioka, S. and Bruner, D. W. (1963). *Cornell Vet.* **53**, 41–53.
Namioka, S. and Murata, M. (1961). *Cornell Vet.* **51**, 522–528.
Namioka, S. and Murata, M. (1964). *Cornell Vet.* **54**, 520–534.
Rimler, R. B. (1978). *J. Clin. Microbiol.* **8**, 214–218.
Sawada, T., Rimler, R. B. and Rhoades, K. R. (1982). *J. Clin. Microbiol.* **15**, 752–756.

10

Serology and Epidemiology of *Plesiomonas shigelloides*

R. SAKAZAKI

Enterobacteriology Laboratories, National Institute of Health, Tokyo, Japan

I.	Introduction	259
II.	Cultural and biochemical characteristics	260
III.	Serology	260
	A. Preparation of diagnostic antiserum	260
	B. Dermination of antigens	263
	C. Antigenic schema of *P. shigelloides*	265
IV.	Epidemiology and ecology	265
	References	268

I. Introduction

Plesiomonas shigelloides was first described by Ferguson and Henderson (1947) and called strain C27 or "paracolon" C27 because of its slow lactose-fermentative properties. Early investigators were confident that the C27 organism was a member of the family *Enterobacteriaceae* and Cowan (1956) suggested a new species *Escherichia sonnei* for this organism together with *Shigella sonnei*. Bader (1954), who was unaware of previous papers on the C27 organism, studied a culture of this organism isolated from a dog and found that the culture possessed lophotrichated flagella. He classified it into the genus *Pseudomonas* as *Pseudomonas shigelloides*. Because of its ability to ferment glucose and to give a positive reaction in indophenol oxidase test, Ewing *et al.* (1961) later transferred the organism to the genus *Aeromonas*. The generic name *Plesiomonas* was given by Habs and Schubert (1962) who suggested that the organism did not exhibit some key characters of the genus *Aeromonas*. This suggestion was supported by the DNA study of Sebald and Véron (1963). They demonstrated that the (G + C) content of its DNA was 51 mole % as opposed to that of *Aeromonas* which ranged from 57 to 63 mole % and of *Vibrio* which ranged from 40 to 50 mole %. The genus *Plesiomonas* was also justified by a numerical taxonomic study by Eddy and Carpenter (1964).

The genus *Plesiomonas* is classified into the family *Vibrionaceae* and only a single species *Plesiomonas shigelloides* is included in this genus. The type strain of the species is ATCC 14029.

II. Cultural and biochemical characteristics

P. shigelloides is a Gram-negative, facultatively anaerobic rod with polar tuft flagella (1–5). The organism grows well not only on/in ordinary media but also on selective plating media for enterobacteria, such as *Salmonella–Shigella* agar and deoxycholate citrate agar. Although some investigators reported haemolysis of a few strains of *P. shigelloides* (Aldova *et al.*, 1966; Zajc-Satler *et al.*, 1972), none of the strains were included in a large collection of Ewing *et al.* (1961) nor in our collection.

Physiological and biochemical characteristics of *P. shigelloides* based on 280 strains are summarized in Table I.

III. Serology

P. shigelloides first attracted attention when it was said to possess the identical somatic antigen with that of *Shigella sonnei* (Ferguson and Henderson, 1947; Vandepitt *et al.*, 1957; Bader, 1954; Schmidt *et al.*, 1954). However, Sakazaki *et al.* (1959) who studied 29 strains biochemically identical with the C27 organism, distinguished five somatic (O) and four flagellar (H) antigens without designating antigenic symbols. Later, Quincke (1967) studied 57 strains and divided them into 16 O-antigen groups without taking account of their H-antigens. Aldova and Geizer (1968) divided their 14 strains into five O-antigen groups and distinguished five H-antigens among the strains. Independently of the former investigators, Shimada and Sakazaki (1978) established an antigenic schema that included 40 serovars of 30 O-antigen groups and 11 H-antigens.

A. Preparation of diagnostic antiserum

1. *O-antiserum*

After plating an ordinary agar medium, selected smooth colonies are inoculated into infusion broth of pH 7.6–7.8. After overnight incubation at 35°C by shaking in a water bath, the cultures are heated at 100°C for 2 h. The

TABLE I
Biochemical characteristics of *Plesiomonas shigelloides* based on 280 strain

Test (Substrate)	Sign	% +	%(+)^a
Oxidase	+	100	
Indole	+	100	
Voges–Proskauer (25°C)	−	0	
Citrate (Simmons)	−	0	
Nitrate to nitrite	+	100	
H$_2$S (Kligler)	−	0	
Urease (Christensen)	−	0	
Gelatinase (Kohn)	−	0	
Phenylalanine deaminase	−	0	
Lysine decarboxylase	+	99.3	0.7
Arginine dihydrolase	+	97.8	1.7
Ornithine decarboxylase	+	91.4	2.8
Gas from glucose	−	0	
Acid from			
Arabinose	−	0	
Glucose	+	100	
Lactose	d	40.0	52.8
Maltose	d	62.8	
Raffinose	−	0	
Rhamnose	−	0	
Sucrose	−	0	2.8
Trehalose	+	100	
Xylose	−	0	
Adonitol	−	0	
Dulcitol	−	0	
Mannitol	−	0	
Sorbitol	−	0	
Salicin	d	11.7	10.3
Inositol	+	98.2	1.8
Esculin hydrolysis	−	0	
β-Galactosidase	+	100	
Tween 80 hydrolysis	−	0	
Deoxyribonuclease	−	0	

Signs: +, 90% or more positive within one or two days; −, 90% or more negative; d, 11–89% positive.

^{a} Figure in parentheses indicates percentage of delayed reaction (three days or more).

heated organisms are then washed twice by centrifugation and finally the packed cells are resuspended in PBS. The density of the suspension is adjusted to that of the original broth culture.

Rabbits were injected with 0.5-ml, 1.0-ml, 2.0-ml and 4.0-ml amounts of the suspension at intervals of three to four days. The animals are bled on the sixth day after the final injection and the serum is then collected. Four injections usually result in antisera of satisfactory titres of 1:1000 or more against the homologous antigen. If the titre is lower on test bleeding, an additional one or two injections of 4.0-ml amounts of the antigen may be given.

2. H-antisera

Since *P. shigelloides* have only a few polar flagella, it is rather difficult to produce H-antisera of satisfactory titre with broth cultures. Crude flagellar preparation may be appropriate for the production of H-antisera. Actively motile cultures, obtained by several passages of the organism through semi-solid medium, are inoculated onto 10–20 plates of soft agar medium enriched with 0.5% yeast extract and 0.2% glucose. After overnight incubation at 30°C, the growth is harvested in PBS and centrifuged at 15 000 r/min for 30 min. The packed cells are resuspended in PBS and deflagellated in a blender at 8000 r/min for 10 min. The cells are removed by centrifugation at 15 000 r/min for 15 min and supernatant fluid then centrifuged at 50 000 × $\mathbf{g}$ for 60 min; the sediment is washed twice in PBS. The flagellar fragment obtained from cultures of approximately 15 plates is finally suspended in 15 ml of PBS.

Rabbits are immunized with the flagellar fragment suspension in the same manner as that for O-antiserum production. Four injections usually result in antisera of satisfactory H-antibody titres of 1:10 000 or more. Although the resulting antisera still give O-titre in response to immunization with crude flagellar preparation, a higher working dilution of antisera of satisfactory H-titre is enough to avoid O-agglutination. If necessary O-antibody in H-antisera is removed by absorption with homologous cultures heated at 100°C for 2 h.

3. *Absorption of antiserum*

Agglutinin absorption may be necessary in certain antisera in which some reciprocal or unilateral reactions to heterologous antigens are recognized. For absorption of O-antisera, a certain amount of appropriately diluted O-antiserum are mixed with packed cells derived from culture harvested from growth on plates of infusion agar enriched with 0.5% yeast extract and heated at 100°C for 1 h. Growth on plates of brain heart infusion medium containing 0.8% agar and 0.5% yeast extract, which is washed off in formalin-treated saline by centrifugation, may be appropriate for absorption of H-antibodies. In all instances, the mixture is incubated for 2 h in a 50°C water bath followed by standing overnight at 4°C and then centrifuged.

B. Determination of antigens

1. *O-antigen*

For the determination of O-antigens of *P. shigelloides*, slide agglutination is the method of choice. Slide agglutination is performed by the usual techniques using overnight agar cultures and the working dilution of O-antisera. Some cultures of *P. shigelloides* may not be agglutinated by homologous O-antisera. Because of this, O-antigen determination of *P. shigelloides* is carried out with cell suspensions heated at 100°C for 2 h followed by washing. The such antigen may give the highest titres with homologous O-antisera. O-antisera are customarily diluted to 1:10 for slide O-agglutinations.

Thirty O-antigen groups have been recognized in *P. shigelloides* by Shimada and Sakazaki (1978). In their study, no actual cross-reactions were observed among any of the 30 O-antigen groups, but some minor unilateral reactions occurred with antisera of O:3, O:8, O:19, O:21, O:25 and O:27 groups as shown in Table II. Absorption of those antisera in which only the such minor reactions occur is not necessary in the working dilution of the antisera.

2. *H-antigen*

An overnight culture of actively motile organisms in brain heart infusion medium (pH 6.8–7.0) is satisfactory for the determination of H-antigen of *P. shigelloides*. The actively motile organisms may be obtained by the same method as described in the H-antiserum preparation. After overnight incubation at 30°C, an equal volume of PBS containing 0.2% thimerosal is added to the broth culture. The formalized broth cultures which are usually applied in H-agglutination of enterobacteria may not be satisfactory in some instances. In such cases, live culture should be used.

The determination of H-antigens is best performed by qualitative tube agglutination. The working dilution of H-antisera is usually at 1:1000. 0.1 ml of a 1:100 dilution of each antiserum is distributed into a small tube and 1.0 ml of the broth cultures mentioned above is added. The test is incubated in a water bath at 50°C for 4 h and read. H-agglutination of *P. shigelloides* cultures is flocculant in nature resembling that of enterobacteria, but it may usually appear after incubation at 50°C for 4 h or more.

Shimada and Sakazaki (1978) recognized 11 H-antigens among strains of *P. shigelloides* studied. Although most of the 11 H-antisera they used gave no significant reaction with any of the heterologous H-antigens, antisera of H:5 and H:8 agglutinated unilaterally H:1- and H:5-antigens, respectively, as shown in Table III. Those antisera with extensive heterologous reactions

TABLE II
O-antigenic relationships

O-antiserum	Homologous titre	Titre with other O-antigen	
1	1280		
2	640		
3	640	7:80	
4	2560		
5	2560		
6	1280		
7	640		
8	640	6:160	
9	1280		
10	640		
11	1280		
12	640		
13	640		
14	2560		
15	640		
16	320		
17	320		
18	640		
19	1280	6:160	
20	640		
21	1280	6:160	8:40
22	1280		
23	1280		
24	640		
25	1280	22:40	
26	640		
27	1280	4:80	
28	640		
29	1280	25:80	
30	1280		

TABLE III
H-antigenic relationships

H-antiserum	Homologous titre	Titre with other H-antigen	
1a,1b	8000	*1a,1c:2000*	
1a,1c	8000	*1a,1b:2000*	
2	8000		
3	8000		
4	16000		
5	16000	1a,1b:2000	1a,1c:2000
6	16000		
7	8000		
8	4000	5:1000	
9	16000		
10	8000		
11	16000		

Italic indicates reciprocal relationships.

should be absorbed to remove minor antibodies. In H:1-antigen two antigenic subgroups, designated 1a,1b and 1a,1c, are distinguished. The factor sera of H:1b and H:1c are obtained by cross-absorption of 1a,1b and 1a,1c antisera.

C. Antigenic schema of *P. shigelloides*

An antigenic schema for *P. shigelloides* consisting of 30 O-antigen groups and 11 H-antigens giving a total of 40 serovars was established by Shimada and Sakazaki (1978) (Table IV). As mentioned before, the majority strains of *P. shigelloides* are O-inagglutinable, but the capsular K or masked antigens of this organism have not been considered in the antigenic schema.

Quincke (1967) reported 16 O-antigen groups within 57 strains of *P. shigelloides*, taking no account of H-antigens. Comparison of O-antigen groups described here with those of Quincke has not been carried out, since the test strains used by Quincke are no longer available.

Aldová and Geizer (1968) studied 14 strains of *P. shigelloides*. They divided their strains into five O-antigen groups distinguishing five H-antigens. A comparative study with their strains is now in progress.

Extra-generic relationships of P. shigelloides *antigens.*   Since *P. shigelloides* was first recognized as the C27 organism by Ferguson and Henderson (1947), it has attracted attention because of its antigenic relationship to *Shigella sonnei* (Vandepitt *et al.*, 1957; Bader, 1954; Schmidt *et al.*, 1954). Hori *et al.* (1966) also reported the O-antigenic relationship between their isolates of *P. shigelloides* and *Shigella dysenteriae* 7. These two O-antigen groups of *P. shigelloides* concerned were designated O:17 and O:22, respectively, in the schema by Shimada and Sakazaki (1978). The O:17-antigen of this organism is identical with that of *Shigella sonnei*, and the O:22-antigen relates to that of *Shigella dysenteriae* 7 in an a,b–a,c type of relationship. In addition to these, Shimada and Sakazaki (1978) demonstrated an a,b–a,c type of relationship between *P. shigelloides* O-11 and *Shigella dysenteriae* 8 and between *P. shigelloides* O-23 and *Shigella boydii* 13.

No significant relationships have been demonstrated between O-antigens of *P. shigelloides* and *Vibrio cholerae* (Shimada and Sakazaki, 1978). Whang *et al.* (1972) recognized the presence of the common antigen of *Enterobacteriaceae* in strains of *P. shigelloides*.

IV. Epidemiology and ecology

P. shigelloides has long been noticed as a possible enteropathogen by many investigators (Schmidt *et al.*, 1954; Osada and Shibata, 1956; Vandepitt *et al.*,

TABLE IV

Antigenic schema of *Plesiomonas shigelloides*

O-group	O-antigen	H-antigen	O-group	O-antigen	H-antigen
1	1	1a,1b	16	16	5
2	2	1a,1c	17	17	2
3	3	2		17	6
4	4	3	18	18	2
5	5	4	19	19	2
6	6	3	20	20	2
7	7	2	21	21	7
	7	—[a]		21	8
8	8	3	22	22	3
	8	5		22	5
9	9	2		22	8
10	10	11	23	23	1a,1c
11	11	2	24	24	5
	11	5		24	8
12	12	2	25	25	3
	12	3	26	26	1a,1c
	12	9	27	27	3
13	13	2	28	28	3
14	14	4	29	29	2
	14	5	30	30	1a,1c
15	15	10			

[a] Non-motile.

1957, 1974; Sakazaki *et al.*, 1959, 1971; Ewing *et al.*, 1961; Ueda *et al.*, 1963; Eddy and Carpenter, 1964; Aldová *et al.*, 1966; Geizer *et al.*, 1966; Hori *et al.*, 1966; Cooper and Brown, 1968; Paucova and Fukalova, 1968; Winton, 1968; Zajc-Salter *et al.*, 1972; Fourquet *et al.*, 1973; Jandl and Lee, 1976; Tsukamoto *et al.*, 1978; Arai *et al.*, 1980). Most intestinal disorders from which *P. shigelloides* is isolated are gastro-enteritis or watery diarrhoea, but dysentery-like symptoms were also reported (Osada and Shibata, 1956). Although the majority of those incidents are sporadic cases, outbreaks or epidemics of acute gastro-enteritis or diarrhoea in which *P. shigelloides* was suggested as the possible causative agent have been reported by Hori *et al.* (1966) and Tsukamoto *et al.* (1978).

On the other hand, *P. shigelloides* is occasionally isolated from healthy persons who have no clinical history. Bhat *et al.* (1974) described that the prevalence of *P. shigelloides* in patients with diarrhoea was temporally unrelated to the presence of diarrhoea. It has been reported by Catsaras and Butiaux (1965), Nakanishi *et al.* (1969) and Vandepitt *et al.* (1974), however, that the isolation of *P. shigelloides* from faecal specimens of healthy persons is rare compared with that from patients with diarrhoea.

Enteropathogenicity of *P. shigelloides* has not been also conclusively proved. N. Kosakai (personal communication) carried out feeding experiments with human volunteers using a strain possessing the O:17-antigen, which is identical with that of *Shigella sonnei*, but no clinical symptoms developed in any cases. Ligated ileal loop tests in rabbits with strains of this organism have so far given negative reaction and they do not produce an enterotoxin (K. Tamura and R. Sakazaki, unpublished data; Sanyal *et al.*, 1975). However, Sanyal *et al.* (1980) reported recently that live cells and culture filtrates of six strains of *P. shigelloides* caused fluid accumulation in ileal loops in rabbits.

P. shigelloides may be found as an opportunistic pathogen in patients with certain underlying diseases. A case of fatal septicaemia caused by this organism which developed in a patient with cellulitis who suffered from sickel cell anaemia was reported by Ellner and MacCarthy (1973). Another case of cellulitis caused by *P. shigelloides* was described by Von Graevenitz and Mensch (1968).

The isolation of *P. shigelloides* from animals including monkeys, goats, sheep, cows, dogs and cats has also been reported (Bader, 1954; Vandepitt *et al.*, 1957; Sakazaki *et al.*, 1959; Ewing *et al.*, 1961; Habs and Schubert, 1962; Arai *et al.*, 1980).

Aldová *et al.* (1966), Geizer *et al.* (1966), Cooper and Brown (1968), Zakhariev (1971) and Vandepitt *et al.* (1974) suggested that *P. shigelloides* is water-borne. This has been confirmed by Tsukamoto *et al.* (1978). From the results of ecological studies by Tsukamoto *et al.* (1978), it seems that the optimum place for survival and growth of *P. shigelloides* is in the mud at the bottom of rivers and ponds which presumably contains sufficient nutrient. The presence of the organisms in surface water is more frequent in the warmer season than in the winter season. This may explain the fact that most outbreaks of infection with *P. shigelloides* occur during the summer season.

Fish is incriminated as the causative food in the majority of cases of diarrhoea due to *P. shigelloides*. It has been found in some patients in association with *Vibrio parahaemolyticus* (Ueda *et al.*, 1963), as well as *Shigella, Salmonella* and *Vibrio cholerae* (Vandepitt *et al.*, 1957; Aldova *et al.*, 1966; Sakazaki *et al.*, 1971). Although they are sometimes found in seawater in the coastal area, it is probable that the sea environment allows only a brief life span of *P. shigelloides* as suggested by Zakhariev (1971).

No difference of distribution of *P. shigelloides* serovars has been found between human patients and the environment. The value of serotyping of *P.*

shigelloides as an epidemiological tool has been confirmed in the outbreaks of
P. shigelloides infection described by Tsukamoto *et al.* (1978).

References

Aldová, E. and Geizer, E. (1968). *Zbl. Bakt. Hyg., I. Abt. Orig.* **207**, 35–40.
Aldová, E., Bakovsky, J. and Chovancová, A. (1966). *J. Hyg. Epidemiol. Microbiol. Immunol.* **10**, 470–482.
Arai, T., Ikejima, N., Itoh, T., Sakai, S., Shimada, T. and Sakazaki, R. (1980). *J. Hyg.* **84**, 203–311.
Bader, R. E. (1954). *Z. Hyg.* **140**, 450–456.
Bhat, P., Shanthakumari, S. and Rajan, D. (1974). *Indian J. Med. Res.* **62**, 1051–1060.
Catsaras, M. and Buttiaux, R. (1965). *Ann. Inst. Pasteur* **16**, 85–88.
Cooper, R. G. and Brown, G. W. (1968). *J. Clin. Pathol.* **21**, 715–718.
Cowan, S. T. (1956). *J. Gen. Microbiol.* **15**, 345–348.
Eddy, B. P. and Carpenter, K. P. (1964). *J. Appl. Bacteriol.* **27**, 96–109.
Ellner, P. D. and MacCarthy, L. R. (1973). *Am. J. Clin. Pathol.* **59**, 216–218.
Fauckova, V. and Fukalova, A. (1968). *Zbl. Bakt. Hyg., I. Abt. Orig.* **216**, 212–216.
Ferguson, W. W. and Henderson, N. D. (1947). *J. Bacteriol.* **54**, 179–181.
Fourquet, R., Couturier, Y., Jamet, A. and Griffet, P. (1973). *Arch. Inst. Pasteur Madagascar* **42**, 61–68.
Geizer, E., Kopecky, K. and Aldová, E. (1966). *J. Hyg. Epidemiol. Microbiol. Immunol.* **10**, 23–26.
Habs, H. and Schubert, R. H. W. (1962). *Zbl. Bakt. Hyg., I. Abt. Orig.* **186**, 316–327.
Hori, M., Hayashi, K., Maeshima, K., Miyamoto, T., Yeneda, Y. and Hagihara, Y. (1966). *J. Jpn. Assoc. Infect. Dis.* **39**, 441–448.
Jandl, G. and Linke, K. (1976). *Zbl. Bakt. Hyg., I. Abt. Orig. Reihe A* **236**, 136–140.
Nakanishi, H., Leistner, L. and Hechelmann, H. (1969). *Fleischwirtschaft* **49**, 1501.
Osada, A. and Shibata, I. (1956). *Acta Pediatr. Jpn.* **60**, 739–742.
Quincke, G. (1967). *Arch. Hyg.* **151**, 525–529.
Sakazaki, R., Namioka, S., Nakaya, R. and Fukumi, H. (1959). *Jpn. J. Med. Sci. Biol.* **12**, 355–363.
Sakazaki, R., Tamura, K., Prescott, L. M., Bencic, A., Sanyal, S. C. and Sinha, R. (1971). *Indian J. Med. Res.* **59**, 1025–1034.
Sanyal, S. C., Singh, S. J. and Sen, P. C. (1975). *J. Med. Microbiol.* **8**, 195–198.
Sanyal, S. C., Saraswathi, B. and Sharma, P. (1980). *J. Med. Microbiol.* **13**, 401–409.
Schmidt, E. E., Velaudapillai, T. and Niles, G. R. (1954). *J. Bacteriol.* **68**, 50–52.
Sebald, M. and Veron, M. (1963). *Ann. Inst. Pasteur* **105**, 897–910.
Shimada, T. and Sakazaki, R. (1978). *Jpn. J. Med. Sci. Biol.* **31**, 135–142.
Tsukamoto, T., Kinoshita, Y., Shimada, T. and Sakazaki, R. (1978). *J. Hyg.* **80**, 275–280.
Ueda, S., Yamazaki, S. and Hori, M. (1963). *Jpn. J. Public Health* **10**, 67–70.
Vandepitt, J., Ghysels, G., Goethem, V. H. and Marrecau, N. (1957). *Ann. Soc. Bel. Med. Trop.* **37**, 737–742.
Vandepitt, J., Makulu, A. and Gatti, F. (1974). *Ann. Soc. Bel. Med. Trop.* **54**, 503–513.
Von Graevenitz, A. and Mensch, A. H. (1968). *N. Engl. J. Med.* **278**, 245–249.

Weing, W. H., Hugh, R. and Johnson, J. G. (1961). Studies on the Aeromonas group. U.S. Department of Health, Education and Welfare, Public Health Service. Communicable Disease Center, Atlanta, Georgia.

Whang, H. Y., Heller, M. E. and Neter, E. (1972). *J. Bacteriol.* **110**, 161–164.

Winton, F. W. (1968). *J. Pathol. Bacteriol.* **95**, 562–567.

Zajc-Satler, J., Dragasm, A. Z. and Kumelj, M. (1972). *Zbl. Bakt. Hyg., I. Abt. Orig. Reihe A* **219**, 514–521.

Zakhariev, Z. A. (1971). *J. Hyg. Epidemiol. Microbiol. Immunol.* **15**, 402–404.

11

Serology and Epidemiology of *Vibrio cholerae* and *Vibrio mimicus*

R. SAKAZAKI[1] AND T. J. DONOVAN[2]

[1] *Enterobacteriology Laboratories, National Institute of Health, Tokyo, Japan*
and [2] *Public Health Laboratory, Maidstone, Kent, England*

I.	Introduction	271
II.	Cultural and biochemical characteristics	272
III.	Serology	274
	A. H-antigen of *Vibrio cholerae*.	275
	B. O-antigen of *Vibrio cholerae*.	278
	C. O-antigen variation in 01 *Vibrio cholerae* . . .	281
	D. Comparison of serovars with other antigenic schemata .	284
	E. Serology of *Vibrio mimicus*	284
IV.	Epidemiology of *Vibrio cholerae* and *Vibrio mimicus* . .	285
	References	288

I. Introduction

Since Robert Koch first isolated the cholera vibrio in 1884 during the fifth pandemic, it has been known as the causative agent of cholera under the name of *Vibrio cholerae*, a name which was originally given by Pacini (1854). Cholera vibrio is now divided into two biovars, classical (or *V. cholerae* biovar *classical*) and eltor (or *V. cholerae* biovar *eltor*). As well as the cholera vibrio a large number of vibrios which are biochemically similar to cholera vibrio but which are not agglutinated by cholera antisera have been recognized in cholera epidemic areas for many years. They have been referred to in the past as non-agglutinable (NAG) vibrios or as non-cholera vibrios (NCV) because no scientific name was given them. Sakazaki *et al.* (1967) carried out a taxonomic study on these vibrios and concluded that they should be classified as *V. cholerae*. The International Subcommittee on Taxonomy of Vibrios (1972) therefore recommended that the species *V. cholerae* should no longer be restricted to the cholera vibrio. Gardner and Venkatraman (1935) studied serologically these vibrios under the name

METHODS IN MICROBIOLOGY VOL. 16
ISBN 0-12-521516-9

"cholera group of vibrios" and divided them into six O-subgroups in which O-subgroup 1 was assigned to cholera vibrios. Sakazaki *et al.* (1970), Shimada and Sakazaki (1977) and T. Shimada, T. J. Donovan and R. Sakazaki (unpublished data) extended the serological study and established an antigenic schema consisting of 83 O-groups (serovars). Currently cholera vibrio is referred to as O1 *V. cholerae* and other strains non-agglutinable by O1 *V. cholerae* antiserum as non-O1 *V. cholerae*.

Although most strains of *V. cholerae* ferment sucrose, some sucrose-negative strains were also included in this species (Sakazaki *et al.*, 1967; Hugh and Sakazaki, 1972). On the basis of DNA hybridization these strains which do not ferment sucrose have recently been placed in a separate species and the name *Vibrio mimicus* has been proposed for the species by Davis *et al.* (1981). Because of the serological identity and probably similar clinical significance of *V. mimicus* and *V. cholerae*, the former is also reviewed in this chapter.

II. Cultural and biochemical characteristics

Strains of *V. cholerae* and *V. mimicus* are gram-negative, straight or slightly curved rods with a single polar flagellum. They are facultatively anaerobic and can grow readily on or in ordinary media. Growth occurs between pH 6.0 and 9.6, but is optimal between pH 7.6 and 8.6. Colonies on nutrient agar are usually translucent, amorphous, but sometimes wrinkled or rugose colonies may occur. Broth cultures of these vibrios show moderate turbidity and sometimes pellicle formation, especially in an alkaline broth (pH 8.0–9.0). They do not require more than trace amounts of NaCl for their growth, and on this basis can be differentiated from other *Vibrio* species.

The physiological and biochemical characteristics of *V. cholerae* and *V. mimicus* based on 751 and 85 strains, respectively, are summarized in Table I. The most useful characters for identifying these species include positive reaction in tests for oxidase, and lysine and ornithine decarboxylases but negative reaction in the arginine dihydrolase test; fermentation of glucose with no gas production; acid production from mannitol but not from arabinose and inositol; no hydrogen sulphide produced in TSI or Kligler iron agar. All strains of O1 *V. cholerae* ferment mannose, but the reaction is variable in non-O1 *V. cholerae*.

V. mimicus is very similar to *V. cholerae* (Table I). The only clear distinguishing characters are failure to ferment sucrose and lack of amylase production. *V. mimicus* in addition is sensitive to Polymyxin B (50 unit disc), Voges–Proskauer negative, chick cell agglutination negative and does not

TABLE I

Biochemical characteristics of *Vibrio cholerae* and *Vibrio mimicus*

Test or substrate	V. cholerae			V. mimicus		
	Reaction	% +	%(+)	Reaction	% +	%(+)
Oxidase	+	100		+	100	
Indole	+	100		+	100	
Voges–Proskauer	d	89.5		−	0	
Citrate (Simmons)	(+)	5.0	85.0	(+)	5.2	90.0
H₂S (TSI)	−	0		−	0	
Urease (Christensen)	−	0		−	0	
Phenylalanine deaminase	−	0		−	0	
Lysine decarboxylase	+	99.3		+	98.8	1.2
Arginine dihydrolase	−	0		−	0	
Ornithine decarboxylase	+	94.4	4.3	+	100	
Amylase	+	94		−	4	
Chick cell agglutination	d	54		−	0	
Polymyxin B 50 units	R	69		S	88	
Gelatinase	+	100		+	100	
DNAase	+	100		+	100	
Tween 80 hydrolase	+	100		+	100	
Growth in peptone water						
Without NaCl	+	98.6		+	100	
With 8% NaCl	−	0		−	0	
Growth at 42°C	+	100		+	100	
Esculin hydrolysis	−	0		−	0	
β-Galactosidase	+	100		+	100	
Fermentation						
Glucose, acid	+	100		+	100	
Glucose, gas	−	0		−	0	
Arabinose	−	0		−	0	
Lactose	(+)	0	98.0	(+)	0	85.5
Mannose	d	75.6		+	100	
Raffinose	−	0		−	0	
Rhamnose	−	0		−	0	
Sucrose	+	100		−	0	
Trehalose	+	100		+	100	
Xylose	−	0		−	0	
Adonitol	−	0		−	0	
Dulcitol	−	0		−	0	
Inositol	−	0		−	0	
Mannitol	+	97.8		+	100	
Sorbitol	−	0		−	0	
Salicin	−	0		−	0	

Symbols: +, 90% or more positive; −, 90% or more negative; d, different reaction (11–89% positive); (+), delayed positive reactions; R, resistant; S, sensitive.

grow at 45°C. All these characters are positive with the majority of *V. cholerae* cultures (Furniss *et al.*, 1978).

Reichelt *et al.* (1976) demonstrated that a luminous vibrio designated *Vibrio albensis* (Lehmann and Neumann, 1896) was indistinguishable from *V. cholerae* by DNA homology. The similarity between the two species has also been confirmed on the basis of nutritional characterization by Desmarchelier and Reichelt (1981) using additional *V. albensis* strains isolated from fresh water.

III. Serology

V. cholerae possesses two principal antigenic components, O-(somatic) and H-(flagellar) antigens. The O-antigen is thermostable and not destroyed by heating at 100°C for 2 h, but short periods of heat at 100°C, for example 30 min, usually reduce the O-agglutinability of suspensions (Sakazaki *et al.*, 1970; Donovan, 1982). The H-antigen is inactivated by heating the cultures at 100°C for a few minutes. This means that suspensions of *V. cholerae* rapidly lose their ability to be agglutinated by H-antisera but they require further heating for up to 2 h at 100°C before they lose their ability to produce H-agglutinins when used as an immunizing agent in animals. Some live cultures of *V. cholerae* may be inagglutinable or only weakly agglutinable by the homologous O-antiserum. Such inagglutinability of live cultures by O-antiserum may be due to the presence of surface mucoid substance (M-antigen). A culture which showed reduced O-agglutination in the living and enhanced O-agglutination after heating was examined by electron microscopy and showed the presence of a partially detached structure resembling a micro-capsule (Donovan, 1982). So far as studied by Sakazaki *et al.* (1967) the M-antigen of *V. cholerae* is identical in each strain and has no diagnostic value.

Apart from the cholera vibrio, the serology of *V. cholerae* was first studied by Gardner and Venkatraman (1935) who divided the "cholera group of vibrios" into six O-subgroups, as mentioned before, in which O-subgroup 1 was assigned to the cholera vibrio and non-choleragenic vibrios were included in O-subgroups 2 to 6. Gallut (1962) studied 47 strains of non-cholera vibrio and found 24 O-subgroups among those strains. However, identification of those strains studied by him was not based on the recent classification of *V. cholerae*. Sakazaki and co-workers studied extensively the serology of *V. cholerae* using a large number of strains which had been identified as *V. cholerae* by strict taxonomic criteria; some strains were sucrose-negative and would now be recognized as *V. mimicus*, and in all more than 80 serovars were recognized (Sakazaki *et al.*, 1970; Shimada and

Sakazaki, 1977; T. Shimada, T. J. Donovan and R. Sakazaki, unpublished data). The International Subcommittee on Taxonomy of Vibrios (1979) stated that more than two laboratories should undertake the serotyping of *V. cholerae* and further development of such a schema; these laboratories should regularly interchange strains. As a result of this all serovar strains were mutually studied in Tokyo and Maidstone.

All *V. cholerae* strains share a common H-antigen (Gardner and Venkatraman, 1935; Sakazaki *et al.*, 1970; Bhattacharyya, 1977). Because of this H-antigen determination was originally considered of little value for serotyping *V. cholerae* and these O-antigen groups are referred to as serovars, although the serovar should be principally expressed by an assortment of O- and H-antigens. However, the determination of the H-antigen may be useful for recognition of *V. cholerae* strains, particularly those non-O1 strains which cannot be recognized by agglutination with any commercially available antisera.

A. H-antigen of *Vibrio cholerae*

Flagellins from the polar flagellum of *V. cholerae* share a common antigenic determinant, as demonstrated by immunodiffusion, with other species of the genus *Vibrio* such as *V. alginolyticus, V. anguillarum, V. campbelli, V. fischeri, V. harveyi, V. metschnikovii, V. nereis* and *V. parahaemolyticus* (Shinoda *et al.*, 1976). *V. cholerae* H-agglutinating sera agglutinates all motile strains of *V. cholerae* and *V. mimicus, V. metschnikovii, V. fluvialis* and *V. anguillarum*. The sera can be rendered species specific for *V. cholerae* and *V. mimicus* by absorption with *V. fluvialis*. Use of both unabsorbed and absorbed species-specific *V. cholerae/mimicus* sera is useful to distinguish *V. cholerae/mimicus* from other vibrios (Table IA).

1. *H-antiserum*

Crude flagellar preparation is appropriate for the production of H-antiserum. Cultures for use as vaccines for H-antiserum production are inoculated into semi-solid medium in a Craigie tube. After several passages through the medium, actively motile organisms are inoculated onto 30 or more plates of nutrient agar (pH 7.0–7.2). After overnight incubation at 30°C, the growth is harvested into buffered saline containing 0.6% formalin. The suspension is centrifuged at 15 000 r/min for 30 min. The packed cells are resuspended in buffered saline and deflagellated in a blender at 8000 r/min for 5–10 min. Precautions should be taken to cool the suspension during deflagellation by packing the vessel in an ice bath. The cells are removed by centrifugation at 10 000 r/min for 15 min and the supernatant fluid then

TABLE IA
Relationship of H-antigens

Serum	Absorbed with	Antigens			
		V. cholerae H	V. metschnikovii H	V. fluvialis H	V. anguillarum H
V. cholerae H	Unabsorbed	10 000	2500	2500	160
V. cholerae H	V. cholerae	80	<20	<20	<20
V. cholerae H	V. metschnikovii	2 500	<20	80	<20
V. cholerae H	V. fluvialis	5 000	<20	<20	<20
V. cholerae H	V. anguillarum	5 000	20	80	<20
V. metschnikovii H	Unabsorbed	20	5 000	80	<20
V. metschnikovii H	V. metschnikovii	<20	80	20	<20
V. metschnikovii H	V. cholerae	<20	2 500	20	<20
V. metschnikovii H	V. fluvialis	<20	5 000	<20	<20
V. metschnikovii H	V. anguillarum	<20	5 000	40	<20
V. fluvialis H	Unabsorbed	320	640	10 000	160
V. fluvialis H	V. fluvialis	40	80	1 280	20
V. fluvialis H	V. cholerae	40	160	5 000	80
V. fluvialis H	V. metschnikovii	160	20	5 000	80
V. fluvialis H	V. anguillarum	160	320	5 000	<20
V. anguillarum H	Unabsorbed	80	80	80	640
V. anguillarum H	V. anguillarum	<20	20	<20	40
V. anguillarum H	V. cholerae	<20	<20	40	320
V. anguillarum H	V. metschnikovii	20	<20	20	80
V. anguillarum H	V. fluvialis	40	<20	<20	320

Results shown as reciprocal of the last dilution showing positive agglutination.

centrifuged at $70\,000 \times g$ for 60 min. The sediment is washed twice in phosphate-buffered saline (PBS). The resulting sediment from cultures of about 30–50 plates is finally suspended in 20 ml of PBS. It is advisable to examine the deposit by electron microscopy to estimate the amount of flagellar filaments and the degree of cellular debris.

Rabbits receive four to six intravenous injections of the crude flagellar preparation at intervals of four days. The amounts given are 0.5 ml, 1.0 ml, 2.0 ml and 3.0 ml. The total protein content of the flagellar suspension can be estimated using the method of Lowry *et al.* (1951) and immunizing doses of 100, 200, 300, 400, 500 and 600 μg of flagellar protein used. The animals are bled on the sixth day after the final injection, and the sera are then collected. Four to six injections usually result in antisera of satisfactory titres of 1:10 000 or more. Since the resulting H-antisera still give O-antigen titres in response to immunization even with crude flagellar preparations, O-antibodies in the H-antisera should be removed by absorption with homologous cultures heated at 100°C for 2 h.

Alternatively, Sakazaki found that a method described by Fey and Westzstein (1974) to obtain high titre *Salmonella* H-antiserum using the detached flagella for immunization was also satisfactory for production of *V. cholerae* H-antiserum with high titre of H-antibody and only small amounts of O-antibody.

Although any strains of *V. cholerae* may be employed for H-antiserum production. *V. cholerae* NIH 4 or *V. cholerae* O2 NCTC 4711 are recommended for this purpose.

2. H-agglutination test

Because *V. cholerae* possesses only a single flagellum and the flagellum is covered by a sheath that may prevent access to the H-antigenic sites on the flagellum filament, it is difficult to detect the H-antigen. It is possible to remove the sheath by treatment with phenol, formalin and by ageing of the cultures (Bhattacharyya, 1977).

(a) *Tube agglutination test.* Antigen for the tube agglutination test is prepared by using 48 h cultures on firm dry agar. The cultures are harvested in PBS containing 0.3% formalin and kept for 48 h in the cold. The density of the suspension is adjusted to about one-quarter of an overnight broth culture. The H-agglutination test is carried out by mixing an equal volume of antigen with serum of appropriate dilution and incubated in a water bath at 50°C. The results are read after 18 h.

(b) *Slide agglutination test* (Sil and Bhattacharyya, 1979). For the routine procedure to detect H-antigen of *V. cholerae* a modification of their slide agglutination test using phenolized suspension may be more appropriate than tube agglutination. Growth from an overnight culture at 30°C on nutrient agar is suspended in 0.5–1.0 ml of PBS containing 0.5% formalin and kept for 20 min to allow clumps to sediment out. Two separate loopfuls of the suspension are placed on a slide, and one loopful of 3% phenol is added to each of the two suspensions. After allowing to stand for 1 min, a loopful of H-antiserum diluted to 1:5 or 1:10 is added to the first suspension and the slide tilted back and forth. The H-agglutination may take up to 2 min to develop and appears finely granular, stringy or floccular. The control suspension should remain uniform.

B. O-antigen of *Vibrio cholerae*

As mentioned above, *V. cholerae* is subdivided into serovars on the basis of its O-antigens. So far 83 O-antigen groups (serovar) of *V. cholerae* have now been recognized. The quantitative chemical analysis of the sugar composition of the O-lipopolysaccharides has recently been studied by Hisatsune *et al.* (1980) and Raziuddin (1980). Unlike the majority of other Gram-negative rods, *V. cholerae* (and also some other *Vibrio* species) does not possess 2-keto-3 deoxyoctonic acid in the polysaccharide core of its lipopolysaccharide (Hisatsune *et al.*, 1980). Raziuddin (1980) stated that the majority of identifiable components in the core polysaccharide of *V. cholerae* were phosphorus, glucose, heptose, fructose and ethanalanine phosphate, with traces of mannose, rhamnose and D-perosamine, and those in the O-specific side chain polysaccharide were glucose, fructose, mannose, rhamnose, glucosamine, D-quinosamine and D-perosamine.

R-antigens of all strains of *V. cholerae* are serologically identical regardless of their O-antigen groups (Shimada and Sakazaki, 1973). The R form is indistinguishable from the parent S form in colonial morphology, but may be distinguished by agglutination with anti-R serum and 0.1% acriflavine (trypaflavine). Most O-antisera against *V. cholerae* contain some quantity of the R-antibody which may cause overlapping reactions in the determination of O-antigen groups.

1. *O-antiserum*

In order to obtain antigen for the preparation of O-antiserum, an overnight nutrient broth (pH 7.8) culture incubated in a water bath at 35°C by a rocking culture method is heated at 100°C for 2 h. When an agar culture is used a sterile swab is recommended to remove the growth from slant or plate in order to yield smooth suspensions. The heated culture is centrifuged and the packed cells are resuspended in 1% saline. The density of the suspension is adjusted to that of an overnight broth culture.

A rabbit is injected with 0.5 ml, 1.0 ml, 2.0 ml and 4.0 ml of the suspension at intervals of four days. Six days after several injections with 4.0 ml of the antigen, the animal is bled and the antiserum collected. Two to three injections of 4.0-ml amounts of the antigen may result in O-antisera of satisfactory titres of 1:1000 or more.

It should be emphasized that all O-antisera, even O1 *V. cholerae* antiserum must be absorbed with an R-culture of *V. cholerae* (strain CA385) for practical use.

2. *O-agglutination test*

(a) *Slide agglutination.* Slide agglutination is used for the determination of the O-antigen groups of *V. cholerae*. Growth from an overnight culture at 30–35°C on a nutrient agar slant is harvested into 1 ml of 0.3% formalized PBS using a sterile swab to give a smooth suspension and is kept for 20–30 min at room temperature. A loopful of the suspension is added to a loopful of appropriately diluted O-antiserum on a slide, and the slide tilted back and forth for 30 s. The suspension is usually satisfactory for O-antigen group determination, but sometimes it may be inagglutinable by the homologous O-antiserum because of the presence of the M-antigen. For this reason when agglutination is not recognizable with any of the O-antisera, the test should be repeated with antigen heated at 100°C for 2 h. The heated organisms are washed twice with PBS, centrifuged and the packed cells are resuspended in PBS. The antisera are diluted with phenolized saline. The dilution employed, usually 1:5 or 1:10, is dependent on the strength of the individual antiserum. A newly prepared antiserum should be titrated with its homologous antigen and used in the highest dilution in which it gives a strongly positive slide agglutination within 30 s.

For practical purposes O-antigen suspensions prepared from unknown cultures of non-O1 *V. cholerae* are tested first in O-antiserum pools in which the final dilution of each component O-antiserum is 1 to 10. If a culture is agglutinated by one of the pools the tests are repeated with each of the individual O-antisera that comprise the pool in which a reaction occurred.

(b) *Tube agglutination.* Determination of O-antigens may also be carried out by single tube agglutinations. The dilution used may be different with individual O-antisera, the final dilution is usually 1 in 100 (1 in 10 of the titre with the homologous suspension). 0.1 ml of the appropriate dilution of antiserum is transferred into a small tube and 0.9 ml of heated culture mentioned above is added. The test is incubated in a water bath at 50°C for

18 h. For recognition of the antigens by tube agglutination, washing bacterial cells after heating is particularly important in order to yield a satisfactory reaction.

(c) *Micro-titre agglutination.* Alternatively O-agglutinations may be performed using U-well micro-titre trays. This method is useful for large numbers of cultures. Test suspensions are harvested by the method described under slide agglutination and a further dilution in 5 ml of 0.3% formalized PBS prepared to an opacity of approximately 5×10^8 organisms per millilitre is employed. O-sera are titrated with their homologous antigen suspensions using 25-μl volumes of serum dilutions and 25-μl volumes of antigen suspension in micro-titre trays. The trays are covered and enclosed in a moist sealed container and incubated in an air incubator at 37°C for 18 h. Negative results are shown as a smooth button of cells and positive results by a dispersed agglutinated button of cells. Sera are used at dilutions that give clear-cut agglutination under these conditions. It is recommended that dilutions of the range of 1 in 40 should be used. Sera may be pooled to reduce preliminary testing. It is recommended that only four to five sera are used in a single pool. Strains are tested using unheated and heated suspensions. Positive reactions with pooled sera may be further tested by in use dilutions of single component sera by either the micro-titre tray method or by slide agglutination. A multi-reagent dispenser (Dynadrop-Dynatech Laboratories Ltd) may be used to simultaneously dispense 12 different pools of sera to a row of a micro-titre tray when large numbers of strains are being serotyped. This method is based on a method for serotyping *Salmonella* (Shipp and Rowe, 1980).

Because of the clinical importance of O1 *V. cholerae* it is useful to test an organism suspected of being *V. cholerae* first of all with O1 antiserum, and then if the reaction is negative the organism can be further studied biochemically and serologically.

Non-O1 *V. cholerae* strains are not agglutinated by the O1-antiserum, but they are agglutinated by their specific O-antiserum. Table II shows the test strains of *V. cholerae* for each O-antigen group with which diagnostic antisera should be prepared. Although most of the O-antigen groups show no significant relationship to any other O-antigen groups, a reciprocal (a,b–a,c) reaction may be recognized between O:2 and O:9, O:13 and O:29, O:15 and O:25, O:23 and O:73, O:32 and O:68, O:34 and O:75, and O:79 and O:83 respectively, as shown in Table III. To determine these O-antigen groups, therefore, absorbed antisera should be used. On the other hand unilateral reactions may be seen in some O-antisera, but the reactions are usually minor or very weak when the antisera are appropriately diluted for use in O-antigen group determination.

TABLE II
Test strains of *Vibrio cholerae* O-antigens

O-antigen	Strain	O-antigen	Strain	O-antigen	Strain
1 (Ogawa)	NIH 41	28	12530-62	56	475-75
1 (Inaba)	NIH 35A3	29	161-68	57	1463-76
2	NCTC 4711	30	12795-62	58	1162-74
3	NCTC 4715	31	5473-62	59	1333-74
4	NCTC 4716	32	171-68	60	195-75
5	B4202-64	33	151-68	61	12-74
6	7007-62	34	152-68	62	1-76
7	8394-62	35	1311-69	63	19-76
8	10317-62	36	1321-69	64	1280-75
9	112-68	37	1322-69	65	981-75
10	218-68	38	215-68	66	993-75
11	10843-62	39	225-68	67	121-79
12	211-72	40	212-72	68	293-78
13	11416-62	41	284-73	69	1861-79
14	B8645-64	42	104-73	70	1111-77
15	103-79	43	108-73	71	162-78
16	316-71	44	112-73	72	431-79
17	110-68	45	122-73	73	113-79
18	B5257-64	46	128-73	74	428-79
19	139-68	47	131-73	75	429-79
20	10332-62[a]	48	133-73	76	1158-76
21	109-68	49	1154-74	77	8-76
22	169-68	50	190-75	78	27-76
23	317-71	51	198-73	79	1103-76
24	14438-62	52	207-73	80	1421-77
25	14821-62	53	1157-74	81	318-78
26	334-72	54	1175-74	82	355-80
27	10432-62	55	197-75	83	1042-73

[a] Sucrose-negative (*V. mimicus*).

C. O-antigen variation in O1 *Vibrio cholerae*

A variation in the amount of partial O-antigens may sometimes occur in *V. cholerae*. Among O-antigen groups of *V. cholerae* O-antigen variants of serovar 1 (true cholera vibrio) have been known as Ogawa, Inaba and Hikojuma, although there are some discrepancies in the antigenic formulas of these variants which are often referred to as "serotypes" or "subtypes". Nobechi (1923) described the antigenic formulas of the two variants of O1 *V. cholerae* Inaba and Ogawa as A, B and C and A, B and X, respectively. Sholtens (1933), Heiberg (1935) and Kauffmann (1950) concluded that the antigenic difference between the Ogawa and the Inaba variants were quanti-

TABLE III

Intra-relationship of *Vibrio cholerae* O-antigen

O-antiserum[a]	Homologous titre	Other O-antigens[b]	Titre
2	1000	*9*	500
4	1000	15	100
7	1000	47	50
8	1000	23	100
		57	50
		73	50
		74	50
		75	50
9	1000	*2*	250
13	2000	*29*	500
		40	100
14	1000	19	100
15	1000	*25*	250
18	500	32	100
23	500	24	100
		73	100
25	1000	*15*	100
29	1000	*13*	250
32	1000	68	100
34	500	*75*	100
35	1000	42	100
41	1000	67	50
54	1000	60	50
65	1000	*80*	100
68	1000	*32*	100
73	500	23	50
75	1000	*34*	100
79	1000	*83*	200
80	1000	*65*	100
83	1000	*79*	50

[a] All antisera employed were absorbed by strain 353A to remove R-antibody.

[b] Reciprocal relationships are in italic.

tative but not qualitative. Sakazaki and Tamura (1971) who analysed O-antigens of the two variants by cross-agglutination and by agglutinin–absorption tests, confirmed that the Inaba-specific antigen as described by Nobechi (1923) was not found in any of the Inaba strains, whereas the Ogawa specific antigen was demonstrated in all of the Ogawa strains. They also observed that the Inaba variant was frequently dissociated from the Ogawa cultures, whereas none of the Ogawa variants was isolated from the Inaba strains. These observations may support the hypothesis that the Inaba variants are descended from an Ogawa strain by losing an O-antigen factor. The results of the studies by these authors mentioned above are interpreted to indicate that O1 *V. cholerae* strains produce three O-antigen fractions designated a, b and c. These antigen fractions are arranged hypothetically in Fig. 1 to represent the Ogawa, Inaba and Hikojima (intermediate form). The fractions a and c are common to the Ogawa and Inaba variants; the fraction b, which is absent in the Inaba variant, can be used as the specific fraction for recognizing the Ogawa variant. The Inaba variants possess a large quantity of the c fraction because of lack of the b fraction and are strongly agglutinated by the c fraction antiserum. Although the variation from the Ogawa to the Inaba is irreversible, an intermediate variant (Hikojima) possessing little fraction b and consequently a large quantity of fraction c may be misdiagnosed as an Inaba variant and such a falsely identified Inaba variant may sometimes appear to revert to the Ogawa form.

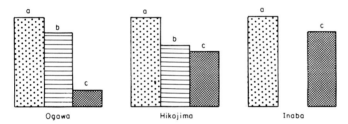

Fig. 1.   Hypothetical picture of antigenic formula of O1 *Vibrio cholerae*.

In the identification of O1 *V. cholerae* differentiation of the two variants is usually performed by using two absorbed antisera, the Ogawa-specific and the Inaba-specific sera. The Ogawa-specific antiserum is easily obtainable by absorbing the Ogawa antiserum with a sufficient quantity of the Inaba culture (growth from 10 to 15 plates per millilitre of the antiserum) and the titre of the Ogawa agglutinin in the absorbed antiserum may not be diminished. However, it is very difficult to obtain a pure Inaba-specific antiserum. All of the agglutinins in the Inaba antiserum are removed by the Ogawa culture when the absorption is done with an excess quantity of growth. To prepare the "Inaba-specific" antiserum therefore, the quantity of the Ogawa culture to be added to the Inaba antiserum must be adjusted very carefully. In such absorbed antiserum if absorption of the Ogawa agglutinin is incomplete agglutination of an Inaba culture occurs rapidly and an Ogawa culture is only slowly and more weakly agglutinated. In a completely absorbed Inaba serum there is no reaction with Ogawa suspensions and the reaction with Inaba suspensions is slow and weak.

Apparent Hikojima reactions should always be investigated using quantitative dilution methods with absorbed Inaba and Ogawa sera. The use of known control cultures and tests for the non-specific R-antigen are recommended in these studies. Genuine Hikojima subtypes are rare amongst *V. cholerae* O1 strains and care is needed with some commercial Inaba and Ogawa sera which may lack specificity and sensitivity.

Bhaskaran and Sinha (1971) demonstrated that O-antigen specificity was transferable from the O:6 strain to the O:1 strain of *V. cholerae* by chromosomal hybridization. Bhattacharji and Bose (1964) suggested that transformation of O-antigen of O1 *V. cholerae* into non-O1 *V. cholerae* may occur. However, no other investigators have been able to reproduce such changes and no such changes have been known to occur in nature.

D. Comparison of serovars with other antigenic schemata

Independently of Sakazaki and co-workers, Smith (1979) has recently reported a serotyping system of non-O1 *V. cholerae*. A comparative study of serovars described here and those by Smith has been carried out at NIH, Tokyo (T. Shimada and R. Sakazaki, unpublished data). Antisera for test strains of all Smith's serovars were prepared at NIH and cross-agglutination and agglutinin absorption tests were carried out between the two systems. The results of the study are summarized in Table IV.

In the system of serotyping of *V. cholerae* described here, serovars 2, 3, 4 and 7 correspond to O-subgroups III, V, VI, II of Gardner and Venkatraman (1935).

E. Serology of *Vibrio mimicus*

As mentioned before, strains which do not ferment sucrose which were formerly included in *V. cholerae*, have recently been separated from *V. cholerae* as *V. mimicus* on the basis of DNA hybridization study (Davis *et al.*, 1981). The two species, however, appear to be indistinguishable by serological methods. The H-antigen of *V. mimicus* is identical with that of *V. cholerae*. The test strain of serovar 20 of *V. cholerae* (10332-62) has recently been found to be *V. mimicus*. So far the majority of strains of *V. mimicus* have been serogrouped with O-antisera of *V. cholerae*. O-antigen groups of *V. mimicus* cover a wide range of O-antigen groups of *V. cholerae*. Therefore, a single serotyping system may be applicable for both species.

TABLE IV
Comparison of serovars of Sakazaki (SA) and of Smith (SM)

SA	SM	SA	SM	SA	SM
2	17 (5047)[a]	24	56 (5053)	51	40 (5079)
3	62 (5059)		22 (5060)	52	21 (5487)
4	20 (5042)	27	37 (5078)	53	59 (5092)
6	14 (5408,7951)[b]		312 (6970)	54	308 (7349)
7	13 (5064)	28	75 (5681,6361)	57	32 (7888)
8	22 (6301)	30	110 (7555)	60	28 (6541)
	176 (5180)	31	320 (7995)	63	309 (6313)
9	343 (8497)	34	42 (5411)	64	148 (7963)
10	27 (6345,7443)	35	15 (6707)	65	340 (8025,8585)
11	102 (5811)	36	30 (5128)	66	38 (5052)
	180 (7647)		83 (5697)		94 (5051)
12	33 (7920)	37	23 (5072)	67	24 (5066)
13	29 (6696)	39	68 (5028)	68	12 (5162)
14	31 (5410,7447)		44 (5046)		43 (5165)
16	48 (5068)		329 (8531)	69	77 (5694)
17	57 (6355)	40	11 (6305)	70	64 (7449)
18	25 (5163)	41	106 (5379,5803)	71	111 (7556)
20	113 (6814)[b]	44	61 (5029)	72	74 (7900)
21	44 (5037,6535)	46	76 (6335)	73	45 (5009,6308)
23	19 (6701,7586)		79 (5444)	74	342 (8462)
	115 (7902)	47	175 (5043)	75	332 (8536)
		48	60 (6326)	76	18 (7977)
		49	69 (5086)		

[a] Numbers in parentheses = test strains.

[b] Of three strains of SM serovars 14, two (5408 and 7951) were *V. cholerae* and the remaining one (5008) was not *V. cholerae* based on biochemical characteristics.

SA serovars 5, 15, 19, 22, 25, 29, 32, 33, 38, 42, 43, 45, 50, 55, 56, 58, 59, 62, 77, 78, 79, 80, 81, 82 and 83 do not correspond to any of the SM serovars.

Unilateral reactions were observed between SA O:5-antigen and SM O:12 serum, SA O:16-antigen and SM O:344-serum, and SA O:15-antigen and SM O:321-serum.

IV. Epidemiology of *Vibrio cholerae* and *Vibrio mimicus*

Cholera has been endemic in the delta of the Ganges River in eastern India and Bangladesh for centuries. It extended beyond these areas in six pandemics between 1818 and 1923. Those pandemics were presumably caused by the classical biovars of O1 *V. cholerae*. On the other hand, the present seventh pandemic of cholera is caused by the eltor biovar of O1 *V. cholerae*, and originating in the Southwest Pacific in 1961. It subsequently spread to Asia, Europe and Africa and still continues to spread. Thus, cholera has become a great public health problem, not only in the developing countries but also in industrial countries. Although cases as severe as those due to the classical vibrio resulting even in death, without adequate therapy, do occur, many patients have only mild diarrhoea and many symptomless excreters may be discovered.

Epidemiologists usually attach much importance to the Ogawa and Inaba variants of O1 *V. cholerae* in following cholera epidemics, but Inaba variants often occur in the stool specimens from patients during cholera epidemics caused by the Ogawa form. This clearly indicates that to distinguish the two variants of O1 *V. cholerae* is not of value in an epidemiological study of cholera.

Non-O1 *V. cholerae* is also a potential agent of cholera-like disease. Outbreaks of a cholera-like syndrome caused by non-O1 *V. cholerae* have been reported by many investigators (Bäck *et al.*, 1974; Chatterjee and Neogy, 1971; Gaines *et al.*, 1964; Ko *et al.*, 1973; McIntyre *et al.*, 1965; Nacescu *et al.*, 1974). Gastro-enteritis type disease associated with non-O1 *V. cholerae* has also been reported (Aldová *et al.*, 1968; Dakin *et al.*, 1974; Zakhariev *et al.*, 1976). Gastro-intestinal disease caused by non-O1 *V. cholerae* may be frequent in warmer countries. The vibrios have been more frequently found than O1 *V. cholerae* in the stool specimens of travellers returning from warmer countries.

Apart from imported cases of *V. cholerae* infection, small localized outbreaks and sporadic cases of diarrhoeal disease associated with *V. cholerae* which are domestically acquired may occur in industrial countries, although the proportion of those diarrhoeal diseases in these countries is unknown. Those patients have frequently a history of consumption of seafoods, especially crustaceans such as crabs, shrimps and lobsters, and oysters. Crustaceans and shellfish imported from warmer countries are very often contaminated with *V. cholerae*. There is no conclusive proof that diarrhoeal disease, even cholera, is spread by direct contact. The most important mode of spread is through water and food. Although *V. cholerae* has only a limited ability to survive in fresh water in laboratory experiments (Pandit *et al.*, 1967; Pesigan *et al.*, 1967), it will persist for months or years in the natural environment, such as rivers, lakes, shallows and estuaries. It may survive in the sediment. Early in the 1960s when Sakazaki had begun to study so-called "NAG vibrios", *V. cholerae* could not be found in the stools in cases of human diarrhoea nor in the natural environment in Japan. Recent investigations, however, indicate that the vibrios are commonly found in rivers and estuaries throughout Japan, especially in the warmer season. A similar situation to this has been reported in England, the USA and Germany (Bashford *et al.*, 1979; Kaper *et al.*, 1979; Müller, 1977). The ecology and pathogenicity of *V. mimicus* appears to be similar to *V. cholerae*.

It has been recognized that an exotoxin is involved in the pathogenesis of cholera. Non-O1 *V. cholerae* has also been found to produce a cholera-like exotoxin (Zinnaka and Carpenter, 1972; Ohashi *et al.*, 1972). However strains of *V. cholerae* are not always toxigenic. Shimada *et al.* (Sakazaki, 1981) reported, as shown in Table V, that 67% of 335 strains of O1 *V. cholerae* isolated from patients with clinical cholera or with inapparent symptoms produced cholera toxin, whereas about 30% of those isolates from seafoods and the natural environment were toxigenic. On the other hand, only 6% of non-O1 *V. cholerae* found in patients with diarrhoea and 9–20% of isolates from seafoods and the environment were demonstrated to produce exotoxin which could be neutralized with cholera antitoxin. When 72 strains of *V. mimicus* were studied, approximately 20% of the strains were toxigenic. The production of a toxin different from cholera or cholera-like toxin in non-O1 *V. cholerae* has also been reported (Spira *et al.*, 1979; Ohashi *et al.*, 1972) but its significance for human disease is unknown.

After examination of more than 2000 strains of *V. cholerae* no clear relationship between any particular serovars, areas and sources was recognized. There was no correlation between serovars of *V. cholerae* and their ability to produce toxin (Sakazaki and Shimada, 1977; T. Shimada and R. Sakazaki, unpublished data; Donovan, 1982).

TABLE V

Toxigenicity of strains of *Vibrio cholerae* and *Vibrio mimicus* isolated from various sources

Organism	Source	No. of strains tested	Toxin	
			Positive	Negative
O1 *V. cholerae*	Patient or contact	300	202 (67%)	98 (33%)
	Seafish	15	6 (40%)	9 (60%)
	Environment	20	6 (30%)	14 (70%)
	Subtotal	335	214 (64%)	121 (36%)
Non-O1 *V. cholerae*	Patient or contact	48	3 (6%)	45 (94%)
	Seafish	165	15 (9%)	150 (91%)
	Environment	166	33 (20%)	133 (80%)
	Subtotal	402	64 (16%)	338 (84%)
V. mimicus	Patient or contact	53	13 (25%)	40 (75%)
	Seafish	1	0	1
	Environment	18	1 (6%)	17 (94%)
	Subtotal	72	14 (20%)	58 (80%)
Total		809	292 (36%)	517 (64%)

References

Aldová, E., Láznic̆ková, K., Štĕpánková, E. and Lietavá, J. (1968). *J. Infect. Dis.* **118**, 25–31.

Bäck, E., Ljunggren, H. and Smith, H. L., Jr (1974). *Lancet* **1**, 723–724.

Bashford, D. J., Donovan, T. J., Furniss, A. L. and Lee, J. V. (1979). *Lancet* **1**, 436–437.

Bhaskaran, K. and Sinha, V. B. (1971). *Indian J. Exp. Biol.* **9**, 119–120.

Bhattacharji, L. M. and Bose, B. (1964). *Indian J. Med. Res.* **52**, 777–786.

Bhattacharyya, F. K. (1977). *Jpn. J. Med. Sci. Biol.* **30**, 259–268.

Chatterjee, B. D. and Neogy, K. N. (1971). *Indian J. Med. Res.* **57**, 95.

Davis, B. R., Fanning, G. R., Madden, J. M., Steigerwalt, A. G., Bradford, H. B., Jr, Smith, H. L., Jr and Brenner, D. J. (1981). *J. Clin. Microbiol.* **14**, 631–639.

Dakin, W. P. H., Howell, D. J., Sutton, R. G. A. and O'Keefe, M. F. (1974). *Med. J. Aust.* **2**, 487–490.

Desmarchelier, P. M. and Reichelt, J. L. (1981). *Curr. Microbiol.* **5**, 127–130.

Donovan, T. J. (1982) (in press).

Fey, H. and Wetzstein, H. P. (1974). *Pathol. Microbiol.* **41**, 164–165.

Furniss, A. L., Lee, J. V. and Donovan, T. J. (1978). Public Health Laboratory Service Monograph No. 11.

Gaines, S., Duangmani, C., Noyes, H. E. and Occeno, T. (1964). *Thailand J. Microbiol. Soc. Thailand* **8–10**, 6.17.

Gardner, A. D. and Venkatraman, K. V. (1935). *J. Hyg.* **35**, 262–282.

Heiberg, B. (1935). "On the Classification of *Vibrio cholerae* and the Cholera-like Vibrios". Busk. Copenhagen.

Hisatune, K., Kondo, S., Iguchi, T. and Takeya, K. (1980). Proc. 15th Jt. Conf. US–Japan Coop. Med. Sci. Prog., 1979, pp. 148–165.

Hugh, R. and Sakazaki, R. (1972). *J. Conf. Public Health Lab. Directions* **30**, 133–137.

International Subcommittee on Taxonomy of Vibrios (1972). *Int. J. Syst. Bacteriol.* **22**, 123.

International Subcommittee on Taxonomy of Vibrios (1979). *Int. J. Syst. Bacteriol.* **29**, 170–171.

Kaper, J., Lockman, H., Colwell, R. R. and Joseph, S. W. (1979). *Appl. Environ. Microbiol.* **37**, 91–103.

Kauffmann, F. (1950). *Acta Pathol. Microbiol. Scand.* **27**, 283–299.

Ko, H. L., Lutticken, R. and Pulverer, G. (1973). *Dtsch. Med. Wochenschr.* **98**, 1494–1499.

Lehmann, K. B. and Neumann, R. (1896). *Atlas und Grundriss der Bakteriologie und Lehrbuch der speziellen bakteriologischen Diagnostik. Teil II.* pp. 1–448.

Lowry, O. H., Rosebrough, N. J., Farr, A. L. and Randall, R. J. (1951). *J. Biol. Chem.* **193**, 265–275.

McIntyre, O. R., Feeley, J. C., Greenough, W. B. III, Benenson, A. S., Hassan, S. I. and Saad, A. (1965). *Am. J. Trop. Med. Hyg.* **14**, 412–418.

Müller, G. (1977). *Zbl. Bakt. Hyg., I. Abt. Orig. Reihe B* **165**, 487–497.

Nacescu, N., Ciufecu, C., Nicoara, I., Florescu, D. and Konrad, I. (1974). *Zbl. Bakt. Hyg., I.Abt. Orig. Reihe A* **229**, 209–215.

Nobechi, K. (1923). *Sci. Rep. Gov. Inst. Infect. Dis. Tokyo Imp. Univ.* **2**, 43–88.

Ohashi, M., Shimada, T. and Fukumi, H. (1972). *Jpn. J. Med. Sci. Biol.* **25**, 179–194.

Pacini, F. (1854). *Gazz. Med. Ital.* **6**, 405–412.

Pandit, C. G., Pal. S. C., Murti, G. V. S., Misra, B. S., Murty, D. K. and Shrivastava, J. B. (1967). *Bull. W. H. O.* **37**, 681–685.

Pesigan, T. P., Plantilla, J. and Rolda, M. (1967). *Bull. W. H. O.* **37**, 779–786.

Raziuddin, S. (1980). *Infect. Immun.* **27**, 211–215.

Reichelt, J. L., Baumann, P. and Baumann, L. (1976). *Arch. Mikrobiol.* **110**, 101–120.

Sakazaki, R. (1982). Proc. 4th Int. Symp. Tox. Org. US-Japan Conf. Develop. Util. Nat. Res. 1981, pp. 177–184.

Sakazaki, R. and Shimada, T. (1977). *Jpn. J. Med. Sci. Biol.* **30**, 279–282.

Sakazaki, R. and Tamura, K. (1971). *Jpn. J. Med. Sci. Biol.* **24**, 93–100.

Sakazaki, R., Gomez, C. Z. and Sebald, M. (1967). *Jpn. J. Med. Sci. Biol.* **20**, 265–280.

Sakazaki, R., Tamura, K., Gomez, C. Z. and Sen. R. (1970). *Jpn. J. Med. Sci. Biol.* **23**, 13–20.

Shimada, T. and Sakazaki, R. (1973). *Jpn. J. Med. Sci. Biol.* **26**, 155–160.

Shimada, T. and Sakazaki, R. (1977). *Jpn. J. Med. Sci. Biol.* **30**, 275–277.

Shinoda, S., Kariyama, R., Ogawa, M., Takeda, Y. and Miwatani, T. (1976). *Int. J. Syst. Bacteriol.* **26**, 97–101.

Shipp, C. R. and Rowe, B. (1980). *J. Clin. Pathol.* **33**, 595–597.

Sholtens, R. Th. (1933). *C. R. Soc. Biol.* **114**, 422–424.

Sil, J. and Bhattacharyya, K. (1979). *J. Med. Microbiol.* **19**, 63–00.

Smith, H. L., Jr (1979). *J. Clin. Microbiol.* **10**, 85–90.

Spira, W. M., Daniel, R. R., Ahmed, Q. S., Huq, A., Yusuf, A. and Sack, D. A. (1979). Proc. 14th Jt. Conf. US–Japan. Coop. Med. Sci. Prog., 1978, pp. 137–153.

Zakhariev, Z., Tyujekchiev, T., Valkov, V. and Todeva, M. (1976). *J. Hyg. Epidemiol. Microbiol. Immunol.* **20**, 150–156.

Zinnaka, Y. and Carpenter, C. C. J., Jr (1972). *Johns Hopkins Med. J.* **131**, 403–411.

12

Identification of *Bacillus* Species

R. C. W. BERKELEY,[1] N. A. LOGAN,[2] L. A. SHUTE[3]
and A. G. CAPEY[4]

[1] *Department of Microbiology, University of Bristol, The Medical School, Bristol, UK,* [2] *Department of Biological Sciences, Glasgow College of Technology, Glasgow, UK* [3] *Department of Microbiology, University of Bristol, The Medical School, Bristol, UK* [4] *Department of Microbiology, University of Bristol, The Medical School, Bristol, UK*

I.	Introduction	292
II.	Classical identification methods	293
	A. Morphological and biochemical characterization	293
	B. Bacteriophage typing	293
	C. Serotyping	299
III.	Identification using the API system	303
	A. Introduction	303
	B. Test reproducibility	303
	C. Advantages and disadvantages of classical and miniaturized methods	308
	D. Morphology	309
	E. Taxonomy	309
	F. Identification	310
	G. Microcomputer-assisted identification	314
IV.	Recent developments in instrumented approaches to identification	315
	A. Introduction	315
	B. Pyrolysis	315
	C. Pyrolysis gas–liquid chromatography	317
	D. Standardization problems peculiar to Py-GC	318
	E. Pyrolysis mass spectrometry	318
	F. Analysis of pyrolysis data	319
	G. Problems of standardization concerning both Py-GC and Py-MS	319
	H. Use of Py-GC in identification of *Bacillus* species	320
	I. Use of Py-MS in identification of *Bacillus* species	321
	J. Laser pyrolysis	322
	K. Flow cytometry	322
	L. Overview	323
V.	Summary	323
	References	323

METHODS IN MICROBIOLOGY VOL. 16
ISBN 0–12–521516–9

I. Introduction

Aerobic endospore-forming rods, that is members of the genus *Bacillus*, are ubiquitous and because of the resistance of their spores and the ability of the vegetative organisms to degrade many organic materials, the genus is probably the most important group of spoilage and contaminating bacteria affecting the industrial and medical activities of man. Yet many, or possibly, even most, bacteriologists consider the identification of these organisms too difficult to carry out properly or even at all. It is clear, however, from the increasing number of strains received in this laboratory from other workers that there is a growing need for accurate identification of these organisms.

The process of identification presupposes the existence of a classification scheme containing taxa with which unknown strains may be identified. For the genus *Bacillus* there have been three main proposals: those of Krasil'nikov (1949), Prèvot (1961) and Gordon and colleagues (1973).

Krasil'nikov's scheme depends on the extent to which the rod is swollen by the endospore. This character is one about which there is so much conflict of interpretation, not surprisingly since some strains produce both swollen and non-swollen sporangia (Bonde, 1981), that it is difficult to equate his nomenclature with that of other systems (Skerman, 1967). Prèvot arranged the organisms into four genera: *Bacillus, Bacteridium, Inominatus* and *Clostridium*. The classification of Gordon and colleagues, based on a meticulous study of 1134 strains, is essentially similar to that in the eighth edition of "Bergey's Manual of Determinative Bacteriology" (Gibson and Gordon, 1974). The chief differences are the allocation of species rank to *B. anthracis* and *B. thuringiensis* and the inclusion of 26 species of arguable status in a section separate from the 22 more widely accepted types.

Of the three schemes, that of Gordon and colleagues is pre-eminent and it is this classification which underpins the list of *Bacillus* species included in the Approved List of Bacterial Names published by the International Committee for Systematic Bacteriology (Skerman *et al.*, 1980). It is, however, recognized that this classification is not entirely satisfactory. Evidence from studies of DNA composition suggests that the genus *Bacillus* is the equivalent to some bacterial families (Priest, 1981) and numerical analyses of the data published by Gordon and colleagues (1973), by Priest and co-workers (1981) and of that obtained by using the API System, by Logan and Berkeley (1981), both indicate that splitting of the genus into perhaps six new genera is called for. Such an operation would, however, be premature at this time as sufficient data to define satisfactorily all the new groups are not yet at hand although work towards this goal is currently proceeding in several laboratories. In this chapter long established means of identifying *Bacillus* strains are described briefly. There is then a fuller description of the newly developed method based on API materials. Finally, there is a section dealing with recent work on instrumented approaches to identification of members of this genus.

II. Classical identification methods

A. Morphological and biochemical characterization

The prime authority on the cellular morphology and biochemical characterization of *Bacillus* strains is the monograph by Gordon (1973). In this publication is presented a simplified key to typical strains of many species in the genus. As a result of an international inter-laboratory reproducibility trial a modified version of this key has been published (Norris *et al.*, 1981) in which two new tests (nitrate reduction and rod diameter 1 μm or greater) have replaced the two (growth at pH 5.7 and citrate utilization) which were shown to produce disagreement even in the hands of experts and given rigorous test standardization. These methods (details in Gordon *et al.* 1973) and this key (Table I) should enable the tentative identification of typical strains of common species. The tentative identification should be checked using all the characters tested against the summary in Table II. Finally, a number of known strains should be tested in parallel with the unknown. Sources of type strains for each species in the Approved List of Bacterial Names (Skerman *et al.*, 1980) are shown in Table III.

Provided careful attention is paid to details, especially of medium composition and test procedure, this method should result in the proper identification of many unknown strains. Similar results may be obtained using tables of test results (Cowan and Steel, 1965, 1974) rather than a dichotomous key, in the first instance. Less common examples may be named by further reference to Gordon and co-workers (1973). But difficulty may still be encountered with what prove to be atypical or intermediate strains. The API system described in Section III enables identification of a greater number of species. Before describing this approach the applications of phage typing and serology to identification of *Bacillus* is discussed.

B. Bacteriophage typing

1. *Introduction*

Bacteriophage typing of *Bacillus* species is not widely undertaken, but has been used in certain areas. Smith *et al.* (1952) used this technique to a limited extent in the early part of their study but as their work progressed they found

TABLE I

Key for the tentative identification of typical strains of common *Bacillus* species[a]

---

1. Catalase: positive . . . 2
        negative . . . 16
2. Voges–Proskauer: positive . . . 3
        negative . . . 9
3. Growth in anaerobic agar: positive . . . 4
        negative . . . 8
4. Growth at 50°C: positive . . . 5
        negative . . . 6
5. Growth in 7% NaCl: positive . . . *B. licheniformis*
        negative . . . *B. coagulans*
6. Acid and gas from glucose (inorganic N): positive . . . *B. polymyxa*
        negative . . . 7
7. Reduction of NO_3 to NO_2: positive . . . *B. cereus*
        negative . . . *B. alvei*
8. Hydrolysis of starch: positive . . . *B. subtilis*
        negative . . . *B. pumilus*
9. Growth at 65°C: positive . . . *B. stearothermophilus*
        negative . . . 10
10. Hydrolysis of starch: positive . . . 11
        negative . . . 14
11. Acid and gas from glucose (inorganic N): positive . . . *B. macerans*
        negative . . . 12
12. Width 1.0 μm or greater: positive . . . *B. megaterium*
        negative . . . 13
13. pH in V-P broth 6.0: positive . . . *B. circulans*
        negative . . . 15
14. Growth in anaerobic agar: positive . . . *B. laterosporus*
        negative . . . 15
15. Acid from glucose (inorganic N): positive . . . *B. brevis*
        negative . . . *B. sphaericus*
16. Growth at 65°C: positive . . . *B. stearothermophilus*
        negative . . . 17
17. Decomposition of casein: positive . . . *B. larvae*
        negative . . . 18
18. Parasporal body in sporangium: positive . . . *B. popilliae*
        negative . . . *B. lentimorbus*

---

[a] Modified from Gordon et al. (1973).
  Numbers on the right indicate the number (on the left) of the next test to be applied until the right-hand number is replaced by a species name.

that other reactions were simpler and more dependable. They do, however, describe their methods. A more general reference, in which the principles of bacteriophage typing are to be found, is Adams (1959). In general, a lawn of the bacterial strain under investigation is spotted with bacteriophage dilutions and the spots examined for lysis, after the plate has been incubated for a

TABLE II

Summary of the characters used in the key for typical *Bacillus* species[a]

	Catalase	V-P reaction	Growth in anaerobic agar	Growth at 50°C	Growth in 7% NaCl	Acid and gas in glucose	NO₃ reduced to NO₂	Starch hydrolysed	Growth at 65°C	Rods 1.0 μm wide or wider	pH in V-P medium <6.0	Acid from glucose	Hydrolysis of casein	Parasporal bodies
B. megaterium	+	−	−	−	+	−	o	+	−	+	o	+	+	−
B. cereus	+	+	+	−	+	−	+	+	−	+	+	+	+	o
B. thuringiensis	+	+	+	−	+	−	+	+	−	+	+	+	+	+
B. licheniformis	+	+	+	+	+	−	+	+	−	−	o	+	+	−
B. subtilis	+	+	−	+	+	−	+	+	−	−	o	+	+	−
B. pumilis	+	+	−	+	+	−	−	−	−	−	+	+	+	−
B. firmus	+	−	−	−	+	−	+	+	−	−	−	+	+	−
B. coagulans	+	+	+	+	−	−	o	+	−	o	+	+	o	−
B. polymyxa	+	+	+	−	−	+	+	+	−	−	o	+	+	−
B. macerans	+	−	+	+	−	+	+	+	−	−	−	+	−	−
B. circulans	+	−	o	+	o	−	o	+	−	−	o	+	o	−
B. stearothermophilus	o	−	−	+	−	−	o	+	+	o	+	+	o	−
B. alvei	+	+	+	−	−	−	−	+	−	o	+	+	+	−
B. laterosporus	+	−	+	+	−	−	+	−	−	−	−	+	+	+
B. brevis	+	−	−	+	−	−	o	−	−	−	−	+	+	−
B. larvae	−	−	+	−	+[b]	−	o	−	−	−	−	+	+	−
B. popilliae	−	−	+	−	+[b]	−	−	−	−	−	−	+	−	+
B. lentimorbus	−	−	+	−	−	−	−	−	−	−	−	+	−	−
B. sphaericus	+	−	−	−	o	−	−	−	−	o	−	−	o	−

[a] +, greater than 85% of strains examined by Gordon *et al.* (1973) positive; −, greater than 85% of strains negative; o, variable character.

[b] Growth on 2% NaCl.

suitable period. From the lysed and unlysed areas of the plate, the bacteriophage sensitivities of the strain can be determined and the resulting data used in taxonomic analysis.

The areas of the genus where work with bacteriophage has been reported are discussed below under the names of the susceptible organisms.

2. B. cereus

Bacteriophage in *B. cereus* strains have been investigated and used by a

TABLE III
Sources of type strains of valid *Bacillus* species

	Culture collection[a] and number			
	ATCC	DSM	NCIB	NCTC
B. acidocaldarius	27009	446	11725	
B. alcalophilus	27647	485	10436	4553
B. alvei	6344	29	9371	6352
B. anthracis	14578		9388	10340
B. badius	14574	23	9364	10333
B. brevis	8246	30	9372	2611
B. cereus	14579	31	9373	2599
B. circulans	4513	11	9374	2610
B. coagulans	7050	1	9365	10334
B. fastidiosus	29604	91	11326	
B. firmus	14575	12	9366	10335
B. globisporus	23301	4	11434	
B. insolitus	23299	5	11433	
B. larvae	9545			
B. laterosporus	64	25	9367	6357
B. lentimorbus	14707	2049	11202	
B. lentus	10840	9	8773	4824
B. licheniformis	14580	13	9375	10341
B. macerans	8244	24	9368	6355
B. macquariensis	23464	2	9934	10419
B. megaterium	14581	32	9376	10342
B. mycoides	6462	2048		
B. pantothenticus	14576	26	8775	8162
B. pasteurii	11859	33	8841	4822
B. polymyxa	842	36	8158	10343
B. popilliae	14706	2047		
B. pumilus	7061	27	9369	10337
B. schlegelii		2000		
B. sphaericus	14577	28	9370	10338
B. stearothermophilus	12980	22	8923	10339
B. subtilis	6051	10	3610	3610
B. thuringiensis	10792	2046	9134	

[a] ATCC, American Type Culture Collection; DSM, Deutsche Sammlung von Mikro-organismen; NCIB, National Collection of Industrial Bacteria; NCTC, National Collection of Type Cultures.

number of workers. McCloy (1951, 1958) studied bacteriophage derived from *B. cereus* strain W. Strain W was found to be lysogenic with a temperate bacteriophage W which appeared to mutate under different conditions to produce a virulent bacteriophage W, which caused lysis of W cultures. Wy, another non-lysogenizing bacteriophage, was also isolated, but this did not cause lysis in strain W. These bacteriophage were, however, not considered to be taxonomically useful.

Bacteriophage isolated from soil, against the streptomycin-resistant strain *B. cereus* 569 (Földes *et al.*, 1961) showed that *B. cereus* bacteriophage show patterns of specificity at the subspecies level, as the bacteriophage attacked only certain *B. cereus* strains. It also discriminated between most *B. cereus* and *B. anthracis* but not those *B. cereus* whose autolysate reacted with antibody to *B. anthracis* polysaccharide. This discrimination between *B. cereus* and *B. anthracis* was, therefore, not useful as an identification method.

Norris (1961) described the isolation of four bacteriophages from the *B. cereus* group, which were able to discriminate between strongly lecithinase-positive strains, which were resistant to lysis, and lecithinase-negative or weakly positive strains, which were sensitive to lysis. However, the plaque morphologies were not considered to be characteristic enough to be of taxonomic value.

Gordon (1973) reported having used a bacteriophage against *B. cereus* NRS 201, which was isolated from soil for discriminating *B. cereus* strains. All the *B. cereus* cultures held in her 1952 collection were sensitive to lysis by the bacteriophage. The rhizoid form of *B. mycoides* and *"B. praussnitzi"* were resistant but the variants of these two species were sensitive. *B. cereus* cultures were also found to be lysed by another bacteriophage developed against *B. mycoides* NRS 319.

3. B. anthracis

The discovery of a phage lysing *B. anthracis*, gamma phage (Brown and Cherry, 1955), a variant of phage W, led to hopes of a specific means of identification of this organism. However only 85% of the *B. anthracis* cultures tested were lysed by this phage. If a single phage was used, 108, which lyses 91%, might be better (Buck *et al.*, 1963). Thus this test has to be used in conjunction with others to achieve definitive identification.

4. B. thuringiensis

de Barjac *et al.* (1974) tried to differentiate *B. thuringiensis* strains using bacteriophage. In their study, 12 different bacteriophages were used, but the subspecies of *B. thuringiensis* recognized by other methods such as H-serotyping or biochemical tests could not be differentiated by bacteriophage sensitivities. Thus it was concluded that bacteriophage typing of *B. thuringiensis* strains is not viable at the present time.

5. B. sphaericus

In contrast to the experience with *B. thuringiensis* (Section II.B.4), de Barjac (1981) reports that phage typing of *B. sphaericus* gives a grouping in perfect agreement with that obtained by H-serotyping. The strains toxic for mosquitoes divided into three different serotypes, each of which reacts specifically with a number of bacteriophages.

6. B. stearothermophilus

Sharp *et al.* (1979) used bacteriophage typing as an aid to characterizing some thermophilic species of *Bacillus*. Bacteriophage sensitivities coupled with data from biochemical tests, antibiotic and bacteriocin sensitivities, esterase patterns and DNA base composition allowed them to compare the three caldoactive strains isolated by Heinen and Heinen (1972) with eight strains of *B. stearothermophilus*. Analysis of their results enabled them to place the strains into the three main taxonomic groups suggested by Walker and Wolf (1971). Genotypic and phenotypic data produced slightly different groupings. The caldoactive strains proved to be more sensitive to infection by selected bacteriophage than the *B. stearothermophilus* strains. Among the caldoactive strains, *B. caldolyticus* showed a slightly lower level of infection and it was thought this might be due to its possession of the restriction enzyme BcII (Bingham *et al.*, 1978).

7. *Alkalophilic strains*

Gordon *et al.* (1977) report having used bacteriophage as an aid to discriminating between 27 variants of alkalophilic strains that were able to grow at pH 7.0, some morphologically similar strains and strains of *B. firmus*.

Tests with a bacteriophage whose host is *B. firmus* ATCC 14575 showed that all the alkalophilic strains were lysed on infection with the bacteriophage. Strains of *B. firmus* and some of the *B. firmus-B. lentus* intermediates also proved to be sensitive, but 162 morphologically similar strains of *B. licheniformis*, *B. subtilis* and *B. pumilus* were resistant.

These results, coupled with data from other work (Gordon, 1981) were later used to promote the acceptance of the 27 alkalophilic strains as strains of *B. firmus*.

8. *Overview*

Thus bacteriophage typing has not been extensively used for *Bacillus* species.

Where use of bacteriophage has been made, identification of strains tends to be based upon combinations of bacteriophage sensitivities with other types of data, rather than on sensitivity patterns alone.

C. Serotyping

1. *Introduction*

Unlike the situation found in the Enterobacteriaceae and *Salmonella* species, where serological methods are used extensively in identification, serological identification of *Bacillus* species is not normal practice. It is, however, used with success in a number of areas. The use of spore antigens, H-antigens, O-antigens and enterotoxins in the identification of *Bacillus* species are discussed separately below.

Henriksen (1978) gives a comprehensive introduction to the subject of serotyping of bacteria and also discusses the types of serological reaction that may be used in detecting antigens.

2. *Spore antigens*

Spores are characteristic features of *Bacillus* species and as would be expected, many investigations into the nature of spore antigens and their potential use in identification of *Bacillus* species have been made. Norris (1962) reviewed the earliest investigations of spore antigens and the problems encountered in using spore antigens.

The earliest paper dealing with spore antigens was that of Defalle (1902) who recognized that if living spores were used as antigens, they might germinate on inoculation into animals and give rise to antisera containing antibodies against vegetative cell antigens as well as spore antigens. This was confirmed to be the case by Teale and Bach (1919).

Another problem was that the fairly long incubation period of spores with their antisera in the normal agglutination test may again promote germination. Noble (1919) circumvented this by proposing a rapid method agglutination test which was used by later workers satisfactorily.

Due to the hydrophobic surface of spores (Lamanna and Eisler, 1960) it was found that they tend to auto-agglutinate. Noble (1927) developed a technique for producing homogeneous suspensions of *Bacillus anthracis* for agglutination reactions. This method was successfully used by Norris and Wolf (1961) and by other workers, but even using this method some bacterial spores still show clumping in the controls, and it is impossible to study these by simple agglutination tests.

To rid spore suspensions of contaminating vegetative cell debris two main approaches have been used. The first is to grow organisms on media designed to produce complete sporulation; see, for example, Howie and Cruickshank (1940), Davies (1951). The second is the removal of debris by autolysis using thiomersalate (Delpy and Chamsy, 1949; Norris and Wolf, 1961) or lysozyme (Walker, 1959; Tomcsik and Baumann-Grace, 1959).

Autoclaving of spore suspensions can also be used as spore antigens are heat resistant (Howie and Cruickshank, 1940; Bekker, 1944), and completely independent of vegetative cell antigens (Norris and Wolf, 1961). Treatment of spores with KOH has also been used, but this alters their serological properties (Doak and Lamanna, 1948).

As long as the special problems of using spore antigens are catered for, they can be used quite successfully in serological investigations, as evidenced by the following reports.

As early as 1902 Defalle showed cross-agglutination between spores of *Bacillus mycoides* and *Bacillus anthracis*, organisms which are now known to be closely related. Lamanna (1940a) placed the "small-celled" *Bacillus* species into three serological groups, which were later identified as *B. licheniformis*, *B. subtilis* and *B. brevis*. Within the "large-celled" species he was able to separate *B. cereus* strains from the others, and also found that some strains of *B. mycoides* cross-reacted with *B. cereus* strains and it is now thought to be a variant of this species (Lamanna, 1940b).

Davies (1951) tested spores of 39 strains of *B. polymyxa* using seven spore antisera. He found all the *B. polymyxa* strains agglutinated with the antisera and none of 81 strains of 15 other *Bacillus* species did so.

Norris and Wolf (1961) showed that spore agglutinations were species specific in some organisms, e.g. *B. subtilis* and *B. licheniformis*, although this was not the case in all species. For example, Lamanna and Eisler (1960) tried to separate *B. anthracis* and *B. cereus* using spore agglutinogens, but were unable to do so.

Further work with *B. cereus* spore antigens was carried out by Kim and Goepfert (1972) using a fluorescent antibody technique for confirming the tentative identification of food poisoning organisms as *B. cereus*. This technique was used instead of simple agglutination tests, because problems are encountered due to clumping of the cells even in the absence of any antibody (Norris, 1961). The initial identification was base on the growth of the organisms from food on the KG medium of Kim and Goepfert (1971) which relies on the lecithinase reaction to differentiate *B. cereus* from other *Bacillus* species. This must be confirmed, however, as some isolates will be lecithinase negative and also a few other species produce lecithinase. These workers concluded that the fluorescent antibody technique would prove useful for this, but would not be completely effective because of its failure to detect asporogenic mutants and its inability to differentiate between *B.*

thuringiensis and *B. cereus* species. Although *B. thuringiensis* strains can be identified by their production of crystals in culture, in practice no typing scheme based on this method has yet been used.

Walker and Wolf (1971) used spore agglutination tests to substantiate their classification of *B. stearothermophilus* strains into three groups. Chowhery and Wolf (reported in Wolf and Sharp, 1981) extended this work, and their results were in accordance with the groups suggested by Walker and Wolf.

3. H-antigens

H-antigens have been used in serological studies of *Bacillus* species more extensively than the other antigens. This is probably due to the fact that although the spore antigens confer the highest species specificity, the H-antigens provide the highest strain specificity (Norris and Wolf, 1961). Also, the problems encountered with spore antigens promoted the use of H-antigens.

The most successful use of H-antigens has been in the construction of a serotyping scheme for *B. cereus* strains by Taylor and Gilbert (1975) and Gilbert and Parry (1977). This is used in the Food Hygiene Laboratory for investigating food poisoning outbreaks (Gilbert, 1979). It has also been used for serotyping non-food related clinical isolates (Fitzpatrick *et al.*, 1979).

The work of Taylor and Gilbert (1975) followed on from that of Le Mille and colleagues (1969) who reported 17 serotypes among 33 cultures of *B. cereus*. Taylor and Gilbert (1975) used 18 serotypes of *B. cereus*. The scheme was later extended in Japan by Terayama *et al.* (1978) who, with additional serotypes, typed *B. cereus* strains from various foods.

As well as providing a rapid identification means the scheme has provided useful epidemiological information about *B. cereus* food poisoning. For example, Gilbert and Parry (1977) observed that more isolates from cooked rice were serotype 1 (23%) than from uncooked rice (3%) and that serotype 17 was found frequently in uncooked rice (15%) and rarely in cooked rice (1%). They later confirmed that type 1 strain spores are more heat resistant than type 17 spores (Parry and Gilbert, 1980).

B. thuringiensis strains are also classified using a combination of antigen agglutinations and biochemical tests. Initially de Barjac and Bonnefoi (1962) recognized different serotypes of *B. thuringiensis* by the flagella antigens and later recognized 14 serotypes (de Barjac, 1978), some of which can now be subdivided according to the presence of H-antigenic subfactors.

B. sphaericus has also been successfully characterized using a similar H-serotyping method (de Barjac, 1981).

Siman *et al.* (1977) undertook an extensive study of *B. subtilis* H-antigens and were able to define at least five distinct serological groups. They found difficulties in carrying out simple agglutination due to spontaneous clumping of the organisms, and so used double diffusion in agar methods or complement fixation. Even though distinct serological groups could be recognized, cross-reactions between strains were strong and made it impractical to produce a serotyping scheme, although it was suggested that flagellar antigenicity could be a useful character in identification.

4. *O-Antigens*

These have been very little used in identifying *Bacillus* strains. *B. cereus* which has received a lot of attention due to its pathogenic role has not yet been successfully typed using O-antigens.

Walker and Wolf (1971) attempted to classify *B. stearothermophilus* strains using O-antigens and found this approach to be of limited application although it did add weight to the existence of the three major thermophile subgroups based on biochemical and physiological characteristics (Wolf and Sharp, 1981). Evidence from agglutination reactions using O-antigens of *Sporosarcina ureae* suggests it can be subdivided into several antigenic groups (MacDonald and MacDonald, 1962).

5. *Enterotoxins*

The only instance of enterotoxins being used successfully as antigens for identification purposes is found in *B. cereus*. Several different enterotoxins appear to be involved in *B. cereus* food poisoning (Melling *et al.*, 1976; Turnbull, 1976; Turnbull *et al.*, 1977). Gorina *et al.* (1975) purified an enterotoxin that caused vomiting in cats and successfully detected it in foods using an aggregate-haemagglutination method. This method was very sensitive and detected as little as $0.004 \, \mu g \, ml^{-1}$.

6. *Overview*

Thus serological methods are of use in the identification of *Bacillus* species, though like phage typing, tend to be used in conjunction with other taxonomic data. The most widely investigated area of the genus using serological methods is that of the *B. cereus* and some of its close relatives reflecting their importance in the medical field and as insect pathogens.

III. Identification using the API system

A. Introduction

Recognition of the difficulty in standardizing and having available the media used classically in the identification of *Bacillus* strains, and the time taken in using them, provoked the search for an identification system that might be widely employed and which would be reproducible and rapid (Logan and Berkeley, 1981).

API tests* are produced as ready-to-use micro-tube systems developed from Buissière's (1972) modification of the Ivan Hall tube, and contain dehydrated substrates for performing standard biochemical tests. The two systems finally chosen for *Bacillus* identification were the API 20 Enterobacteriaceae (API 20E) strip and the API 50 Carbohydrate (API 50CH) gallery. For the original work, the API 50 Enterobacteriaceae (API 50E) gallery was used; it comprised 49 tests and one control, with 39 of the tests carbohydrates and the remaining 10 miscellaneous. This gallery has now been withdrawn and the API 50CH, containing 49 carbohydrate tests and one control, is its replacement. The API 50CH gallery contains only the carbohydrate substrates so that different basal media and indicators may be used for suspension and inoculation of the organisms. The API 50CH Enterobacteriaceae (API 50CHE) and the API 50CH *Bacillus* (API 50CHB) are available and both use the API 50CHE/B medium.

The tests of the API 20E that are used for *Bacillus* and the tests in the API 50CH are listed in Table IV.

Before describing the identification system it is convenient to outline the shortcomings of the classical test methods, notably lack of reproducibility, already referred to briefly in Section II, and to give a short account of investigations made to ascertain the efficacy of API tests in solving these problems.

B. Test reproducibility

In 1974 the Sub-Committee of the Taxonomy of the Genus *Bacillus* (*Bacillus* Sub-Committee or BSC) of the International Committee on Systematic Bacteriology of the International Association of Microbiological Societies prepared a list of standard methods for performing the tests classically used in the taxonomy and identification of members of the genus *Bacillus*; these were largely drawn from the monograph of Gordon (1973). In 1975 an international reproducibility trial of these tests was initiated by the BSC. Few

---

*API Laboratory Products Ltd, Basingstoke, Hampshire, UK.

TABLE IV

A percentage positive results matrix for *Bacillus* species

Species	*B. cereus*	*B. mycoides*	*B. thuringiensis*	*B. cereus* (emetic)[a]	*B. anthracis*	*B. firmus*	*B. lentus*	*B. laterosporus*	*B. alvei*	"*B. thiaminolyticus*"	"*B. psychrosaccharolyticus*"	*B. insolitus*	"*B. carotarum*"	*B. badius*	*B. fastidiosus*	"*B. freudenreichii*"	*B. brevis*
Strains studied	119	25	55	30	37	44	27	10	12	11	2	2	47	2	5	3	18
Morphological and Supplementary Tests																	
Cell width (μm)[c]	1.4	1.3	1.4	1.4	1.3	0.8	0.8	0.9	0.8	0.7	1.0	0.9	1.1	0.9	1.3	0.9	0.9
Chains of cells	96	100	100	100	100	29	70	[d]	25			50	100	100	100	100	5
Motility	96		100	100		100	100	100	100	100	100	100	42	100	100	100	100
Spores round							15					50					
Spores ellipsoidal	100	100	100	100	100	100	96	100	100	100	100	50	91	100	100	100	100
Spores cylindrical	22		20			29			17			50	26				
Spores central/ paracentral	70	84	6	30		75	52	100	58		100	100	62	50	80	100	81
Spores subterminal	100	100	100	100	100	100	100	90	100	100	100	100	86	100	80	67	100
Spores terminal						3						50	4	50	20		
Sporangia swollen						51	30	100	92		91	100				33	100
Parasporal bodies								100									
Crystalline inclusions			94														
Vacuoles	83	100	100	100	100												
Gas from carbohydrates																	
API 20E Tests																	
ONPG		32				23	89		42	100	100	100	2				22
ADH	60	36	87	17		2									100		
LDC																	
ODC																	
Citrate (Simmons')	86	60	93	100		34	15	50		100		100	36	100	80	33	44
H₂S										82							
Urease	22	20	13				30		50	91		50	21		100	100	
TDA																	
Indole									100	100							
V-P	92	92	98	100	100	84	55	100	92	27		100	91				72
Gelatin	100	100	100	100	70	95	44	100	100	100	50		76	100			33
Nitrate	80	76	92	87	100	50	15	90		18		50	66			100	77

[a] Strains of serotypes 1, 3, 5 and 8 which include strains isolated in connection with outbreaks of emetic type food poisoning.
[b] Groups of Walker and Wolf (1971).
[c] Average cell width.
[d] Where the value is '0' this has been omitted for clarity.

B. pasteurii	B. sphaericus	B. globisporus	"B. psychrophilus"	B. subtilis	"B. amyloliquefaciens"	B. licheniformis	B. pumilus	B. megaterium	B. circulans	B. macerans	B. polymyxa	B. macquariensis	"B. laevolacticus"	"B. racemilacticus"	B. coagulans	B. stearothermophilus 1 [b]	B. stearothermophilus 2	B. stearothermophilus 3	"B. caldolyticus group"	B. pantothenticus
6	54	3	2	131	52	81	63	33	44	15	15	3	5	4	20	38	4	32	3	18
0.7	1.0	0.9	0.9	0.8	0.8	0.8	0.7	1.5	0.8	0.7	0.9	0.6	0.7	0.7	0.8	0.8	0.7	0.9	0.8	0.6
100		33	50	22	84	80	5	97	23	7	13		20	50	35	66		40		83
100	100	100	100	95	100	100	100	97	100	100	100	100	40	100	100	100	100	100	100	100
100	100	100	100					69												78
	20			100	100	100	61	100	100	100	100	100	100	100	100	87	100	100	100	78
				8					20	7	7				5	37		9		
				40	7	49	40	40	29		40					10	75	3		
33	79	100	100	98	100	100	100	100	84	100	100	100			100	100	97	100	100	22
100	74	67	100	15	36	12			45	33		100	100	100	55	39		53	100	100
100	100	100	50	5		2			84	100	100	100	100	100	85	68		100	100	100
								90												
										100	100									
	7	100	100	90	67	100	100	100	100	100	100	100	60	50	70	26			12	78
						95							20							
	85		100	98	92	99	89	85	4											22
100	65	100	100			16			7				20	25						
	11		100	100	100	100	98	97	73	67	93		100	100	100	81	100	69	33	50
	4	67	100	100	100	100	98	100	38	40	93		20		30	95	100	100	100	94
100	14		100	95	88	96		15	13	47	100		20			60	25	16		50

Continued overleaf

TABLE IV—*continued*

Species	B. cereus	B. mycoides	B. thuringiensis	B. cereus (emetic)[d]	B. anthracis	B. firmus	B. lentus	B. laterosporus	B. alvei	"B. thiaminolyticus"	"B. psychrosaccharolyticus"	B. insolitus	"B. carotarum"	B. badius	B. fastidiosus	"B. freudenreichii"	B. brevis
Strains studied	119	25	55	30	37	44	27	10	12	11	2	2	47	2	5	3	18
API 50CHB Tests																	
Glycerol	92	96	92	70		98	55	100	100	100	100	100	55				50
Erythritol																	
D-Arabinose											100		4				
L-Arabinose						4	70				100	100	57				
Ribose	97	76	98	93	100	34	89	100	100	100	100	100	68				16
D-Xylose						4	18				100	100	15				
L-Xylose																	
Adonitol									100	9							
β-Methyl-xyloside							33						50				
Galactose	6	32		6		11	70		58	100			50				8
D-Glucose	100	100	100	100	100	100	100	100	100	100	100	100	98				11
D-Fructose	98	84	100	100	59	52	100	100		100	100	100	98				22
D-Mannose		8	41	3		23	96	100	42	100	100	100					
L-Sorbose																	
Rhamnose							55						50				19
Dulcitol							15										
Inositol	4	3					22		75	91	100	50	38				11
Mannitol						84	96	100			100	100	70				11
Sorbitol						7	52				50	50	21				
a-Methyl-D-mannoside						2	15			91			50				
a-Methyl-D-glucoside	2	12	4		3	2	52		67	100	100	100					
N-Acetyl glucosamine	99	100	100	100	100	88	92	100	100	100	100	100	47				
Amygdalin	8	24	2				70	70	50	100	100	100	4				
Arbutin	91	84	100	60	32	13	81	100	67	100	100	100	13				
Aesculin	100	100	98	100	97	73	100	100	100	100	100	100	34				16
Salicin	87	80	83			13	78	100	42	100	100	100	19				
Cellobiose	84	60	72	43		7	81	90	50	91	100	100	81				
Maltose	98	100	100	100	100	100	100	100	58	100	100	100	62				5
Lactose	8	8				2	67		25	91	100	50	2				4
Melibiose						2	78		50	100			50				4
Sucrose	47	64	55	83	100	91	100		75	100	100	100	68				
Trehalose	98	92	100	100	100	75	89	100	58	100	100	100	89				
Inulin						7	11						50				40
Melezitose	1		4			2	48		17	100			50				2
D-Raffinose			1			7	89		50	100	100	100	50				42
Starch	96	100	94	6	97	91	100	50	100	100	100	50	36				5
Glycogen	92	100	92	10	92	54	70		67	91	100	50	36				5
Xylitol				3			11										5
β-Gentiobiose	18	3	4			7	63	90	83	100	100	100	15				
D-Turanose	15	24	11	6	3	13	63		67	100	100	100	2				
D-Lyxose							7						50				
D-Tagatose						2	15						50				
D-Fucose																	
L-Fucose			2	3			4		8	100			17				
D-Arabitol						4	15			9			50				
L-Arabitol																	
Gluconate	29	12	6	40	3		26		8	100	50						
2 Keto-gluconate																	
5 Keto-gluconate							4		17	100							

[a] Strains of serotypes 1, 3, 5 and 8 which include strains isolated in connection with outbreaks of emetic type food poisoning.

[b] Groups of Walker and Wolf (1971).

[c] Average cell width.

[d] Where the value is '0' this has been omitted for clarity.

B. pasteurii	B. sphaericus	B. globisporus	"B psy-'trophilus"	B. subtilis	"B amyloliquefaciens"	B. licheniformis	B. pumilus	B. megaterium	B. circulans	B. macerans	B. polymyxa	B. macquariensis	"B laevolacticus"	"B. racemilacticus"	B. coagulans	B. stearothermophilus 1[b]	B. stearothermophilus 2	B. stearothermophilus 3	"B. caldolyticus group"	B. pantothenticus
6	54	3	2	131	52	81	63	33	44	15	15	3	5	4	20	38	4	32	3	18
				97	96	100	98	100	86	100	100		80	100	100	100	100	97	100	100
									23	93	7			25	10					83
				98	82	100	98	97	91	100	100				100	75	71	25	9	
				99	100	100	100	100	98	100	100	100		20	100	80	89	25	44	100
	7			89	71	99	98	91	98	100	100	100			100	70	66	75		
																3	50			
									93	100	100	100				3				
			100	30	44	100	100	97	100	100	100	100	100	100	95	60	50	84	100	100
			100	100	100	100	100	100	100	100	100	100	100	100	100	100	100	100	100	100
			100	100	100	100	100	100	100	100	100	100	100	100	100	100	100	100	100	100
				94	63	99	97	15	98	100	100	33	100	100	95	89	75	94	100	100
				2		7	5						100	75	5	16	25	62		
				2		83	5	3	45	100	33		60	50	50	8				100
									2								75			
				95	86	76	28	100	41	40						40	26	75		28
				96	100	90	100	100	88	100	100	100	100	100	40	79	100		100	50
				88	88	95	17	54	25	73			100	100	35	21	100	9		78
							92	6	18	6	40	67	80	100	20			44		
	54	33	100	99	98	100	55	60	95	100	100	100	100	100	85	60	100	87	100	100
	54	33	100	22	19	89	100	97	79	20	20	100	100	100	95	47		6		100
				99	100	100	97	100	98	100	100	100	100	100	70	18	100	16	67	100
				100	100	100	100	100	98	100	100	100	100	100	65	55	100	3	67	100
	2	33		100	100	100	100	100	100	100	100	100		75	95	76	100	22	100	100
				100	100	100	100	100	100	100	100	100	100	100	65	71	100	34	100	100
				100	100	100	100	98	100	100	100	100	100	100	70	63	100	37	100	94
	4			100	98	100	46	100	100	100	100	100	100	100	100	100	100	93	100	100
				49	84	89	76	100	100	100	100	100	100	100	85	13	25	50		50
				80	56	45	23	100	100	100	100	100	100	100	90	58		87	100	5
				100	100	100	100	100	98	100	100	67	100	100	85	97	100	100	100	89
	4			100	82	100	100	100	98	100	100	100	100	100	100	97	100	90	100	100
				83	11	68	1	85	70	100	87		100	75	5					
						1		69	52	100	40		100	75	5	81	25	100	100	
				90	92	79	89	100	98	100	100	100	100	100	95	60	25	97	100	
	4			98	98	99		100	95	100	100	100	100	100	95	89		100	100	100
				98	94	96		100	93	100	100	100	60	50	35	47		90	100	17
						1		30	4							3	100			
				96	88	89	100	97	98	100	100	100		25	70	37	100	56	100	55
				97	79	100	54	100	100	100	100	100	7	60	75	95	95	100	100	100
			100			96	65	3	2				100	50	45	79	75	87	100	100
								3						25						
								3	43	87							100			78
								60	7	80	7			25	55		100			
											7					3				
				4		2		15	61	93	80	67		50	75					72
									2				20	30						
				2				6	79	93	53		60	50	55	89	100	100		

Reproduced, with permission, from Logan and Berkeley, 1984.

tests were found to give really consistent results even in the hands of experts on the genus (Logan, 1980; Logan and Berkeley, 1981).

In the light of this trial it was decided to examine tests in the API system for applicability to *Bacillus* taxonomy and to subject any promising materials to inter- and intra-laboratory reproducibility trials. The inter-laboratory trial was carried out on an international scale as before but some participants were inexperienced with *Bacillus* species or unfamiliar with API materials. It was found that, after problems highlighted in the inter-laboratory trial were largely eliminated in the intra-laboratory trial, the API tests gave more reproducible results than did the classical tests (Logan *et al.*, 1978; Logan, 1980; Logan and Berkeley, 1981).

C. Advantages and disadvantages of classical and miniaturized methods

The difficulties and neglect of *Bacillus* identification already referred to are partially explained by some of the disadvantages of the classical tests; they require special media which are time consuming, expensive and inconvenient to prepare, and some of these media have short shelf lives so that considerable wastage may occur when use is infrequent. Several tests take 14 days or more from pure culture to final reading and many people are unwilling to wait this long. The difficulty of media standardization, a problem which is, of course, not confined to *Bacillus* tests, means that inconsistent, and therefore misleading, results may be obtained (Logan and Berkeley, 1981).

The API tests were investigated because they offered the possibility of overcoming the problems of media preparation and standardization, shelf life and test duration. The quality control measures taken during the manufacture of API materials result in highly standardized test media; this should significantly improve test consistency (Janin, 1976). In addition, a shelf life of 18 months, rapidity (results 48 h after obtaining a pure culture), applicability to a wide range of organisms and comparative cheapness (Cox *et al.*, 1976; Miller and Lu, 1976; Robertson *et al.*, 1976) make such materials very attractive, especially to the occasional user.

As intimated earlier (Section II), identification schemes using classical *Bacillus* tests, and the dichotomous scheme of Gordon and colleagues (1973) or the tables of Cowan and Steel (1965, 1974) do not allow the identification of all of the species included in the Approved List of Bacterial Names (Skerman *et al.*, 1980) and are usually unable to facilitate the identification of atypical or intermediate strains. The API test based system described here enables the identification of 38 taxa of which 31 probably merit species status; 26 of the 31 species included in the Approved List of Bacterial Names can be identified by the system and the remaining five, which have special media requirements, can be easily identified by other methods. The system also enables the identification of atypical and intermediate strains by presenting the results of many tests as a table of percentage positive reactions; such recognition of, and allowance for, strains difficult to assign to a taxon should encourage the confidence of users of the system.

D. Morphology

Colony morphology and other such cultural characteristics are of limited use for the classification and identification of *Bacillus* species (Gibson, 1944; Wolf and Barker, 1968; Bonde, 1975), though to the experienced worker they may be valuable. For experienced and inexperienced workers alike, observations on the morphology of vegetative cells and sporangia (Wolf and Barker, 1968; Gordon *et al.*, 1973; Wolf and Sharp, 1981) are more useful.

Whatever the identification system employed, it must be confirmed that any strain suspected of being a *Bacillus* species is indeed an aerobic endospore-forming rod. This is most conveniently done by the examination of smears of live cultures, young and sporulated, by phase contrast microscopy; features observed at this time may usefully be taken into account in the subsequent identification. In the identification scheme described below the information generated by the API tests is supplemented with details of cell width, arrangement and morphology and sporangial morphology. Some of these characters gave inconsistent results in the BSC reproducibility trial and it was believed that stricter definitions of descriptive terms would improve consistency (Logan, 1980; Logan and Berkeley, 1981; Sneath and Collins, 1974).

E. Taxonomy

To confirm the usefulness of the 139 tests originally chosen [119 API tests, comprising 12 API 20E, 49 API 50E and 58 enzyme tests, 39 of which are not available commercially (Logan and Berkeley, 1981) and 20 morphological and miscellaneous tests], it was necessary to examine a large number of strains.

The results of a computer taxonomy of 600 *Bacillus* strains using all 139 tests were described by Logan and Berkeley (1981); further work (which was published earlier), in which the tests of greatest reproducibility and discriminatory value were used for a taxonomy of the same strains, was described by Logan (1980). In later work (Logan and Berkeley, 1984), 1075 strains (including the original 600) were characterized using the API 50CH gallery instead of the API 50E. A full taxonomy based upon these data awaits completion, but the tests in the API 50CH gallery are, in all but 11 cases, equivalent to those in the API 50E and the species pattern of results shown by the 11 new tests support the taxa defined previously.

The results of the taxonomies based upon API tests were found to be largely in agreement with the results of studies based upon a wide range of different characterization and analytical methods (Berkeley and Goodfellow, 1981); in addition some areas of confusion in the taxonomy of the genus have been elucidated (Logan, 1980; Logan and Berkeley, 1981) and the existence of new groups revealed (Logan *et al.*, 1979; Logan, 1980).

The taxonomy described by Logan (1980), based as it was upon reproducible tests of high separation values, provided a strong foundation for a diagnostic system.

F. Identification

1. *General considerations*

An identification system intended for routine use should satisfy several requirements, namely accuracy, rapidity, availability, convenience and cheapness. Rapidity, convenience and cheapness are greatly affected by the number of tests used; a reduction in the number of tests, from the 80 used by Logan (1980), was considered desirable for these reasons. It had to be remembered, however, that when the number of tests employed falls there is a concomitant loss in the accuracy of the identification system.

A problem with most commercially available identification kits is that individual tests cannot be purchased separately. As a result, although identification may be possible using just a few of the API 20E and API 50CH tests, the results of all the tests might as well be recorded as there is no saving in money, and very little in time, if these less useful tests are omitted.

The scheme described here, therefore, employs a large number of tests; it must be borne in mind, however, that the genus *Bacillus* is very wide [in terms of DNA homology the equivalent of some bacterial families (Priest, 1981)], and that the system recognizes 38 taxa. Logan (1980) found that a scheme using only six API 20E tests, 18 API 50E tests and 11 morphological tests, 35 tests in all, gave a misidentification rate of 15%; this unacceptably high figure was largely owing to the fact that the data base of 100 strains was too small.

This raises two important points: firstly, as Smith *et al.* (1952) observed, a taxonomist is naturally limited by his collection (an obvious consequence is that any identification system developed by the taxonomist will be likewise limited) and second, as Bonde (1975) pointed out, an identification system based too heavily upon strains from collections, and not giving adequate attention to new isolates, will often leave many fresh isolates unidentified. Both of these problems were experienced by Willemse-Collinet *et al.* (1980).

2. *Media and cultivation*

The majority of strains grew satisfactorily on nutrient agar (Difco); such strains used in the study were maintained on nutrient agar containing 5 mg l^{-1} $MnSO_4.4H_2O$ (maintenance medium). The manganese sulphate enhances sporulation (DSM, 1977) and strains were allowed to sporulate prior to being stored at 4°C in the dark.

For morphological studies strains were grown on the appropriate maintenance medium at 30°C for 48 h (or 20°C for 96 h for psychrophilic strains, 37°C for 48 h for *B. coagulans* and 55°C for 24 h for thermophilic strains). For API tests strains were grown overnight (or, for psychrophiles, 48 h) on plates of nutrient agar at the appropriate temperature. For slow growing organisms, such as *B. lentus*, two plates were prepared.

For strains not growing satisfactorily on nutrient agar or maintenance medium, special media had to be used:

Allantoin Mineral medium (DSM, 1977) for *B. fastidiosus* contained (g l^{-1}): K_2HPO_4, 0.8; KH_2PO_4, 0.2; $MgSO_4.7H_2O$, 0.5; $CaCl_2.2H_2O$, 0.05; $FeSO_4.7H_2O$, 0.01; $MnSO_4.4H_2O$, 0.001 (increased to 0.005 for maintenance); allantoin (Sigma) 20; agar, 15; at pH 6.8.

Bacillus pasteurii medium (Gibson and Gordon, 1974): to nutrient agar or maintenance medium 10 g l^{-1} NH_4Cl was added and the reaction adjusted to pH 9 prior to autoclaving.

"Bacillus racemilacticus" medium (DSM, 1977) for *"B. laevolacticus"* and *"B. racemilacticus"* contained (g l^{-1}): glucose, 5; peptone, 5; yeast extract, 5; $CaCO_3$, 5; agar, 15; at pH 6.8. The maintenance form of the medium contained 5 mg l^{-1} $MnSO_4.4H_2O$.

Strains growing only on such special media are unlikely to be isolated fortuitously and their growth requirements give some idea of their identities; furthermore in the cases of *B. fastidiosus* and *B. pasteurii* the results of tests in the API 50CH gallery are likely to be of little value. In such circumstances the API, morphological and supplementary tests are used merely to confirm or challenge the provisional identification.

3. *Morphological and supplementary tests*

Observations on vegetative cell morphology were made on cultures grown overnight or longer, if necessary to obtain visible growth, on a suitable maintenance medium at the appropriate temperature. Using phase contrast microscopy at $1000\times$ magnification, organisms were examined for cell shape, chains of cells and a foamy or vacuolate appearance of the cytoplasm (enhanced by growing on 1% glucose nutrient agar). Cell widths were measured using a Vickers-AEI Image Splitting Eyepiece which offers greater accuracy than an eyepiece graticule (Quesnel, 1971).

For determination of motility, strains were grown on slopes of nutrient agar and after 6 h, or as soon as growth was visible, a loopful of the water of condensation at the base of the slope was examined by phase contrast microscopy at 1000 × magnification. Motility was frequently observed in slides prepared for vegetative cell morphology studies and in such cases special motility examinations were omitted. Spores were sometimes observed in slides prepared from overnight cultures on maintenance media but many strains required incubation for two days or more before sporangial morphology could be observed. Slides were examined for the shapes of spores, their positions in the sporangia, distension of sporangia by mature spores and presence of parasporal bodies and crystals. Categories of spore shape were round, ellipsoidal and cylindrical, oval spores were scored as round and ellipsoidal, bean or kidney shapes as ellipsoidal, and banana shapes as cylindrical; intermediate cases were scored for both of the categories that they lay between. Categories of spore position were terminal, subterminal and central or paracentral; several different positions might be observed in one culture. Sporangial swelling was scored positive only if the distension was appreciable.

The determination of gas production from carbohydrates is described in Section III.F.4.

4. *API tests*

Strains were grown as described above and cells were harvested in 2 ml of sterile normal saline and the dense suspension so produced was used to prepare two further suspensions: (i) in 4 ml of sterile normal saline to correspond to tube No. 3 of the McFarland (1907) series of standard opacities (approximately equivalent to 9×10^8 organisms ml^{-1}) and (ii) in 10 ml of API 50CH E/B medium which contained (g l^{-1}) $(NH_4)_2 SO_4$, 2.0; yeast extract, 0.5; tryptone 1.0; phenol red, 0.18; with mineral base of Cohen-Bazire *et al.* (1957) 10 ml; in a phosphate buffer of pH 7.5 after autoclaving, to correspond to tube No. 3 of the McFarland Scale.

The API test strips were placed in moistened incubation chambers and then inoculated with the appropriate suspensions. Using the 4 ml of saline suspension, the first 12 tests of the API 20E strip were inoculated (the last eight test substrates being carbohydrates which were duplicated by the API 50CHB and had unsuitable indicator in the API 20E); the tubes and cupules of the citrate, Voges–Proskauer, and gelatine liquefaction tests were filled, but in all of the other tests only the tubes were filled with suspension. The tubes of the arginine dihydrolase (ADH), lysine decarboxylase (LDC), ornithine decarboxylase (ODC) and urease tests were sealed by completely filling the cupules with sterile mineral oil. All the tubes of the API 50CHB were filled with the suspension in the API 50CH E/B medium. Any air bubbles forming in tubes during inoculation were pulled out using a sterile loop.

Strips were incubated at 30°C or 37°C for 48 h and read at 24 h and 48 h (20°C for 96 h, reading at 48 h and 96 h for psychrophiles; and 55°C for 24 h reading at 12 h (approximately) and at 24 h for thermophiles). The API 50CHB strips were tilted, bases of tubes uppermost, at about 5° in order to trap any gas evolved.

Results were scored according to the manufacturer's instructions. A test scoring positive at either reading time was considered positive; occasionally, tests appeared positive at the first reading but reverted to negative by the final reading. In the API 50CHB this was due to the production of large quantities of alkali (there being tryptone in the medium) which masked the acid production; it was frequently observed when members of the *B. subtilis* group were tested. The tubes of the API 50CHB strips were examined for gas bubbles.

Studies on strains of *B. anthracis* were carried out in safety cabinets at CAMR, Porton Down, Salisbury, and prophylactic measures were taken beforehand.

5. *Data base*

The results of the tests for the 1075 *Bacillus* strains are expressed as percentage positive results for each species or group of strains in Table IV.

For several species, few representatives were available for study; where less than ten strains were characterized the results presented in Table IV are of limited value for identification. In 25 species or groups, however, large numbers of strains from a wide range of sources were available and the results shown in Table IV give an adequate indication of within-species variation for each test (Sneath, 1978; Gordon, 1981).

A percentage positive results matrix is of greater value than a dichotomous key, or a table giving results as "positive", "negative" and "variable", because not only may it be used as it is for diagnosis (with the indication of within-species variation making the identification of atypical strains possible), it may also be employed as a computer data base for identification by the "taxon-radius" model (Sneath and Sokal, 1973; Gyllenberg and Niemelä, 1975; Sneath, 1978) providing that each species is represented by at least ten strains (Sneath, 1978). An alternative basis is that of Lapage *et al.* (1973) (Section III.G). Another advantage is the ease with which such a matrix or data base is updated as information on further strains becomes available.

The scheme described here, and the data base presented, facilitate the identification of a *Bacillus* strain to species level within 48 h of obtaining a pure culture, in the case of a mesophilic strain; the identification of a psychrophilic strain may take between 48 h and 96 h but that of a thermophile only 18–24 h. With experience, a presumptive identification of a mesophile may be made in 24 h.

It is anticipated that the increasing interest in *Bacillus* identification in the medical, veterinary and industrial fields will be encouraged by the availability of this rapid and convenient diagnostic scheme.

G. Microcomputer-assisted identification

Data derived from results for a new isolate have to be compared with the matrix of results (Table IV). This can be done by hand and eye but a measure of the degree of matching with the values in the matrix cannot rapidly be obtained. A computerized system though, is fast, accurate and can supply statistics on the degree of matching and highlight any unexpected results. A mathematical basis for such a system exists (Lapage *et al.*, 1973). Other methods have been reviewed by Willcox *et al.* (1980). The advent of cheap microcomputers has permitted the use of such programs without the recourse to large machines or the skills of computer scientists.

A program to process results of API tests has been written using Fortran and to run on a NorthStar Advantage under the CP/M operating system. The program is interactive and prompts the user to input the morphological and supplementary characters and the results of the tests at 24 h and 48 h of incubation. An identification score (Lapage *et al.*, 1973) is set above which an identification is accepted as reliable. The identity of the most probable species is displayed on the screen with its corresponding identification score and any test differences between the isolate and the most probable species. If the program fails to identify the isolate, that is there is not a high enough degree of correspondence between the isolate test results and those of any of the groups in the results matrix, then the names of those groups with the highest identification score are displayed, again with test differences. There is the option for a printed version of the results to be output at the same time as the screen display.

IV. Recent developments in instrumented approaches to identification

A. Introduction

Classical methods of identification of *Bacillus* species can be time-consuming and difficult to standardize. Some tests can take up to 14 days. Miniaturized methods such as the API system have the advantages of being quicker and more standardized, but methods allowing earlier identifications, using reproducible techniques, are constantly being investigated.

Rapid identification techniques which are amenable to automation, such as those combining pyrolysis with the analytical techniques of gas-liquid chromatography (Py-GC) or mass spectrometry (Py-MS) are likely to fulfil the requirements of a rapid identification system for *Bacillus* species. Pyrolysis is the thermal degradation of matter in an inert atmosphere to produce a series of volatile lower molecular weight substances characteristic of the original material. These volatile fragments can then be analysed by gas-liquid chromatography or mass spectrometry, both of which require volatile samples for analysis.

The complex analogue trace or pyrogram produced by Py-GC and the mass spectra produced by Py-MS can be examined using multivariate statistical procedures to determine the relationships between and within groups of bacteria. They can also be treated as chemical profiles or "fingerprints" and used for identification purposes, by matching the profile against those held in a data base of known strains.

The first report of Py-MS of biological materials was that of Zemany (1952), who showed that complex materials are degraded reproducibly if the pyrolysis parameters are standardized. The first report of Py-GC was that of Davison *et al.* (1954). At first, little impact was made by these reports, but interest in these techniques was increased after they were used by workers in the US space exploration programme (Wilson *et al.*, 1962; Oyama, 1963). Since then, much work has been done on the identification of micro-organisms using both Py-GC and Py-MS. For reviews of this subject, see Irwin and Slack (1978), Gutteridge and Norris (1979) and Meuzelaar and colleagues (1982).

B. Pyrolysis

Pyrolysis causes substances to rupture at their weakest points, producing characteristic fragments called the pyrolysate (Drucker, 1976). The pyrolysate fragments can undergo secondary reactions which are not completely reproducible and are thus undesirable. It is necessary to minimize and control these reactions by standardizing the parameters that affect them. If this is not done, the final pyrolysis products may vary between successive analyses.

Parameters that affect secondary reactions are mainly those concerned with the transfer of the pyrolysate from the pyrolysis zone through the gas chromatograph or into the mass spectrometer. In Py-GC the pyrolysate is carried through the column using a carrier gas to regulate the flow rate and the column is kept at an optimum temperature. In Py-MS, most inlet systems have inert walls, as do the expansion chambers, and are heated to a temperature (usually 150°C) which minimizes secondary reactions, but prevents pyrolysate condensation. Another approach is to pyrolyse directly in front of the ionization source (direct probe mass spectrometry) so that no transfer problems are involved (Gutteridge and Puckey, 1982). The temperature/time profile of the pyrolysis process can also affect the secondary reactions (Levy, 1967) and to minimize them it is advisable to use high pyrolysis temperatures between 500°C and 700°C, with short temperature rise times (Farre-Ruis and Guiochon, 1968).

The design of the pyrolysis unit (pyrolyser) is important as it will affect the pyrolysate produced (Hall and Bennet, 1973). Pyrolysers may be continuous mode units, where the sample is introduced into a pre-heated environment, or pulse mode units, where the sample is rapidly heated. Only pulse mode units have been commonly used in microbiological studies, as it is difficult to control the temperature/time parameters of continuous mode units.

Pulse mode units are of three main types—filament pyrolysers, Curie-point pyrolysers and laser pyrolysers. Laser pyrolysers will be discussed separately later. Filament pyrolysers induce pyrolysis by resistive heating of platinum ribbons or coils (Levy *et al.*, 1972). Many problems have been encountered with these pyrolysers (Gutteridge and Norris, 1979) but they have been used successfully to pyrolyse micro-organisms (Reiner, 1965; Needleman and Stuchberry, 1977).

Curie-point pyrolysers are essentially special types of filament pyrolysers, but in this case the filament is a ferromagnetic wire. When the filament is placed in the work coil of a high frequency oscillator and the current is switched on, eddy currents cause the filament temperature to rise until it reaches its Curie-point. The wire then becomes paramagnetic and no longer induces heating—the temperature falls until it drops below the Curie-point and the wire recommences heating. It thus acts as a thermostat, the equilibrium temperature stabilizing near to the Curie-point of the wire. The Curie-point is dependent on the wire composition (Dyson and Littlewood, 1968), pure iron having a Curie-point of 770°C and pure nickel of 358°C. Common alloys of the two have Curie-points in between these values. Curie-point pyrolysis has been used in most pyrolysis studies to date.

C. Pyrolysis gas–liquid chromatography

Pyrolysis gas-liquid chromatography (Py-GC) separates mixtures of volatile materials using a gaseous moving phase and a liquid stationary phase. Py-GC systems are of two sorts—high resolution and low resolution systems. Low resolution systems are most commonly used and consist of tubes packed with a granular porous support material coated with a liquid stationary phase. Two columns are used, one for sample analysis and one as a control, to show how much increased current is due to loss of phase (bleed) from the column.

High resolution systems are not often used, but consist of capillary columns, of which there are two types—wall coated open tubular (WCOT) and support coated open tubular (SCOT) columns. In WCOT columns the capillary inner wall is coated directly with the liquid phase—these are used least often. In SCOT columns the inner wall is packed with porous material which is in turn coated with the liquid phase. These columns are used singly, as there is negligible bleed from this type of column.

The aim of the system is to separate the individual components of the pyrolysate and to resolve them into individual peaks on the pyrogram, which is the end-product of the analysis. Separation is due to varying retentions of the pyrolysate fragments by the liquid phase, which is described in terms of polarity. In general, the greater the polarity of the liquid phase, the greater the retention of a polar solute relative to a non-polar solute with a similar boiling point.

Most studies have used Carbowax 20M as the liquid phase, which is polar and thermostable up to 200°C. Some work has been done with other solutes, especially where higher temperatures are needed. Reiner and Kubica (1969), for example, used Carbowax 20M-TPA which is thermostable up to 250°C. French *et al.* (1981) reported a study on the use of solid stationary phases as an alternative to Carbowax 20M, which has problems associated with it of poor baseline resolution, rapid column deterioration, long analysis times and difficulty in reproducing pyrograms from control organisms on new columns. Using these phases they were able to examine four to six samples an hour and found they could reproduce pyrograms on two separate columns. The best discrimination of the bacteria examined was achieved using Chromosorb 104, followed by Chromosorb 101 and Tenax-GC. Thus the use of different stationary phases may well overcome some of the technical problems that have long been associated with Py-GC.

D. Standardization problems peculiar to Py-GC

Although Py-GC has been used successfully to characterize micro-organisms (Reiner *et al.*, 1972; Oxborrow *et al.*, 1977a; Stack *et al.*, 1978) there are problems associated with setting up this technique as a widespread identification method. These are mainly concerned with the reproducibility and standardization of the technique.

Reproducibility may be obtained in the same laboratory using the same column in successive analyses (Meuzelaar *et al.*, 1975) but degeneration of the chromatographic column affects reproducibility (Quinn, 1974; Needleman and Stuchberry, 1977). Replacing the column often gives results that are markedly different, due to difficulties in standardizing column performance (Sekhon and Carmichael, 1973). Some workers have managed to renew columns and retain reproducibility (Haddadin *et al.*, 1973; Stack *et al.*, 1977) but inter-laboratory reproducibility is still poor, although results can be compared by assessing them relative to certain reference strains of bacteria or chemical standards. Another, smaller problem associated with Py-GC is that in most systems the pyrogram baseline has to be fixed by eye and is thus subjective. In Py-MS a fixed baseline is recorded.

As a result of these problems with Py-GC, although it has been used to a limited extent with *Bacillus* species, there has been increasing interest in the use of Py-MS as an alternative method which avoids some of these difficulties.

E. Pyrolysis mass spectrometry

Pyrolysis mass spectrometry (Py-MS) separates the pyrolysate fragments on the basis of their mass/charge ratio (m/z), using a mass spectrometer. It was not until 1973 when Meuzelaar and Kistemaker designed a pyrolysis mass spectrometer specifically for fingerprinting complex biological samples such as bacteria that Py-MS became a real alternative to Py-GC. This first purpose-built Py-MS system was basically a Curie-point pyrolyser coupled to a rapid scanning quadrupole mass spectrometer. Initially, a fixed electron energy of 100 eV was used, with resolution up to m/z 50. This instrument produced reproducible results, but a lot of information was being lost due to the high ionization and the low resolution. Meuzelaar *et al.* (1973) added a quadrupole mass filter allowing resolution up to m/z 250 and a lower ionization energy of 14 eV, to overcome this problem. In 1976, a fully automated version of this instrument was developed, which allowed analysis of a sample every 2–4 min. It has been used for discriminating a wide variety of bacterial types (Meuzelaar *et al.*, 1976; Borst *et al.*, 1978; Boon *et al.*, 1980).

Since then, many developments in this field have occurred. Purpose-built Py-MS systems are now commercially available, for example, from VG Gas Analysis, Middlewich, Cheshire, along with the computer software necessary for data analysis, and offer ever-increasing efficiency, reproducibility and computerized facilities. For further information about Py-MS, see Meuzelaar *et al.* (1982). Py-MS possesses many advantages over Py-GC, the main ones being that it is faster, has fewer problems of reproducibility, and so is more likely to enable data exchange between laboratories, and is more amenable to automation and computerization. It does have the disadvantage of being more expensive than the budget of most microbiology laboratories can accommodate, but there is reason to believe that less expensive Py-MS systems will be developed in the near future. With the recent advances in microcomputers it is also likely that the cost of computerized facilities for instrument control and data analysis will also decrease.

F. Analysis of pyrolysis data

The pyrograms or mass spectra that are produced from pyrolysis studies require the application of special data analysis procedures. It is necessary to be able to pick out characteristic peaks or masses and to be able to compare data from separate analyses to assess similarity. For identification purposes, simple matching of data is sufficient, although recently factor analysis of pyrolysis data to determine the identity of the materials which contribute to spectra differences has become an important area.

Many data processing methods have been applied to pyrolysis data, varying from visual comparisons of peak heights to complex computerized multivariate statistical analysis. The large amounts of data generated by automated systems and the increasing emphasis on the need for quantitative rather than qualitative analysis make the visual method inadequate, and computerized data analysis has now become more of a necessity than an aid.

There now appears to be agreement that multivariate data analysis is the best approach to analysis of pyrolysis data, as evidenced by its use in all major recent studies. The principles and use of multivariate data analysis are complex and are discussed fully elsewhere (Gutteridge *et al.*, 1979; MacFie and Gutteridge, 1978; MacFie *et al.*, 1978; MacFie and Gutteridge, 1982).

G. Problems of standardization concerning both Py-GC and Py-MS

If reproducibility is to be maintained, all pyrolysis parameters must be carefully standardized. Apart from standardization of pyrolysis equipment and conditions, it has been found that variations in sample preparation, growth conditions and in some cases the time of harvest of organisms affect reproducibility. These must be standardized as much as possible and any variations in procedure should be carefully recorded.

In sample preparation, it is necessary to prepare specimens as free from medium contamination as possible. Wherever possible solid media are preferable to liquid media, as organisms can be picked off the plates and applied directly to the Curie-point wire using a disposable loop. Alternatively a slurry in water or another solvent such as methanol can be made and then applied to the wire (Meuzelaar *et al.*, 1976). Where organisms tend to grow into the media, they can be grown on membrane filters placed on the plate surface (Oxborrow *et al.*, 1976). Harvests from liquid media require extensive preparatory washing, see, for example, Wickman (1977).

The problems of inter-laboratory and long-term reproducibility have already been discussed with respect to Py-GC (Section IV.D). In the case of Py MS both short- and long-term reproducibility appear to be good, with figures in the area of 94% for short-term and 92% for long-term reproducibility of averaged pyrolysis mass spectra from glycogen and bovine serum albumin (Windig *et al.*, 1979). Inter-laboratory reproducibility between Py-MS systems of the same basic design looks promising and between the same instrument in different laboratories should be even better (Meuzelaar *et al.*, 1982).

H. Use of Py-GC in identification of *Bacillus* species

Py-GC studies on the genus *Bacillus* were carried out by Oxborrow *et al.* (1976, 1977a,b), but their investigations were concerned more with developing standard methods of sample preparation and with investigating the effect of growth time and growth media on pyrogram reproducibility, than with identification of *Bacillus* species as such.

Due to the comparative newness of the method relative to classical methods of identification and the length of time needed to establish a data base of pyrograms, only a few areas of the genus have been studied. Work has been concentrated on areas of *Bacillus* taxonomy which present clearly defined problems (O'Donnell and Norris, 1981).

O'Donnell *et al.* (1980a) conducted a Py-GC study using non-sporulated cultures of *B. subtilis* and the closely related groups of *B. pumilus, B. licheniformis* and *"B. amyloliquefaciens"*. At present *"B. amyloliquefaciens"* is not regarded as a separate species from *B. subtilis*, but they were able to separate the four groups clearly using Py-GC data, strengthening the case for its being afforded separate species status.

They used non-sporulated cultures as the differences between vegetative cells were thought to be greater than those between spored cells. This was upheld by a Py-MS study on the same group of strains (Section IV.I). A study of sporulating cultures of *B. subtilis*, *B. pumilus*, *B. cereus*, *B. thuringiensis* and *B. megaterium* was carried out (O'Donnell and Norris, 1981) as it was thought that organisms such as *B. thuringiensis* which contain a protein crystal when they are sporulating, might be better separated using sporulated cultures.

They were able to separate the groups using this data and also found it easier to standardize preparations made from sporulating cultures. We, however (Section IV.I), found that sporulated cultures of the strains used in the *B. subtilis* study could not be completely discriminated, so it is necessary to investigate the effects of sporulation on pyrolysis data more fully before discounting the use of either sporulated or non-sporulated cultures in pyrolysis studies. It is certainly obvious that in studies with *Bacillus* species, the growth phase of the cultures used should be rigorously standardized and recorded.

O'Donnell *et al.* (1980b) also investigated the relationships between *B. cereus*, *B. thuringiensis* and *B. mycoides*. They were able to separate them into three distinct groups and found that *B. mycoides* more closely resembled *B. cereus* than did *B. thuringiensis*. *B. cereus* and *B. cereus* food poisoning organisms were also distinguishable (O'Donnell and Norris, 1981) and could be separated from *B. anthracis*. Some of the separation of *B. anthracis* may have been due to the fact that these samples had been autoclaved before analysis. Ten *B. cereus* strains isolated in connection with outbreaks of food poisoning could be further discriminated to produce two groups which corresponded well to the diarrhoeal and vomiting types of food poisoning.

Stern (1981) conducted a study of five *Bacillus* species and was able to classify them with 96% accuracy. The main aim of this study was to examine the possibility of using Py-GC to discriminate *B. cereus* from other *Bacillus* species. Using the statistical technique of stepwise discriminant analysis on the Py-GC data he was able to discriminate it with 100% efficiency.

Thus Py-GC has been used successfully to discriminate between groups of *Bacillus* strains and to investigate their relationships with each other. Rapid identification of *Bacillus* species using Py-GC is a possibility for the future, although it is more likely that Py-MS will fulfil this role.

I. Use of Py-MS in identification of *Bacillus* species

Until recently very little investigation of the genus had been carried out using Py-MS. Most Py-MS studies to date have concentrated on parameters affecting reproducibility and have used organisms that are convenient rather than taxonomically interesting. The major area of bacterial taxonomy that has been investigated with Py-MS is that of the mycobacteria (Wieten *et al.*, 1981, 1982).

Some of the parameter studies have used *Bacillus* species, for example, Boon *et al.* (1980) investigated Py-MS of whole cells, cell walls and cell wall polymers of *"B. subtilis* var. *niger"* WM and changes induced in the mass spectra produced when the cells were grown on different media.

Recent work in this laboratory has included a Py-MS study of *B. subtilis* and the related groups of *B. pumilus, B. licheniformis* and *B. "amyloliquefaciens"*. The same strains that were used in the Py-GC study of O'Donnell *et al.* (1980a) were examined, using both sporulated and non-sporulated cultures. Good separation of all four groups was achieved using data from non-sporulated cultures, but they could not be fully discriminated using data from sporulated cultures. *B. licheniformis*, however, was best discriminated from the other three groups using this data.

A small data base of these *Bacillus* strains was constructed, and when challenged with eight "unknowns", the unknowns were correctly identified using computerized matching procedures. Thus Py-MS identification of *Bacillus* species is a real possibility.

At present, the authors know of Py-MS studies being undertaken in three other areas of the genus—the alkalophilic strains of *B. firmus* and *B. lentus*; thermophilic *Bacillus* strains and on the *B. cereus, B. mycoides, B. megaterium* and *B. thuringiensis* groups. It is hoped that the results of these studies will soon be available and will lead to further work in this promising field.

J. Laser pyrolysis

Laser pyrolysis of micro-organisms allows single organisms to be pyrolysed and then analysed using Py-MS or Py-GC. This reduces the time lapse from receipt of the sample to analysis (as cultures do not have to be set up) and also completely standardizes the sample size as long as the cells chosen for analysis are representative. Due to the unavailability of laser equipment, however, laser pyrolysis has not been used in many studies as yet.

Recently work has been done by Böhm (1981) using a Laser-Microprobe-Mass-Analyser (LAMMA) which is basically a laser microscope coupled to a time of flight mass spectrometer. Using this equipment he pyrolysed single cells of *B. cereus* and *B. anthracis* and produced characteristic spectra that allowed separation of the two groups. This was only a pilot study, but the results are promising and laser pyrolysis may well be used in the future.

K. Flow cytometry

Given supplies of specific antisera an extremely rapid and sensitive new method for detecting (identifying) micro-organisms is flow cytometry. Initial work has used fluorescent antibody preparations to study *Bacillus* spores and vegetative cells (Phillips and Martin, 1983). In many instances, because of the formation of long chains by certain *Bacillus* species, interpretation of the results for vegetative cells will be difficult. For this reason, in these instances, this technique may only be easily applicable to spores. Such a limitation may not be a disadvantage if a material to be examined contains sporulated material rather than vegetative cells; more rapid identification will be possible than if any other approach, except possibly laser pyrolysis mass spectrometry, was used.

L. Overview

Thus both Py-GC and Py-MS have great potential value for the identification of *Bacillus* species. To date Py-GC has been used more extensively than Py-MS, but the advantages of Py-MS and the increasing availability of commercial equipment are likely to cause a reversal of this situation. Although it takes a long time to establish pyrolysis data bases from bacterial strains, it is likely that rapid identification of *Bacillus* species using Py-MS will be achievable in the near future. Looking further ahead, we may well see MS coupled to laser pyrolysis, decreasing the identification time still further.

V. Summary

Classical methods for identification of *Bacillus* are effective if carried out properly but are slow and the reproducibility of certain tests is not good. Using the API system greater reproducibility and more rapid identification may be achieved. Instrumented methods hold out the promise of obtaining reliable results even faster but none is fully developed yet. Pyrolysis mass spectrometry is the technique of this type most fully advanced and, at the moment, at least, of most promise for general application to the genus

References

Adams, M. H. (1959). "Bacteriophages". Wiley (Interscience), New York.
de Barjac, H. (1978). *Entomophaga* 23, 309–319.
de Barjac, H. (1981). *In* "The Aerobic Endospore-forming Bacteria: Classification and Identification" (R. C. W. Berkeley and M. Goodfellow, Eds), pp. 241–250. Academic Press, London and New York.
de Barjac, H. and Bonnefoi, H. (1962). *Entomophaga* 8, 223–229.
de Barjac, H., Sisman, J. and Cosmao Dumanoir, V. (1974). *C. R. Hebd. Séances Acad. Sci. Ser. D* 279, 1939–1942.

Bekker, J. H. (1944). *Antonie van Leeuwenhoek* **10**, 67–70.

Berkeley, R. C. W. and Goodfellow, M. (1981). "The Aerobic Endospore-forming Bacteria: Classification and Identification". Academic Press, London and New York.

Bingham, A. H. A., Atkinson, A., Scialy, D. and Roberts, R. (1978). *Nucleic Acids Res.* **5**, 3457–3467.

Böhm, R. (1981). *Fresenius Z. Anal. Chem.* **308**, 258–259.

Bonde, G. J. (1975). *Dan. Med. Bull.* **22**, 41–61.

Bonde, G. J. (1981). *In* "The Aerobic Endospore-forming Bacteria: Classification and Identification" (R. C. W. Berkeley and M. Goodfellow, Eds), pp. 181–215. Academic Press, London and New York.

Boon, J. J., de Boer, W. W., Kruysen, F. J. and Wouters, J. T. M. (1980). *J. Gen. Microbiol.* **122**, 119–127.

Borst, J., van der Snee-Enkelaar, A. C. and Meuzelaar, H. L. C. (1978). *Antonie van Leuwenhoek* **44**, 253.

Brown, E. R. and Cherry, W. B. (1955). *J. Infect. Dis.* **96**, 34–39.

Buck, C. A., Anacker, R. C., Newman, F. S. and Eisenstadt, A. (1963). *J. Bacteriol.* **85**, 1423–1430.

Buissière, J. (1972). *C. R. Hebd. Séances Acad. Sci. Ser. D* **274**, 1426–1429.

Cohen-Bazire, G., Sistrom, W. R. and Stanier, R. Y. (1957). *J. Cell. Comp. Physiol.* **49**, 25–68.

Cowan, S. T. and Steel, K. J. (1965). "Identification of Medical Bacteria". Cambridge University Press, Cambridge.

Cowan, S. T. and Steel, K. J. (1974). "Identification of Medical Bacteria", 2nd edn. Cambridge University Press, Cambridge.

Cox, N. A., Mercuri, A. J. and McHan, F. (1976). *In* "Rapid Methods and Automation in Microbiology" (H. H. Johnston and S. W. B. Newsom, Eds), pp. 163–168. Learned Information (Europe), Oxford.

Davies, S. N. (1951). *J. Gen. Microbiol.* **5**, 807–816.

Davison, W. H. T., Slaney, S. and Wragg, A. L. (1954). *Chem. Ind. (London)* 1356.

Defalle, W. (1902). *Ann. Inst. Pasteur Paris* **16**, 756–774.

Delpy, L. P. and Chamsy, H. M. (1949). *C. R. Acad. Sci.* **228**, 1077–1973.

Doak, B. W. and Lammana, C. (1948). *J. Bacteriol.* **55**, 373–380.

Drucker, D. B. (1976). *In* "Methods in Microbiology", Vol. 9. (J. R. Norris, Ed.), pp. 51–125. Academic Press, London and New York.

DSM (Deutsche Sammlung von Mikroorganismen) (1977). "Catalogue of Strains" (D. Claus and C. Schaab-Engels, Eds), p. 276. Gesellschaft für Strahlen und Umweltforschung mbH, Munich.

Dyson, N. and Littlewood, A. B. (1968). *J. Gas Chromatogr.* **6**, 449–454.

Farre-Ruis, F. and Guiochon, G. (1968). *Anal. Chem.* **40**, 998–1000.

Fitzpatrick, D. J., Turnbull, P. C. B., Keane, C. T. and English, L. F. (1979). *Br. J. Surg.* **66**, 577–579.

Földes, J., Meretey, K. and Varga, I. (1961). *Nature (London)* **189**, 161–162.

French, G. L., Phillips, I. and Chinn, S. (1981). *J. Gen. Microbiol.* **125**, 347–355.

Gibson, T. (1944). *J. Dairy Res.* **13**, 248–260.

Gibson, T. and Gordon, R. E. (1974). *In* "Bergey's Manual of Determinative Bacteriology" (R. E. Buchanan and N. E. Gibbons, Eds), 8th edn, pp. 529–550. Williams & Wilkins, Baltimore, Maryland.

Gilbert, R. J. and Parry, J. M. (1977). *J. Hyg.* **78**, 69–74.

Gilbert, R. J. (1979). *In* "Foodborne Infections and Intoxications" (H. Reimann and F. L. Bryan, Eds), 2nd edn, pp. 495–518. Academic Press, New York and London.

Gordon, R. E. (1973). The genus *Bacillus*. *In* "Handbook of Microbiology, Vol. 1: Organismic Microbiology" (A. I. Laskin and H. A. Lechevalier, Eds), pp. 71–88. The Chemical Rubber Co. Press, Cleveland, Ohio.

Gordon, R. E. (1981). *In* "The Aerobic Endospore-forming Bacteria: Classification and Identification" (R. C. W. Berkeley and M. Goodfellow, Eds), pp. 1–15. Academic Press, London and New York.

Gordon, R. E., Haynes, W. C. and Pang, C. H-N. (1973). "The Genus *Bacillus*", Agriculture Handbook No. 427. US Department of Agriculture, Washington D.C.

Gordon, R. E., Hyde, J. L. and Moore, J. A. (1977). *Int. J. Syst. Bacteriol.* **27**, 256–262.

Gorina, L. G., Fluer, F. S., Olovnekov, A. M. and Ezepcuk, Y. V. (1975). *Appl. Microbiol.* **29**, 201–204.

Gutteridge, C. S. and Norris, J. R. (1979). *J. Appl. Bacteriol.* **47**, 5–43.

Gutteridge, C. S. and Puckey, D. J. (1982). *J. Gen. Microbiol.* **128**, 721–730.

Gutteridge, C. S., MacFie, H. J. H. and Norris, J. R. (1979). *J. Anal. Appl. Pyrol.* **1**, 67–76.

Gyllenberg, H. G. and Niemelä, T. K. (1975). *In* "Biological Identification with Computers" (R. J. Pankhurst, Ed.), pp. 121–136. Academic Press, London and New York.

Haddadin, J. M., Stirland, R. M., Preston, M. W. and Collard, P. (1973). *Appl. Microbiol.* **25**, 40–43.

Hall, R. C. and Bennet, G. W. (1973). *J. Chromatogr. Sci.* **11**, 439–443.

Heinen, U. J. and Heinen, W. (1972). *Arch. Microbiol.* **82**, 1–23.

Henriksen, S. D. (1978). *In* "Methods in Microbiology", Vol. 12 (T. Bergan and J. R. Norris, Eds), pp. 1–13. Academic Press, London.

Howie, J. W. and Cruickshank, J. (1940). *J. Pathol. Bacteriol.* **50**, 235–242.

Irwin, W. J. and Slack, J. A. (1978). *Analyst* **103**, 673–704.

Janin, P. R. (1976). *In* "Rapid Methods and Automation in Microbiology" (H. H. Johnston and S. W. B. Newsom, Eds), pp. 155–162. Learned Information (Europe), Oxford.

Kim, H. V. and Goepfert, J. M. (1971). *J. Milk Food Technol.* **34**, 12–15.

Kim, H. V. and Goepfert, J. M. (1972). *Appl. Microbiol.* **24**, 708–713.

Krasil'nikov, N. A. (1949). "Guide to the Bacteria and Actinomycetes". Akademii Nauk SSR, Moscow.

Lamanna, C. (1940a). *J. Infect. Dis.* **67**, 193–204.

Lamanna, C. (1940b). *J. Infect. Dis.* **67**, 205–212.

Lamanna, C. and Eisler, D. (1960). *J. Bacteriol.* **79**, 435–441.

Lapage, S. P., Bascomb, S., Willcox, W. R. and Curtis, M. A. (1973). *J. Gen. Microbiol.* **77**, 273–290.

Levy, R. L. (1967). *J. Gas Chromatogr.* **5**, 107–113.

Levy, R. L., Fanter, D. L. and Wolf, C. J. (1972). *Anal. Chem.* **44**, 48–52.

Logan, N. A. (1980). PhD Thesis, University of Bristol.

Logan, N. A. and Berkeley, R. C. W. (1981). *In* "The Aerobic Endospore-forming Bacteria: Classification and Identification" (R. C. W. Berkeley and M. Goodfellow, Eds), pp. 104–140. Academic Press, London and New York.

Logan, N. A. and Berkeley, R. C. W. (1984). *J. Gen. Microbiol.* **130**, 1871–1882.

Logan, N. A., Berkeley, R. C. W. and Norris, J. R. (1978). *J. Appl. Bacteriol.* **45**, xxviii–xxix.

Logan, N. A., Capel, B. J., Melling, J. and Berkeley, R. C. W. (1979). *FEMS Microbiol. Lett.* **5**, 373–375.

MacDonald, R. E. and MacDonald, S. W. (1962). *Can. J. Microbiol.* **8**, 795–808.

MacFie, H. J. H. and Gutteridge, C. S. (1978). *J. Appl. Bacteriol.* **45**, 4–5.

MacFie, H. J. H. and Gutteridge, C. S. (1982). *J. Anal. Appl. Pyrol.* **4**, 175–204.

MacFie, H. J. H., Gutteridge, C. S. and Norris, J. R. (1978). *J. Gen. Microbiol.* **104**, 67–74.

McCloy, E. W. (1951). *J. Hyg.* **49**, 114–125.

McCloy, E. W. (1958). *J. Gen. Microbiol.* **18**, 198–220.

McFarland, J. (1907). *J. Am. Med. Assoc.* **49**, 1176–1178.

Melling, J., Capel, B. J., Turnbull, P. C. B. and Gilbert, R. J. (1976). *J. Clin. Pathol.* **29**, 938–940.

Meuzelaar, H. L. C. and Kistemaker, P. G. (1973). *Anal. Chem.* **45**, 587–590.

Meuzelaar, H. L. C., Kistemaker, P. G., Posthumous, M. A. and Kistemaker, J. (1973). *Anal. Chem.* **45**, 1546–1549.

Meuzelaar, H. L. C., Ficke, H. G. and Den Harink, H. C. (1975). *J. Chromatogr. Sci.* **13**, 12–17.

Meuzelaar, H. L. C., Kistemaker, P. G., Eshuis, W. and Boerboom, H. A. J. (1976). *In* "Advances in Mass Spectrometry" (N. R. Daley, Ed.), Vol. 7B, pp. 1452–1456. Heyden, London.

Meuzelaar, H. L. C., Haverkamp, J. and Hileman, F. D. (1982). "Pyrolysis Mass Spectrometry of Recent and Fossil Biomaterials. Compendium and Atlas". Elsevier, Amsterdam.

Le Mill, F., de Barjac, H. and Bonnefoi, A. (1969). *Ann. Inst. Pasteur Paris* **116**, 808–819.

Miller, R. E. Jr and Lu, L-P. (1976). *Am. J. Med. Technol.* **42**, 238–242.

Needleman, M. and Stuchberry, P. (1977). *In* "Analytical Pyrolysis" (C. E. R. Jones and C. A. Cramers, Eds), pp. 77–88. Elsevier, Amsterdam.

Noble, A. (1919). *J. Bacteriol.* **14**, 287–300.

Noble, A. (1927). *J. Immunol.* **4**, 105–109.

Norris, J. R. (1961). *J. Gen. Microbiol.* **26**, 167–173.

Norris, J. R. (1962). *J. Gen. Microbiol.* **28**, 393–408.

Norris, J. R. and Wolf, J. (1961). *J. Appl. Bacteriol.* **24**, 42–56.

Norris, J. R., Berkeley, R. C. W., Logan, N. A. and O'Donnell, A. G. (1981). *In* "The Prokaryotes" (M. P. Starr, H. Stolp, H. G. Truper, A. Balows and H. G. Schlegel, Eds), pp. 1711–1742. Springer-Verlag, Berlin and New York.

O'Donnell, A. G. and Norris, J. R. (1981). *In* "The Aerobic Endospore-forming Bacteria" (R. C. W. Berkeley and M. Goodfellow, Eds), pp. 141–179. Academic Press, London and New York.

O'Donnell, A. G., Norris, J. R., Berkeley, R. C. W., Claus, D., Kaneko, T., Logan, N. A. and Nozaki, R. (1980a). *Int. J. Syst. Bacteriol.* **30**, 448–459.

O'Donnell, A. G., MacFie, H. J. H. and Norris, J. R. (1980b). *J. Gen. Microbiol.* **119**, 189–194.

Oxborrow, G. S., Fields, N. D. and Puleo, J. R. (1976). *Appl. Environ. Microbiol.* **32**, 306–309.

Oxborrow, G. S., Fields, N. D. and Puleo, J. R. (1977a). *In* "Analytical Pyrolysis" (C. A. Cramers and C. E. R. Jones, Eds), pp. 69–76. Elsevier, Amsterdam.

Oxborrow, G. S., Fields, N. D. and Puleo, J. R. (1977b). *Appl. Environ. Microbiol.* **33**, 865–870.

Oyama, V. I. (1963). *Nature (London)* **200**, 1058–1059.

Parry, J. M. and Gilbert, R. J. (1980). *J. Hyg.* **84**, 77–82.

Phillips, A. P. and Martin, K. L. (1983). *Cytometry* **4**, 123–121.

Prèvot, A.-R. (1961). "Traité de Systématique Bactérienne", Vol. 2. Dunod, Paris.

Priest, F. G. (1981). *In* "The Aerobic Endospore-forming Bacteria: Classification and

Identification" (R. C. W. Berkeley and M. Goodfellow, Eds), pp. 33–57. Academic Press, London and New York.

Priest, F. G., Goodfellow, M. and Todd, C. (1981). *In* "The Aerobic Endospore-forming Bacteria: Classification and Identification" (R. C. W. Berkeley and M. Goodfellow, Eds), pp. 91–103. Academic Press, London and New York.

Quesnel, L. B. (1971). *In* "Methods in Microbiology" (J. R. Norris and D. W. Ribbons, Eds), Vol. 5A, pp. 1–103. Academic Press, London and New York.

Quinn, P. A. (1974). *J. Chromatogr. Sci.* **12**, 796–806.

Reiner, E. (1965). *Nature (London)* **206**, 1272–1274.

Reiner, E. and Kubica, G. P. (1969). *Am. Rev. Respir. Dis.* **99**, 42–49.

Reiner, E., Hicks, J. J., Ball, M. M. and Martin, W. J. (1972). *Anal. Chem.* **44**, 1058–1061.

Robertson, E. A., Macks, G. C. and MacLowry, J. D. (1976). *J. Clin. Microbiol.* **3**, 421–424.

Sekhon, A. S. and Carmichael, J. W. (1973). *Can. J. Microbiol.* **19**, 409–411.

Sharp, R. J., Brown, K. J. and Atkinson, A. (1979). *J. Gen. Microbiol.* **117**, 201–210.

Simon, M. A., Emerson, S. V., Shaper, J. M., Bernard, P. D. and Glazer, A. N. (1977). *J. Bacteriol.* **130**, 200–204.

Skerman, V. B. D. (1967). "A Guide to the Identification of the Genera of Bacteria", 2nd edn. Williams & Wilkins, Baltimore, Maryland.

Skerman, V. B. D., McGowan, V. and Sneath, P. H. A. (1980). *Int. J. Syst. Bacteriol.* **30**, 225–420.

Smith, N. R., Gordon, R. E. and Clark, F. E. (1946). "Aerobic Mesophilic Sporeforming Bacteria". Publication 559. US Department of Agriculture, Washington D.C.

Smith, N. R., Gordon, R. E. and Clark, F. E. (1952). "Aerobic Sporeforming Bacteria". Monograph No. 16. US Department of Agriculture, Washington D.C.

Sneath, P. H. A. (1978). *In* "Essays in Microbiology" (J. R. Norris and M. H. Richmond, Eds), pp. 9/1–9/13. Wiley, Chichester and New York.

Sneath, P. H. A. and Collins, V. G. (1974). *Antonie van Leeuwenhoek* **40**, 481–527.

Sneath, P. H. A. and Sokal, R. R. (1973). "Numerical Taxonomy". Freeman, San Francisco, California.

Stack, M. V., Donoghue, H. D., Tyler, J. E. and Marshall, M. (1977). *In* "Analytical Pyrolysis" (C. E. R. Jones and C. A. Cramers, Eds), pp. 57–68. Elsevier, Amsterdam.

Stack, M. V., Donoghue, H. D. and Tyler, J. E. (1978). *Appl. Environ. Microbiol.* **35**, 45–50.

Stern, N. J. (1981). *J. Food Sci.* **46**, 1427–1429.

Taylor, A. J. and Gilbert, R. J. (1975). *J. Med. Microbiol.* **8**, 543–550.

Teale, F. H. and Bach, E. (1919). *J. Pathol. Bacteriol.* **23**, 315–332.

Terayama, T., Shengaki, M., Yamada, S., Ushida, H., Igarashi, H., Sakai, S. and Zen Hoji, H. (1978). *Food Hyg. Soc. Jpn.* **19**, 98–104.

Tomcsik, F. and Baumann-Grace, J. B. (1959). *J. Gen. Microbiol.* **21**, 666–675.

Turnbull, P. C. B. (1976). *J. Clin. Pathol.* **29**, 941–948.

Turnbull, P. C. B., Nottingham, J. F. and Ghosh, A. C. (1977). *Br. J. Exp. Pathol.* **58**, 273–280.

Walker, P. D. (1959). PhD Thesis, University of Leeds.

Walker, P. D. and Wolf, J. (1971). *In* "Spore Research 1971" (A. N. Barker, G. W. Gould and J. Wolf, Eds), pp. 247–262. Academic Press, London and New York.

Wickman, K. (1977). *Acta Pathol. Microbiol. Scand. Sect. B* **259**, 49–53.

Wieten, G., Haverkamp, J., Meuzelaar, H. L. C., Engel, H. B. W. and Berwald, L. G. (1981). *J. Gen. Microbiol.* **122**, 109–118.

Wieten, G., Haverkamp, J., Berwald, L. G., Groothuis, D. G. and Draper, P. (1982). *Ann. Microbiol.* **133B**, 15–27.

Willcox, W. R., Lapage, S. P. and Holmes, B. (1980). *Antonie van Leeuwenhoek* **46**, 233–299.

Willemse-Collinet, M. F., Tromp, Th. F. J. and Huizinga, T. (1980). *J. Appl. Bacteriol.* **49**, 385–393.

Wilson, M. E., Oyama, I. V. and Varigo, S. P. (1962). "Proc. 3rd Int. Symp. Gas Chromatr.", pp. 329–338. Academic Press, New York and London.

Windig, W., Kistemaker, P. G., Haverkamp, J. and Meuzelaar, H. L. C. (1979). *J. Anal. Appl. Pyrol.* **1**, 39–52.

Wolf, J. and Barker, A. N. (1968). *In* "Identification Methods for Microbiologists" (B. M. Gibbs and D. A. Shapton, Eds), Part B, pp. 93–109. Academic Press, London and New York.

Wolf, J. and Sharp, R. J. (1981). *In* "The Aerobic Endospore-forming Bacteria: Classification and Identification" (R. C. W. Berkeley and M. Goodfellow, Eds), pp. 251–296. Academic Press, London and New York.

Zemany, P. D. (1952). *Anal. Chem.* **24**, 1709–1713.

13

Automated Identification of Bacteria: An Overview and Examples

H. GYLLENBERG

Department of Microbiology, School of Agriculture, University of Helsinki, Finland

I.	Introduction	329
II.	Establishment of the reference system	331
III.	Utilization of numerically arranged information for differentiation	332
IV.	Correction of the reference system	334
V.	Examples of application	334
	A. Streptomycetes	334
	B. Clinical isolates	335
	C. Phage typing pattern of *Pseudomas aeruginosa*	338
	References	339

I. Introduction

Earlier chapters of this volume deal with classification. Identification is sometimes defined as a corollary, a complementary function to classification. Actually there is, as will be shown, a quite obvious close link between classification and identification, but from the operational point of view both processes serve different purposes and are defined by different principles and features.

The first and perhaps most significant difference is that whereas a classification can be produced *de novo* without previous organized knowledge, identification can be performed only with reference to existing organized knowledge. One could say, accordingly, that classification represents production of knowledge, identification implies utilization of knowledge.

Let's try to illustrate the point using a simple example. Suppose that you

Present address: Academy of Finland, Drumsovägen 1, 00200 Helsingfors 20, Finland.

find a postcard among your papers. You notice that it is not in colour, so you may conclude that it is not a very new one, but that's all. However, it happens that a good friend of yours collects postcards. He has arranged and classified his collection with reference to the time from which the cards originate. By comparing the cards of different age your friend has produced a lot of knowledge concerning the development of the reproduction and printing techniques, the preferred motives of the pictures at different periods, maybe even differences in styles between different countries, not to mention the fashions worn by ladies shown on the cards. Now you show your card to your friend. Most probably he will immediately tell you that your card is for example a German one from the early 1930s or an Italian one from the 1950s etc. Since there is no general "postcard classification theory", your friend has produced his own classification in order to be able to extract the information inherent in his collection. By utilization of this previous organized knowledge your single card can be identified.

The "postcard example" shows that classification usually provides the reference of organized knowledge that is unavoidable for successful identification. The close link between classification and identification lies in the fact that classification produces the knowledge that is utilized for identification.

There are also other differences between classification and identification, which follow, however, from the principal distinction already defined above. Even a postcard collector will find, sooner or later, that he is unable to handle and to arrange an unlimited amount of material. Moreover he will certainly find that at a certain point a further quantitative increase of the collection does not give a corresponding increase in its content of information. Accordingly, from the point of view of producing new knowledge there is no existing reason to continue to increase the size of the collection *ad infinitum*. In other words, in classification there is always a limit for the size of the material needed and used, and the process, therefore, can be described as non-continuous.

The reverse is true for identification. Although the postcard collection of your friend is complete enough to allow identification of any postcard whatsoever, billions of postcards are sold and mailed every year. This implies that the potential need for identification has no limits at all, and there occurs a continuous input of objects to be identified.

The conclusions to be drawn from the foregoing discussion are summarized in Table I.

Coming then to the problems related to automated, that is computer-assisted, identification three questions arise. Firstly, what kind of reference knowledge is needed? Second, how should this knowledge be utilized for identification? Third, can we be sure that the reference knowledge has already reached a "state of stability"? If not, how can the outputs of

TABLE I

Main features of and mutual differences between classification and identification

Classification	Identification
Produces new knowledge and is not necessarily dependent on previous organized knowledge	Utilizes knowledge and is, therefore, always dependent on previous organized knowledge
Handles limited materials and is, therefore, a non-continuous process	Handles unlimited materials and is, in consequence, a continuous process

identifications be used to improve the reference knowledge? These three general questions define in fact a "system" flow chart for automated identification as is indicated in Fig. 1.

II. Establishment of the reference system

The prerequisite for identification is previous organized knowledge. As far as automated identification is concerned this knowledge is needed in a numerical form. In conventional, non-numerical taxonomy the reference knowledge is arranged to define a number of groups (taxa) by verbal description of their

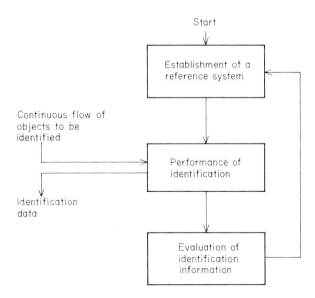

Fig. 1.   System flow chart for automated identification.

characteristics. The same information can easily be given in numerical form by describing, for example, two-state characteristics by the frequency of one of the states. The reference knowledge can then be provided as a matrix which is illustrated by the following simplified example:

Characteristics	Groups			
	A	B	C	D
1	0.99	0.01	0.95	0.05
2	0.05	0.99	0.95	0.05
3	0.05	0.99	0.01	0.95
4	0.95	0.99	0.01	0.99

The same way of numerical description is equally useful for continuous characteristics, the severest difficulty is provided by multi-state characteristics of the nature: colour red, blue, yellow, brown or grey etc. For characteristics of this kind every state has to be handled separately, taking into consideration, however, that the sum of frequencies of the different states must be $= 1$.*

III. Utilization of numerically arranged information for differentiation

This task can be illustrated in a simple way. Let's suppose that we rely on the knowledge given in the matrix presented above. Further suppose that we have to identify an unknown object which shows the following combination of characteristics: $- - + +$. We start by calculating how well these correspond to Group A. We get a probability figure P_A by multiplication of the frequencies of characteristics 1 to 4: $(1 - 0.99) \times (1 - 0.05) \times 0.05 \times 0.95 = 0.00046$. Similarly we obtain $P_B = 0.0097$, $P_C = 0.00000025$ and $P_D = 0.8482$. It is easily seen by comparison with the "reference matrix" that P_D for the "unknown" corresponds to the maximal probability that can be obtained for belonging to Group D on the basis of the given frequencies. At the same time this alternative is exclusive. The share of P_D of the sum $P_A + P_B + P_C + P_D$ is close to 99%. It is then obvious that the "unknown" fulfils two conditions: (1) it shows a very high *absolute affinity* to Group D and (2) it shows a very high *relative affinity* to the same group. Accordingly, it seems well founded to identify the "unknown" as belonging to Group D.

---

* With two-state characteristics it suffices to give the frequency of just one state, because the other is obtained simply by subtraction, e.g. $q = 1 - p$. With quantitative continuous characteristics a transformation of the figures to the range 0 to 1 is necessary.

The absolute and relative affinities were defined as "very high" in the example given. The limits cannot be defined in mathematical terms. As in all kinds of identification the "degree of similarity" expected or required for identification is ultimately a subjective measure. But once defined, and transformed into a numerical expression, identification can be automated.

The requirements for identification were already defined: sufficiently high absolute as well as relative affinity to some group described as reference. However, depending on the reference system it may happen that a very high absolute affinity is shown for two groups, which consequently implies a rather low relative affinity for both. Both groups are hence overlapping and the "unknown" which falls in the intersection of those groups can be described as an *intermediate*. Another alternative is that the absolute affinity to a given group is to some extent in doubt, but the relative affinity is exclusively high. In that case the "unknown" cannot be identified, but can well be described as a *neighbour* to the group in question. Finally, the possibility remains that an "unknown" shows neither absolute nor relative affinity to any one of the reference groups. Hence the "unknown" can be defined as an *outlier* not responding to the reference system. These different "identification states" are illustrated in Fig. 2.

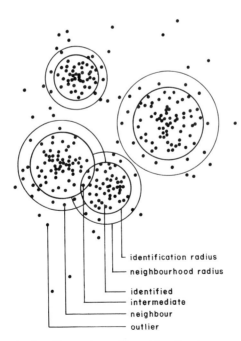

identification radius
neighbourhood radius

identified
intermediate
neighbour
outlier

Fig. 2.   Illustration of four identification states.

IV. Correction of the reference system

Although a proper selection of the reference system is a prerequisite for relevant identifications, it can be assumed that continuous identification may accumulate information to support an adjustment of the reference knowledge. This may imply definition of new groups, but also a lumping together of groups which were originally distinguished. The significant point here is that all necessary calculations and descriptions can be provided by the computer but since the necessary decision criteria hardly can be formulated in mathematical terms, a man–machine interaction is unavoidable. As soon as the investigator comes to the decision that the reference system has to be modified, the machine will do it, but as a corollary of the fact that the subjective element cannot be completely eliminated from identification, the machine remains incapable of evaluating when the reference system becomes irrelevant.

However, human intuition is by no means a definitely correct measure. There occurs in this connection, as has already been mentioned, a theoretically based check, even of man-initiated decisions. Every reclassification, that means every modification of a reference system serving identification, should move towards increased organization and stability. As soon as a modification manifests itself as a step in the direction of disorganization and lability it is certainly based either on wrong information or on poor intuition.

V. Examples of application

Principles remain principles until their practical applicability can be demonstrated. Therefore, a few examples applying automated identification methods to experimental microbiological materials are presented. The materials were distinctly different, and hence problems of different kinds could be elucidated.

A. Streptomycetes

The bacterial genus *Streptomyces* is of particular interest because several of its members are known to produce different antibiotics. Accordingly, there occurs a considerable economic interest related to these organisms as well as an obvious need of exact and objective identifications. The problem here is that much of the efforts to create satisfactory identification schemes based on a reliable classification have largely failed. The defined "species" are mostly rather diffuse and frequently even overlapping. This is at least partly due to the fact that the characters considered for classification and identification have been highly variable, from which it has followed that tests to verify characters show low reproducibility.

To overcome this difficulty a special International Streptomyces Project (ISP) has been conducted, and as a result of this project not less than 448 *Streptomyces* strains have been described in much detail in several different and mutually independent laboratories by standardized methods. The ISP material provides very useful reference information for further work on the streptomycetes.

However, also with the ISP material relevant classifications to serve as a reference system for identifications are difficult to attain. In spite of this problem the present author and his group (Gyllenberg *et al.*, 1975) have tried to use the ISP material for the creation of a model for automatic identification of streptomycetes. For this purpose a special method for the creation of a reference system was elaborated. With the aid of this method, which is described in detail in the original paper (Gyllenberg *et al.*, 1975), the reference matrix given in Table II was obtained.

The material was then identified against this reference matrix. It was found that about only 42% of the material could be exactly identified, whereas more than 30% of the material remained outside the defined reference matrix (outliers). The others were intermediates and neighbours. The low fraction of distinct identifications is due to the special circumstances and difficulties related to *Streptomyces*. As already pointed out, distinct subgroups cannot easily be separated from each other, which makes the identification task hard to fulfil. On the other hand, comparison with certain other classifications indicates that our identification data are largely confirmed. This proves that automatic identification is applicable to streptomycetes within the limits and restrictions that are unavoidable with characters which show that high variability.

B. Clinical isolates

Clinical isolates were selected as an experimental material, since they illustrate very well the continuous function of identification. In fact the experimental material is "open-ended" in contrast to the "closed" material of the *Streptomyces* study. Accordingly, the clinical isolate material was particularly useful for the study of the possibilities to utilize identification data for an adjustment of the reference system (matrix).

In the case of the study of clinical isolates there was no difficulty in obtaining primary reference matrices. The isolates were obtained from a public health laboratory where the isolates were routinely identified into classical taxa (e.g. *Pseudomonas, Proteus, Escherichia coli*). These taxa could be utilized as a reference system, but it was also shown (Gyllenberg, 1976) that a

TABLE II

Reference matrix for identification of the ISP material

Characters	1	2	3	4	5	6	7	8	9	10	11	12	13	14	15
1	.7419	.0050	.9850	.7200	.9850	.9850	.9850	.9850	.9850	.9850	.9850	.0050	.0050	.9850	.9850
2	.1452	.9546	.0050	.2700	.0050	.0050	.0050	.0050	.0050	.0050	.0050	.0050	.0050	.0050	.0050
3	.0323	.0050	.0050	.0050	.0050	.0050	.0050	.0050	.0050	.0050	.0050	.9850	.3300	.0050	.0050
4	.0806	.0354	.0050	.0050	.0050	.0050	.0050	.0050	.0050	.0050	.0050	.0050	.6600	.0050	.0050
5	.9677	.9950	.9950	.9950	.9950	.0050	.9950	.0050	.0050	.0714	.1111	.9950	.9950	.0050	.9950
6	.9838	.9950	.9950	.9950	.9950	.0050	.9950	.0050	.0050	.0050	.9950	.9950	.9950	.6250	.9950
7	.9950	.9950	.1538	.0050	.9950	.0050	.9550	.0050	.0050	.1429	.3883	.1000	.6667	.0050	.9950
8	.9950	.9950	.0050	.9950	.9950	.0050	.7500	.4375	.0050	.9286	.0050	.9950	.9950	.0050	.3333
9	.9950	.9950	.9950	.9950	.9950	.9950	.9950	.9950	.9950	.1429	.9950	.9950	.9950	.2500	.8333
10	.9950	.9950	.9950	.9950	.9950	.9950	.9950	.9950	.0050	.0714	.9950	.9950	.9950	.2500	.9950
11	.8666	.9850	.0050	.9450	.0050	.0050	.4950	.1866	.0050	.0050	.0050	.7960	.0050	.6841	.0050
12	.0802	.0050	.0762	.0050	.2605	.0500	.0050	.1866	.0050	.0050	.0550	.0995	.9850	.1244	.4950
13	.0481	.0050	.9138	.0450	.7295	.9850	.0050	.6219	.0050	.0050	.9350	.0995	.0050	.1866	.4950
14	.0050	.0050	.0050	.0050	.0050	.0050	.4350	.0050	.9850	.9850	.0050	.0050	.0050	.0050	.0050
15	.9950	.9950	.9950	.9950	.9950	.9950	.9950	.9950	.0050	.9950	.7222	.9950	.9950	.8125	.0050
16	.0050	.9950	.0050	.0050	.9950	.1429	.9950	.9950	.0050	.9950	.6667	.9950	.0050	.0050	.9950
17	.9950	.9950	.0050	.0050	.9950	.0050	.9950	.9950	.0050	.0050	.1667	.1000	.0050	.0050	.9950
18	.0050	.3518	.0758	.0895	.7737	.2800	.4875	.1837	.4900	.6332	.1633	.0050	.0050	.4900	.0050
19	.7626	.0050	.2652	.7164	.0050	.1400	.0050	.0050	.2450	.0050	.7078	.3920	.9700	.0050	.1625
20	.1271	.1759	.1137	.0448	.1032	.0050	.4875	.2450	.0050	.1407	.0050	.0050	.0050	.4288	.8125
21	.0794	.0352	.5304	.1343	.1032	.5600	.0050	.5512	.2450	.1407	.1089	.0050	.0050	.0613	.0050
22	.0050	.0050	.0050	.0050	.0050	.0050	.0050	.0050	.0050	.0050	.0050	.0050	.0050	.0050	.0050
23	.0050	.0050	.0050	.0050	.0050	.0050	.0050	.0050	.0050	.0704	.0050	.4900	.0050	.0050	.0050
24	.0159	.4221	.0050	.0050	.0050	.0050	.0050	.0050	.0050	.0050	.0050	.0980	.0050	.0050	.0050

reference system was easily obtained by numerical classification methods. Accordingly, we studied a conventional classification, on the one hand, and a numerical classification, on the other, as alternative reference systems, and we tried to follow the development of both reference systems as adjusted on the basis of new identification data. It was then found that a numerical classification of the 110 first isolates gave rise to a reference system consisting of six groups, whereas a conventional classification of the 150 first isolates indicated the presence of nine "taxa" (groups). The following steps were similar, numerical treatment of the next 223 (183) strains, and necessary readjustment of the reference systems. With identification data from 333 isolates available, there were 13 groups in both reference systems. 303 further isolates were then identified, and the data again used for an adjustment of the reference systems. To the reference system based on numerical treatment from the start, two new groups were added, but one earlier group was dissolved, and the final number of groups was then 14. Again, to the reference system initially based on conventional classification one new group was added. Accordingly, the final result was identical in both cases. Table III shows how the identified isolates (636 altogether) were distributed in the two reference systems. A distinct agreement is obvious. It can therefore be concluded that an automatic procedure which continuously corrects the basic classification gives results which are independent of the classification background.

TABLE III

A comparison of identification results based on reference systems obtained from initial numerical and conventional classification, respectively (numbers of lines and rows, respectively refer to recognized groups; 0 stands for unidentified status)

| | Start with numerical classification (A) | | | | | | | | | | | | | | | |
	0	1	2	3	5	6	7	8	9	10	11	12	13	14	15	$\sum$
0	84	4			2			1					10	1	1	103
1		75														75
2	2		38										10			50
3				115										1		116
4	1				1	80										82
5					70											70
6								17								17
7	15	3							9							27
8	4											7			2	13
9					8											8
10	3						15									18
11	2										18		3			23
12								16								16
13	9				1											10
14														8		8
$\sum$	120	82	38	115	82	80	15	18	16	9	18	7	23	10	3	636

Start with conventional classification (B)

Automatic correction logics develop conventionally based as well as numerically based initial reference systems towards almost similar solutions. This indicates that automated identification methods based on numerical classifications possess general validity.

C. Phage typing patterns of *Pseudomonas aeruginosa*

This material was quite different from those described above. It consisted of 486 strains of *Pseudomonas aeruginosa* which were described solely by bacteriophage sensitivity. Similarly the problem to be elucidated was more specific. Detection of groups with similar phage typing patterns is difficult since variation in lysotype can be postulated. Accordingly, the groups are polythetic in nature, and numerical classification and identification methods seem useful. This study is outlined in full detail by Bergan *et al.* (1975). From the point of view of applicability of automatic identification it was of special interest that, in this case, too, two different and mutually independent approaches to the creation of reference systems together with subsequent identification of further isolates gave well comparable, although not equivalent, results. This is illustrated in Table IV. It was shown that the numerical

TABLE IV

Group correlation of computation approaches A and B

Computation approach A	Computation approach B																				Total
	0^a	1	2	3	4	5	6	7	8	9	10	11	12	13	14	15	16	17	18	19	
0	122^b		5			1		3	1	5	3	7		1	4		5			3	160
1	1	57	5	3														22				88
2	5	5								7						12						29
3	2										3					2						7
4	4	1		5																		10
5					26																	26
6	2						2															4
7	2																					2
8	1						4															5
9	6		6			5																17
10	3									5												8
11	1								1													2
12	7	10				2				6	1		5									31
13						22																22
14	2																					2
15																		4				4
16	2													1								3
17	1										20											21
18	2										2											4
19																				3		3
20	5												33									38
Total	168	73	16	8	26	30	6	3	2	23	29	7	38	2	4	14	27	4	3	3	486

^{a} Group 0 contains "neighbours" and "outliers".

^{b} Number of *Pseudomonas aeruginosa* strains.

grouping and identification of *P. aeruginosa* bacteriophage patterns have epidemiological relevance.

These conclusions provide evidence supporting the readability of automated identification methods of this kind.

References

Bergan, T., Niemelä, T. and Gyllenberg, H. (1975). *Acta Pathol. Microbiol. Scand. Sect. B* **83**, 257–274.

Gyllenberg, H. G. (1976). *Arch. Immunol. Therap. Exp. (Poland)* **24**, 1–19.

Gyllenberg, H. G., Niemelä, T. K. and Niemi, I. S. (1975). "Numerical Taxonomy of Streptomycetes", pp. 83–109. Polish Medical Publishers, Warsaw.

14

Application of Numerical Taxonomy to the Classification and Identification of Microaerophilic Actinomycetes

K. HOLMBERG and C. E. NORD

Section for Medical Mycology, National Bacteriological Laboratory, Stockholm and Department of Oral Microbiology, Huddinge University Hospital, Karolinska Institute, Stockholm, Sweden

I.	The concept of microaerophilic actinomycetes . . .	342
II.	Classical taxonomy and new systematics	342
III.	Numerical taxonomy applied to the microaerophilic actinomycetes	344
IV.	Theoretical basis for the numerical taxonomy . . .	346
	A. Selection of tests strains	346
	B. Selection of test characters	347
	C. Coefficients of resemblance	348
	D. Methods of coding data	348
	E. Test errors	349
	F. Vigour and pattern statistics	350
	G. Clustering analysis	351
	H. Taxonomic structures and ranks	352
	I. Cluster parameters	352
V.	Validity of numerical taxonomy	353
	A. Stability of numerical taxonomy	353
	B. Optimality of numerical taxonomy . . .	354
	C. Significance of natural clusters	355
VI.	Comparison of numerical and classical taxonomy .	355
VII.	Comparison of numerical phenetic classification and serological, genetic and phylogenetic classifications	356
VIII.	Taxonomic criteria in the identification of microaerophilic actinomycetes	357
	A. Construction of diagnostic tables	357
	B. The probability of correct identification . . .	358
	C. The calculated median organisms in computer identification	358
	References	358

METHODS IN MICROBIOLOGY VOL. 16
ISBN 0–12–521516–9

I. The concept of microaerophilic actinomycetes

The use of numerical phenetic methods in the taxonomy of microaerophilic actinomycetes require some introductory comments. The term actinomycetes is widely used to describe Gram-positive, non-sporeforming, non-motile micro-organisms which exhibit pleomorphic morphology. According to comprehensive historical reviews (Gottlieb, 1970; Slack and Gerencser, 1970; Waksman, 1967) the generic name *Actinomyces* was first proposed for these organisms by Harz in 1877, who applied the name *Actinomyces bovis* to microscopically identified organisms in materials isolated from lumpy jaw of cattle. In 1891 Wolff and Israel published their classical paper in which they gave detailed description and photographs of *Actinomyces* isolated from two human cases and implied the anaerobic nature of the organisms.

The generic name *Actinomyces* was quite adequate at a time when there was little appreciation of the extensive and wide distribution of this group of organisms. With an increased interest in and knowledge of the actinomycetic realm new genera were formed. Since the name *Actinomyces* was originally proposed for anaerobic forms it was concluded that this name should be retained for the group of micro-organisms whose metabolism is primarily anaerobic. However, the name *Actinomyces* has given rise not only to the name of the family *Actinomycetaceae*, and the very order *Actinomycetales*, which include these organisms, but also to the common terminology of the entire group of both aerobic and microaerophilic actinomycetes.

II. Classical taxonomy and new systematics

Early schemes for the systematics of the actinomycetes were mainly determinative and made no attempts to mirror a presumed "natural" relationship of these organisms. There was a heavy reliance on morphological criteria. The morphological diversity of the actinomycetes was used for the purpose of defining taxa at the family as well as the genus level. In addition, actinomycetes occasionally produce a rudimentary type of mycelium at some stage of development. It was, therefore, not clear if classification of the actinomycetes should be evolved from the methods of mycology or of bacteriology. The exact taxonomic position of the actinomycetes was, therefore, uncertain. Bacteriologists considered them as bacteria and mycologists generally as fungi. It was not until the cell wall component studies of Cummins (1962) and Cummins and Harris (1958) that the actinomycetes could definitely be shown to be bacteria distinct from fungi. Moreover, they have no nuclear membrane, are sensitive to lysozyme and phages as well as to the common antibacterial agents, and so the actinomycetes are now accepted as bacteria

(Buchanan and Gibbons, 1974; Lechevalier and Lechevalier, 1967; Waksman, 1967).

Classification on the basis of numerical taxonomic criteria (Jones and Sneath, 1970; Jones, 1978; Sneath, 1970) has pointed towards the order *Actinomycetales* as a "natural" phenetic group (Jones and Bradley, 1964; Jones, 1975). It is quite likely, however, that there is no sharp distinction between *Actinomycetales* and coryneform bacteria. The taxonomic relationship between the actinomycetes and the propionibacteria, eubacteria and lactobacilli has also been a vexatious problem (Moore and Holdeman, 1973; Pine and Georg, 1969; Poupard *et al.*, 1973).

The taxonomy and systematics of the order *Actinomycetales* were the subject of consideration in the Jena International Symposium held in 1968 (Slack and Gerencser, 1970). The meeting illustrated the inadequate state of taxonomy and presented conflicting classification. Since then further progress has been made in the classification of these bacteria and alternative taxonomic ideas have been proposed.

Current taxonomic ideas on the actinomycetes were reviewed by Cross and Goodfellow (1973), Goodfellow and Cross (1974) and Goodfellow and Minnikin (1981) who also give full references to previous works on this subject. The order *Actinomycetales* was proposed to accommodate the following families: *Actinomycetaceae, Actinoplanaceae, Dermatophilaceae, Frankiaceae, Micromonosporaceae, Mycobacteriaceae, Nocardiaceae, Streptomycetaceae, Thermoactinomycetaceae, Thermomonosporaceae.* Many of the improvements in the classification of actinomycetes came too late or were not included in the eighth edition of "Bergey's Manual of Determinative Bacteriology" (Buchanan and Gibbons, 1974). All members of the order *Actinomycetales* are aerobic except for genera in the family *Actinomycetaceae* which may be facultatively anaerobic (microaerophilic) or anaerobic. They are Gram-positive, branching and later on fragmenting filaments without aerial hyphae or spores, and possess a fermentative carbohydrate metabolism (Slack, 1974).

A review of the family was given by Pine (1970) who discussed the physiology and structure of members of this family in relation to their classification and phylogenetic relationship. The lines of demarcation between the genera and between them and related taxa outside the family are not always clear. The former subgroup on Taxonomy of Microaerophilic Actinomycetes (1968) (Slack, 1968) International Committee on Bacteriological Nomenclature, gave their definition of the family *Actinomycetaceae* and included three genera: *Actinomyces, Bacterionema* (Gilmour *et al.*, 1961) and *Rothia* (Georg and Brown, 1967). *Arachnia* was accepted as a new genus in 1969 (Buchanan and Pine, 1962; Pine and Georg, 1969). Howell *et al.* (1965) classified an aerobic-facultative and catalase-positive organism isolated from hamster in a new genus *Odontomyces*. The subgroup on Taxonomy of Microaerophilic Actinomycetes recommended later that this organism should be included in the genus *Actinomyces* and Georg *et al.* (1969) proposed that *Odontomyces viscosus* should be reclassified as *Actinomyces viscosus*.

The taxonomic situation of the family *Actinomycetaceae* is further complicated by the opinion of Prévot (1972) that the strictly anaerobic actinomycetes should be placed in a separate genus, *Actinobacterium*.

The validity of this separation has been fortified by results of cell wall analysis but has not gained general acceptance. The genus *Bifidobacterium* (Orla-Jensen, 1924) has also been proposed for the family *Actinomycetaceae*. The exact taxonomic position, however, in the seventh edition of "Bergey's Manual of Determinative Bacteriology" (Breed *et al.*, 1957) is that the bifidobacteria were classified as *Lactobacillus bifidus*. The genus is at present classified in the *Actinomycetaceae* (Cross and Goodfellow, 1973; Pine, 1970; Poupard *et al.*, 1973) and it is recognized as an independent genus of this family in the eighth edition of "Bergey's Manual of Determinative Bacteriology" (Buchanan and Ribbons, 1974).

III. Numerical taxonomy applied to the microaerophilic actinomycetes

In the two studies by Holmberg and Hallander (1973) and Holmberg and Nord (1975) (below referred to as Paper I and Paper II, respectively), attempts were made to present a relationship and propose "natural" (phenons) groupings of Gram-positive, non-sporeforming, non-motile pleomorphic rods on the basis of phenetic data using numerical taxonomic methods.

In the first analysis (Paper I) 76 aerobic and facultative anaerobic isolates of diphtheroidal rods isolated from the human oral cavity, respiratory tracts and urine together with 47 type strains of the genera *Rothia*, *Bacterionema*, *Mycobacterium*, *Corynebacterium* and *Actinomyces*, obtained from various international culture collections were examined by 77 characters, covering a wide range of tests and properties. The strains fell into five main phenons, representing *Corynebacterium*, *Rothia*, *Bacterionema*, and *Actinomyces* and *Nocardia*, respectively. The *Actinomyces* cluster was distinctly separated into two subclusters, reference strains of *A. viscosus* and *A. naeslundii* formed a tight species-group. *Actinomyces* and *Rothia* were more similar to one another than to the other clusters. On the other hand, *Bacterionema* showed a closer phenetic relationship to *Corynebacterium* and *Nocardia* than to the *Actinomycetaceae*.

In the subsequent study (Paper II), 49 facultatively anaerobic and anaero-

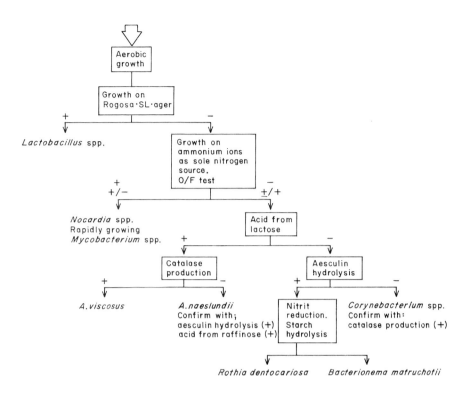

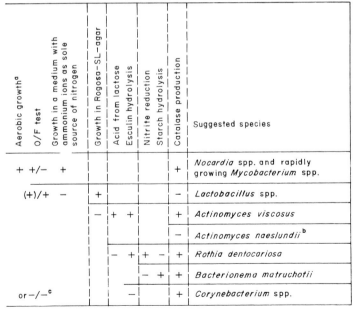

Aerobic growth[a]	O/F test	Growth in a medium with ammonium ions as sole source of nitrogen		Growth in Rogosa-SL—agar	Acid from lactose	Esculin hydrolysis	Nitrite reduction	Starch hydrolysis	Catalase production	Suggested species
+	+/−	+							+	*Nocardia* spp. and rapidly growing *Mycobacterium* spp.
(+)/+	−			+					−	*Lactobacillus* spp.
				−	+	+			+	*Actinomyces viscosus*
									−	*Actinomyces naeslundii*[b]
					−	+	+	−	+	*Rothia dentocariosa*
							−	+	+	*Bacterionema matruchotii*
or −/−[c]					−				+	*Corynebacterium* spp.

[a]On initial isolation   [c]Non-saccharolytic *Corynebacterium* spp.
[b]Optimal growth in the presence of 10% CO_2 [x]According to Holmberg and Hallander (1973)

Fig. 1. Two arrangements of a dichotomous key for *Rothia dentocariosa, Bacterionema matruchotii, Actinomyces naeslundii,* and *A. viscosus.*

bic actinomycetes, isolated from human dental plaques and faeces together with 63 type strains of the genera *Actinomyces, Arachnia, Bifidobacterium, Actinobacterium, Propionibacterium, Eubacterium* and *Lactobacillus* became the subject of a numerical taxonomic study. In all, 90 characters were employed as the taxonomic criteria. The strains fell into six main clusters considered to be equal to genus rank. Two of these represented *Actinomyces*. One cluster contained strains of facultatively anaerobic organisms and type strains of *Actinomyces viscosus, A. naeslundii, A. odontolyticus* and *A. bovis*. The other *Actinomyces* cluster contained anaerobic strains and corresponded to *A. israelii*. The other generic clusters represented *Arachnia, Actinobacterium* and *Propionibacterium*, respectively. The generic clusters of *Actinomyces* had a close phenetic relationship to *Bifidobacterium*, probably corresponding taxonomically to the present family level in the taxonomic hierarchy. It is concluded that strains currently known as *Actinobacterium meyerii* show closer relations to *Actinomyces* than to representative of the other genera examined.

The numerical studies were also involved with the establishment of the range of variabilities and dimension of each of the named taxa and identity calculated hypothetical median organisms (CMOs) of taxa.

IV. Theoretical basis for the numerical taxonomy

Numerical taxonomy operates without *a priori* prejudices regarding the classification of micro-organisms. In practice, numerical taxonomy claims to assess objectively the similarities between strains of bacteria through not introducing any deliberate *a priori* bias in choice of operational taxonomic units (OTUs) and tests and through unbiased mathematical and statistical methods. Recent advances of the philosophy and operational definitions of numerical taxonomy (Lockhart, 1967; Sneath, 1971a,b; Williams and Dale, 1965) have, however, drawn attention to a number of problems in the numerical analytical methods for obtaining a complete objectivity. Theoretical aspects of the numerical analytic methods which had to be taken into account when making the procedural decisions for the studies in Papers I and II have to be commented on.

A. Selection of test strains

No taxonomic study can involve as many strains in as many tests as desirable. The "ideal" classification is based on the reliability of the strains included as representative of the taxa. In practice, not even this may be fully satisfied. While it is desirable to have representative samples of strains—so the study can be probabilistic—this is not always possible, and non-probabilistic studies may be perfectly valid within their limitations.

To obtain a representative sample of microaerophilic actinomycetes for the numerical taxonomic studies morphologically homogeneous populations of fresh isolates and type strains were selected. Isolates from clinical specimens were obtained from different kinds of human habitats, picked on the basis of their Gram-reaction and cell morphology, and collected by a method of random selection. Type strains obtained from various international type culture collections were included to serve as reference strains or "marker strains" of the taxa about which information was sought. Since inter-generic relationships were to be studied, representatives of other genera than those of the family *Actinomycetaceae* were also included in the extent of their known relationship with these actinomycetes (Buchanan *et al.*, 1966). In addition, actinomycetes from a variety of habitats were received from private collections.

B. Selection of test characters

A more critical problem pertained to "the selection of a proper set of tests" which would involve an arbitrary selection among the endless array of attributes characterizing the complete phenotypic expressions of the genomes of the OTUs. The arrays of tests selected were chosen to constitute a suitable testing schedule, balanced with respect to morphology, physiology and biochemistry. Since some kinds of characters are more relevant than others, the recommendations of the Subgroup on the Taxonomy and Microaerophilic Actinomycetes (Slack, 1968) were followed to provide admissible and relevant information of a single property, i.e. unit character of the actinomycetes.

The information derived from this primary choice of test characters was from results in introductory experiments considered for their usefulness to give a good approximation of the total phenetic capabilities of the organisms using the ways devised by Sneath (1968) and Sneath and Collins (1975) to test whether a given selection of tests is adequate for classification purposes. After exclusion of some test characters and a logical subdivision of some unit characters, the number of character states employed in the taxonomic analyses comprised 77 and 90, in respective study. These numbers of character states appear to be appropriate to obtain stable classifications. A minimum of about 60 to 80 characters has been considered imperative (Lockhart and Liston, 1970; Sneath, 1970). Although increasing the number of characters from 50 to 100 leads to considerably narrowing the confidence limits in the similarity coefficient values (Goodall, 1967; Sneath and Johnson, 1972) (Section IV.C).

C. Coefficients of resemblance

The "coefficient of resemblance" was chosen for ease of interpretation, adherence to the Adansonian precept and the ability to express characters in binary form. Moreover, the choice of coefficients was guided by the reliability of the coefficients for data evaluation in the field of microbiology.

Different types of measures for estimating the degree of relationship (similarity or dissimilarity) between a pair of OTUs have been employed for various classificatory objections in biology. An extensive discussion can be found (Sokal and Sneath, 1963). The Jaccard–Sneath similarity index (S_J) (Sneath, 1957) and the simple matching coefficient elaborated by Sokal and Michener (S_{SM}) (1958) were employed in the present studies. The S_J index does not consider matches in negative character states. The philosophy on the significance of negative matches in taxonomic works has been the subject of dispute (Sneath and Sokal, 1973). In effect, the coefficient estimates different aspects of the taxonomic structures. Using the S_{SM} coefficient a pair of organisms can be grouped together principally due to a high proportion of negative correlations. In the present studies the data were, therefore, reexamined with the S_J index and the data obtained for each of the two coefficients of association analysed.

The measure of relationship between organisms may also be affected by the treatment of characters which are present in the whole set of organisms. In the first numerical analysis such invariant characters were included in the computation from the point of view that for obtaining standardized results all tests should be included. Their inclusion increased the absolute values of the resemblance coefficients to some extent. The differences between the strains were diminished introducing an arbitrary element into the value of the coefficient. In the other analysis this was avoided by excluding all characters taking the same value in all organisms from consideration. On the other hand, when this was done, the resemblance of a pair of organisms was not absolute, but dependent upon the set of organisms within which the pair was considered.

D. Methods of coding data

Proper "methods of coding data" into forms suitable for computation is essential for numerical taxonomy since each coding method makes certain assumptions about the manner in which information should be used in the classification (Beers *et al.*, 1962; Lockhart and Liston, 1970; Sneath and Sokal, 1973). The predominant characters employed in the present studies existed in only two mutually exclusive states (recorded as $+$ and $-$) and presented no problem. For the multi-state characters that existed in several different states the situation was more critical. They needed to be adjusted in a coding process with the object of utilizing all of the information in a uniform way. The quantitative multi-state characters such as the amount of anaerobic and aerobic growth in Paper II were converted into two-state characters and arranged in order of magnitude along a one-dimensional axis. These states were scored by the additive coding method (Sneath and Sokal, 1973). This method had the merit of exaggerating dissimilarities due to differences in the overall growth ability and retain the information on the magnitude of differences in these characters. Qualitative multi-state characters such as the colour pattern of a colony were converted into a number of separate two-state characters. This, in effect, means that these characters became weighted. When using the S_{SM} coefficient this coding technique weighted similarities and minimized differences although when using the S_J index this technique weighted differences only. It is, however, important to point out that the weight assigned to a character by the coding systems differed from *a priori* weighting when attempts are made before recording the data to decide which characters are more important than others for the purpose of classification. Meaning that, the weight to be assigned to each character is subjectively determined. Inapplicable characters, such as registering the results from testing acid formation from carbohydrates on non-saccharolytic corynebacteria in Paper I, were designated by the symbol NC, no comparison. This meant that these characters did not contribute to the scoring of the similarities and differences between pairs of organisms, and ignored in calculating similarity coefficients.

E. Test errors

A troublesome problem in microbial taxonomy pertains to "test errors" that affect the coefficient of resemblance. Attempts to master these problems were made by achieving adequate standardization in performing tests by following the standardized conditions in methods for testing cited by Cowan and Steel (1970) and Holdeman and Moore (1973). However, experimental error in performing tests due to lack of test reproducibility is probably greater than is often thought for numerical taxonomic works (Lockhart and Liston, 1970; Sneath, 1971a). This error affects the simple matching coefficient (S_{SM}) based on present/absent test results of which on the average a proportion (p) may give erroneous results (Sneath and Johnson, 1972). The reproducibility of the 90 tests included in the numerical analysis in Paper II was assessed from the results of doing the tests in duplicate on 21 strains at two different occasions. An average proportion of erroneous results taking the majority of results as the true one over the set of tests was calculated to be $p = 3.12\%$. The effect of this error was to shift the value of S that would have been obtained if there had been no errors to a mean value S! The error will introduce scatter, so this expected mean value will not always be the same on every occasion. Following the statistical estimates explored by Sneath and Johnson (1972) both the change from S to S!, and the variance increases stepwise when p is greater than 10%. Thus, in the numerical analysis in Paper II the expected value S!: from comparisons between subcultures of same strains (with $p=0.312$ or 3.1%) was 94%, while its standard error was 2.4%. When, in similar statistical experiments, the average proportion giving erroneous results was assessed in the numerical analysis in Paper I by repeated testing on 50 strains, the expected value of S_{SM} for a pair of identical strains was 96%, while its standard error (based on the 77 tests) was 2.5%. It appears from the findings above that the main source of taxonomic distortion resulting from test error was in both studies mainly due to the sampling errors (Goodall, 1967).

If the variance introduced by experimental error in respective study was added to the expected variance due to the existing sampling error of the 77 tests and the 90 tests employed the total test error on S_{SM} in respective study is achieved (Sneath and Johnson, 1972). These values may be more realistic estimates of the confidence limits of S_{SM} in both studies against which to judge the significance of observed differences between strains or between clusters.

Further helpful information on the reproducibility of the tests used in Paper II was achieved by replicating some of the type strains in the numerical study. The test error was readily estimated by analysis of the variance of the replicates.

F. Vigour and pattern statistics

The variations in growth rates and cultural conditions of micro-organisms may also have an undesirable effect on numerical measurements of relationships. These factors may introduce differences between strains and groups. Analogous to shape and size in biological taxonomy Sneath (1968) suggested the introduction of two new coefficients in the taxonomy of micro-organisms for correcting the influence of these phenetic components in calculating the S_{SM} coefficient. The "concept of vigour" (D_V) expresses the difference between strains in total metabolic activity as measured by the number of positive reactions in a given battery of tests, while the "concept of pattern" (D_p) is the dissimilarity due to a different pattern of reaction. The squares of the D_p and D_V add up to the square of D_T, the total dissimilarity, which equals $1 - S_{SM}$. Since no a priori weighing of characters was carried out in the present study, the potential contribution of dissimilarity from each character could be expected to be the same.

The numerical analysis in Papers I and II delineated phenetically different groups of organisms, in which the differences could plausibly be explained by unequal growth rates of strains. Paper I deals with the taxonomic relationships of saccharolytic and non-saccharolytic corynebacteria, nocardia and facultative anaerobic (microaerophilic) actinomycetes without considerations to any effect of, for example, metabolic activity. In the numerical study in Paper II the treatment of the data was expanded to include vigour and pattern statistics. The values of vigour of the clusters established were fairly uniform and its removal did not alter the resemblances, so that clustering and D_p gave clusterings which well corroborated those obtained by clustering on the similarity values.

A special problem concerns the standard conditions for obtaining reproducible and valid results. The difficulty is that experiments which are optimal for some members of a microbial population studied may be lethal or sublethal for others and characters will change if the experimental conditions change. Thus, the concordance between tests carried out under strict anaerobiosis and aerobic conditions has to be considered. The implication of incongruence in this respect has been amply illustrated in the literature. Melville (1965) discussed this concept on his numerical analysis on some oral actinomycetes. He found the discordance to be remarkably low, which is in accordance with evidence in the present study exemplified by the recovery of a homogeneous taxospecies of *Actinomyces naeslundii* and *A. viscosus* in both Paper I and Paper II on data derived from most similar tests but performed aerobically and strict anaerobically.

G. Clustering analysis

The "clustering analysis" imposes certain constraints on the taxonomic structure of the data matrix of resemblance or dissimilarity. Although the end result of the clustering process is necessarily determined by the inherent structure and relations of the OTUs, it is also appreciably affected by the choice of a clustering method. Bergan (1971) gives a list and discusses a variety of clustering methods.

A common clustering method was employed in the present studies, the single linkage method (SLCA) devised for microbial taxonomy by Sneath (1957). In addition to this method the unweighed pair-group arithmetic average clustering (UPGMA) elaborated by Sokal and Michener (1958) was employed in the numerical analysis in Paper II. The properties and behaviour of these methods have been proved appropriate for use as clustering methods in microbial taxonomy (Sneath, 1971b).

In the computer analysis described in Paper II one OTU joined to a cluster at a time which means that the stems fused only in pairs and recalculation of similarity during each clustering cycle was preceded by an unweighed linkage procedure. The reason for employing this method instead of a weighed method was that unweighed clusterings have been reported to give pheno-grams more closely correlated with the original similarity matrices (Sneath and Sokal, 1973).

‚The pitfalls in the use of a single linkage procedure have been discussed in detail by Bergan (1971), Sokal and Sneath (1963) and Sneath and Sokal (1973). To avoid the extremes in SLCA the adequacy of the clusters formed by SLCA from the matrices of coefficients of association in Paper I was judged by clustering the five strains most similar to each other and preparing a single linkage phenogram of the type described by Gower and Ross (1969).

In additional computer analysis the validity of the SLCA clusters in Paper I was also evaluated by their reappearance in phenograms (dendrograms) derived from the matrix of S_{SM} by UPGMA, by centroid cluster analysis and by complete cluster analysis (Lance and Williams, 1967) (see below).

H. Taxonomic structures and ranks

As an effect of the clustering procedures various types of taxa with different internal taxonomic structure could be extracted from the same matrix of resemblance. Groups at various ranks were taken as equivalent to "natural" taxa. No rigid similarity level for the "rank category of species" is as yet firmly established in numerical classification. Liston *et al.* (1963) suggested that the phenon level of 75% may be a good criterion for the phenetic species level in bacteria. However, differences in test batteries and similarity coeffi-cients employed may cause large differences in the absolute values of numerical measures of resemblance in the number of the clusters and the relative resemblance values among the strains are more significant than the specific numerical values. In the present studies a phenetic level of 90% appeared appropriate for delineating discrete taxa equal to the category of species in the hierarchy tree derived by SLCA. There was also the supporting evidence of serological resemblance that assisted in this consideration.

I. Cluster parameters

By definition, a taxon at a given rank level contained a minimum of five OTUs, and possessed intra-taxon variance less than inter-taxon variance. Among the "cluster parameters" involved in the definition procedure of the taxospecies, the internal phenetic variation had to be taken into account. Even if very complete knowledge of the variation within the taxon was available it is difficult to decide how to compute a mean and its variance for such characters. The coding of the phenetic value of phenetically overlapping characters pose problems. In the present studies, this problem was solved by the following approach. The distribution of positive characters among the strains of a taxon were tabulated, and a point in the taxon representing a "calculated median organism" (CMO) was specified as devised by Liston *et al.* (1963). This mathematical construct possesses all the characters found in more than 50% of the members of the taxospecies. Statistically, the CMOs represent the most typical strain of the taxospecies provided the strains included in the study have been chosen to represent appropriately the variation of strains of a species. Logically, strains designated as the neo-type strains of a defined taxospecies should be the most typical (Tables 5 and 6 in Papers I and II).

A contiguous parameter in the defining procedure of taxospecies is that of the dimension of the taxon. This aspect of taxa was defined by the concept of a taxon radius which could be calculated from the resemblance values of the members of the taxon against its own CMO. If the distribution of the resemblance values around the CMO was plotted as a histogram, a taxon radius could be obtained empirically. The histogram was compared with an appropriate normal curve and the mean of resemblance values of the members of the taxon to the CMO and their standard deviation could be found. Gyllenberg (1965) suggested that twice the intra-cluster standard deviation should be used as a radius. Thus, one would expect 97% of the members of the taxon to be within the radius and the lowest permissible resemblance value for a candidate strain to be included in a taxospecies could be calculated on the basis of a given set of characters.

V. Validity of numerical taxonomy

A. Stability of numerical taxonomy

The application of the numerical methods to a classification of the *Actinomycetaceae* was made by the intention that the phenetic relationships discovered among organisms by this presumptively objective technique would be so stable that essentially similar results will be obtained in other investigations where other kinds of data are used. In effect, the validity of the numerical classification presented in Papers I and II is largely a matter of determining in further comparative studies on the group, the stability of the taxonomic structures recovered by investigating the reappearance of the clusterings under changes in selection of OTUs, unit characters and clustering procedures (Rohlf, 1964).

In a recent numerical taxonomic study of *Actinomycetaceae* using similar unit characters and numerical taxonomic techniques Schofield and Schaal (1981) recognized *Actinomyces israelii* as a well-defined taxon with rather low similarity levels with other species of the genus *Actinomyces*. *A. odontolyticus* and *A. bovis* formed distinct taxa at the generic level. *A. naeslundi* and *A. viscosus* once again grouped together but formed a discrete set of subclusters. The retention of *Arachnia propionica* as a distinct taxon at the generic level was justified on the basis of the numerical data. Because of the lack of representatives of *Actinobacterium meyerii* in the study no phenetic data were obtained on the taxonomic association of *A. meyerii* with the *Actinomycetes*. The *Rothia* strains formed a distinct cluster which was linked to the *Actinomyces* only at low similarity level. A closer phenetic relationship to *Corynebacterium* than to the *Actinomycetaceae* of *Bacterionema* was recognized. Thus confirming the uncertain taxonomic affiliation of genus within the currently defined family *Actinomycetaceae*.

B. Optimality of numerical taxonomy

In Paper II a test on the internal cohesion of the single-link clusters was achieved by the reappearance of the same clusters by the alternative UPGMA clustering procedure. This method to investigate the concordance of clusterings represents a method to define a criterion of optimality for the classification. A classification ought to be optimal if it represents as closely as possible the original similarity matrix among the OTUs. A numerical value on congruence between the resulting taxonomic structures could be obtained by the co-phenetic correlation test introduced by Sokal and Rohlf (1962). By this test, dendrograms can be compared with each other and with the similarity matrix from which they derive. This technique was practised in Paper II to calculate the goodness with which the resulting group structures obtained by SLCA and UPGMA procedure corresponded with each other and reflected the original similarity matrix. In additional studies alternative cluster procedures were also applied to the similarity matrix (S_{SM}) in Paper I. The taxonomic relationships of species derived by a complete linkage, a centroid linkage (Sneath and Sokal, 1973), an average linkage (Sokal and Michener, 1958) and the single linkage clustering procedure using the 77 character descriptions were compared (Fig. 2A–D). The relationships varied according to the analytical method employed. The differences apparent in the dendrograms arising from the various linkage methods indicated that caution should be observed in the interpretation of relationships determined by a single method in numerical analysis. The co-phenetic values calculated from the comparative evaluation of the dendrograms with each other and with the original coefficients of similarity from which they were derived were calculated. The centroid method was found to impose the highest distortion on the taxonomic structure.

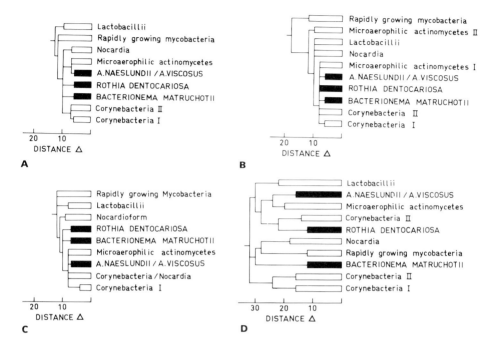

Fig. 2. Taxonomic dendograms formed (A) by single linkage, (B) by complete linkage, (C) by average linkage, (D) by centroid linkage cluster analysis of the 77 test characters of microaerophilic actinomycetes. The abscissa is in distance scale.

C. Significance of the natural clusters

The numerical measures of a method of testing the significance of the clusters selected as "natural" was homogeneity of the clusters. These measurements involved calculation of the inter-group mean similarity values and as a measure of relationship between clusters, the inter-group mean similarity values (Lockhart and Liston, 1970). These measures gave considerable guidance about how tightly linked the clusters were internally and how well isolated they were from the other clusters.

VI. Comparison of numerical and classical taxonomy

The data obtained on the taxonomic structures of the *Actinomycetaceae* based on overall phenetic similarity using numerical methods agreed with the accepted classification of the *Actinomycetaceae* (Slack, 1974) but provided a refined and quantitative description of pre-existing taxa. In most instances, a high degree of concordance has been demonstrated between classifications of bacteria established by numerical taxonomy and those derived by more classical methods (Sneath, 1971a).

VII. Comparison of numerical phenetic classification and serological, genetic and phylogenetic classifications

There are several observations that show the close concordance in the microaerophilic actinomycetes between phenetic resemblance and serological resemblance as measured by immunofluorescence (Slack and Gerencser, 1970). For taxonomic interpretation it is difficult to combine the two estimates. There exists no criterion for allocating weights to them in combining them with the coefficient of phenetic similarity of micro-organisms. Data on the serological affinities obtained by the concomitant use of immunofluorescence testing were, therefore, accumulated for the strains in order to provide a means of additional testing of the clustering of strains obtained by numerical analysis.

Current with the development in numerical taxonomy are the recent advances in molecular genetics in classification. The taxonomic relationships amongst bacteria are being subjected to examination by genetic data by measuring genome size, the overall base composition and the arrangement of the bases within the DNA molecule. There is supporting evidence for the close concordance between bacterial phenetic resemblance and genomic resemblance as measured by DNA-pairing and GC (moles % of guanine-+cytosine) ratios (Jones and Sneath, 1970). Information from analyses of microaerophilic actinomycetes at the genetic level is sparse. Slack and Gerencser (1975) summarized recently the DNA base composition (GC ratios) found by different investigators. These data tend to support (within the limitations of the method) the phenetic concept of the *Actinomyces* as a separate group. A recent DNA homology study (Coykendall and Munzen-maier, 1979) support the separation of *A. naeslundii* and *A. viscosus* at the subspecies level.

The phylogenetic classification pertains to the evolutionary data on micro-organisms. It is obvious that the phenetic studies in Papers I and II have no phylogenetic implications since the test characters used cannot logically be weighted on phylogenetic grounds. Therefore any attempts at evolutionary species definition for the actinomycetes on the basis of the phenetic taxonomic data in Papers I and II would be speculative and largely intuitive.

VIII. Taxonomic criteria in the identification of microaerophilic actinomycetes

A. Construction of diagnostic tables

One of the advantages of numerical taxonomy is that clusters which are defined have a high information content. In the extensive phenotypic characterization of the test strains in Papers I and II, several taxonomic criteria were found to have a descriptive and discriminating value. There was sufficient consistency within clusters and difference between them for some criteria to be weighted for identification. This was readily apparent when the frequency of each test which was positive in each of the clusters had been calculated. A simultaneous strategy of identification (Sneath and Sokal, 1973) was practised in the construction of the tables in Papers I and II. No infinite weight was given to the characters chosen for these tables. Several characters together have to be considered. The test results for a new isolate have to be matched with the tables. The best match is chosen. These tables are therefore operationally polythetic and can accommodate a degree of strain variability, thereby reducing the chances of variants of a species being defined as a new species. Failure to identify using these diagnostic tables means that the results of the identification tests on the unknown isolate are faulty in some respect, or that the species is outside the range for which the tables have been prepared.

Another method for the construction of diagnostic tables to species level selects those characters that were invariant and mutually exclusive within the taxa and then use them as sequential keys (Willcox *et al.*, 1973; Sneath and Sokal, 1973). In succeeding steps these keys will allocate an unknown organism to one of the taxa based upon a unique character. Thus, infinite weight is given to single characters by ignoring all the others. The characters chosen are weighted according to their importance in distinguishing among the taxa.

For the selection of characters for the diagnostic tables, other aspects of the test also have to be considered (Gower and Payne, 1975). These should be well known, easy to determine, give highly repeatable results and not be unduly sensitive to small difference in technique. It should be stressed since this simple requirement is often overlooked that the tables were constructed on the basis of the results obtained by the techniques and media devised in Papers I and II. Other results will undoubtedly be obtained if other techniques and media are used.

B. The probability of correct identification

It is of interest to estimate the probability that the identification of a new isolate is correct by employing the diagnostic tables. This aspect was described in a subsequent paper (Holmberg, 1975) where the diagnostic tables were thoroughly tested with micro-organisms that were not used in their construction. Additional experience of the effectiveness of the tables has also been achieved by their application in routine laboratory use. A figure that expresses the probability that an identification is correct can be worked out by the use of an appropriate conditional probability method (Möller, 1962; Sneath, 1971b; Sneath and Sokal, 1973) and would also make the identification strategy probabilistic. A more extensive discussion of the general problems of probabilistic identification of bacteria has been given by Gyllenberg (1965) and Hill (1974).

C. The calculated median organisms in computer identification

Numerical methods in identification studies have been comparatively neglected (Bascomb et al., 1973; Lapage et al., 1973; Gyllenberg and Niemelä, 1975; Willcox et al., 1980). Computer-generated identification of new isolates by calculating their numerical similarity in the 77 and/or 90 character states to the established calculated median organisms (CMOs) for the homogeneous taxa defined in Papers I and II appears to be an alternative method for identification. This approach would also enable identification of facultative and anaerobic actinomycetes with the following taxonomic aims in mind: (i) to verify the CMOs of clusters that were found homogeneous and probably warranting of species status and (ii) to ascertain which clusters are in need of more critical examination by conventional $n \times n$ analysis.

References

Bascomb, S., Lapage, S. P., Curtis, M. A. and Willcox, W. R. (1973). J. Gen. Microbiol. 77, 291–315.
Beers, R. J., Fisher, J., Megraw, S. and Lockhart, W. R. (1962). J. Gen. Microbiol. 28, 641–652.
Bergan, T. (1971). Bacteriol. Rev. 35, 379–389.
Breed, R. S., Murray, E. G. D. and Smith, N. R. (Eds) (1957). "Bergey's Manual of Determinative Bacteriology", 7th edn. Williams & Wilkins, Baltimore, Maryland.
Buchanan, R. E. and Gibbons, N. E. (Eds) (1974). "Bergey's Manual of Determinative Bacteriology", 8th edn. Williams & Wilkins, Baltimore, Maryland.
Buchanan, B. B. and Pine, L. (1962). J. Gen. Microbiol. 28, 305–323.
Buchanan, R. E., Holt, J. G. and Lessel, E. F. (1966). "Index Bergeyana". Livingstone, Edinburgh.

Cowan, S. T. and Steel, K. J. (1970). "Manual for the Identification of Medical Bacteria". Cambridge University Press, Cambridge.

Coykendall, A. L. and Munzenmaier, T. (1979). *Int. J. Syst. Bacteriol.* **29**, 234–240.

Cross, T. and Goodfellow, M. (1973). "Actinomycetales: Characteristics and Practical Importance (G. Sykes and F. A. Skinner, Eds), pp. 11–112. Academic Press, London and New York.

Cummins, C. S. (1962). *J. Gen. Microbiol.* **28**, 35–50.

Cummins, C. S. and Harris, H. (1958). *J. Gen. Microbiol.* **18**, 173–189.

Georg, L. K. and Brown, J. M. (1967). *Int. J. Syst. Bacteriol.* **17**, 79–88.

Georg, L. K., Pine, L. and Gerencser, M. A. (1969). *Int. J. Syst. Bacteriol.* **19**, 291–293.

Gilmour, M. N., Howell, A., Jr and Bibby, B. G. (1961a). *Bacteriol. Rev.* **25**, 131–141.

Gilmour, M. N., Howell, A., Jr and Bibby, B. G. (1961b). *Int. Bull. Bacteriol. Nomencl. Taxon.* **11**, 161–163.

Goodall, D. W. (1967). *Biometrics* **23**, 54–64.

Goodfellow, M. and Cross, T. (1974). *In* "Actinomycetales" (C. H. Dickinson and G. J. F. Pugh, Eds), pp. 269–302. Academic Press, London and New York.

Goodfellow, M. and Minnikin, D. E. (1981). *In* "Actinomycetes" (K. P. Schaal and G. Pulverer, Eds). Zbl. Bakt. Suppl. 11, pp. 7–16.

Gottlieb, D. (1970). "The Actinomycetales" (H. Prauser, Ed.), pp. 67–77. Fischer, Jena.

Gower, J. C. and Ross, G. J. S. (1969). *Appl. Statist.* **18**, 54–64.

Gower, J. C. and Payne, R. W. (1975). *Biometrika* **62**, 665–672.

Gyllenberg, H. G. (1965). *J. Gen. Microbiol.* **39**, 401–404.

Gyllenberg, H. G. and Niemelä, T. K. (1975). *In* "Biological Identification with Computers" (R. J. Pankhurst, Ed.). Syst. Assoc. Spec Vol. No. 7 pp. 121–136. Academic Press, London and New York.

Hill, L. R. (1974). *Int. J. Syst. Bacteriol.* **24**, 494–499.

Holdeman, L. V. and Moore, W. E. C. (Eds) (1973). "Anaerobe Laboratory Manual". Virginia Polytechnic Institute and State University, Blacksburg, Virginia.

Holmberg, K. and Hallander, H. O. (1973). *J. Gen. Microbiol.* **76**, 43–63.

Holmberg, K. and Nord, C. E. (1975). *J. Gen. Microbiol.* **91**, 17–44.

Holmberg, K. (1975). Studies on the *Actinomycetaceae* by means of numerical taxonomy immunofluorescence and crossed immunoelectrophoresis. Thesis, Karolinska Institute, Stockholm.

Howell, A., Jordan, H. V., Georg, L. K. and Pine, L. (1965). *Sabouraudia* **4**, 65–68.

Jones, D. (1975). *J. Gen. Microbiol.* **87**, 52–96.

Jones, D. (1978). *In* "Coryneform bacteria" (I. J. Bousfield and A. G. Callely, Eds), pp. 13–46 Academic Press, London and New York.

Jones, L. A. and Bradley, S. G. (1964). *Dev. Ind. Microbiol.* **5**, 267–272.

Jones, D. and Sneath, P. H. A. (1970). *Bacteriol. Rev.* **34**, 40–81.

Lance, G. N. and Williams, W. T. (1967). *Computer J.* **10**, 271–277.

Lapage, S. P., Bascomb, S., Wilcox, W. R. and Curtis, M. A. (1973). *J. Gen. Microbiol.* **77**, 273–290.

Lechevalier, H. A. and Lechevalier, M. P. (1967). *Annu. Rev. Microbiol.* **21**, 71–100.

Liston, J., Weibe, W. and Colwell, R. R. (1963). *J. Bacteriol.* **85**, 1061–1070.

Lockhart, W. R. (1967). *J. Bacteriol.* **94**, 826–831.

Lockhart, W. R. and Liston, J. (1970). "Methods for Numerical Taxonomy". American Society for Microbiology, Bethesda, Maryland.

Melville, T. H. (1965). *J. Gen. Microbiol.* **40**, 309–315.

Moore, W. E. C. and Holdeman, L. V. (1973). *Int. J. Syst. Bacteriol.* **23**, 69–74.

Möller, F. (1962). *G. Microbiol.* **10**, 29–47.

Orla-Jensen, S. (1924). *Lait* **4**, 468–474.

Pine, L. (1970). *Int. J. Syst. Bacteriol.* **20**, 445–474.

Pine, L. and Georg, L. K. (1965). *Int. Bull. Bacteriol. Nomencl. Taxon.* **15**, 143–163.

Pine, L. and Georg, L. K. (1969). *Int. J. Syst. Bacteriol.* **19**, 267–272.

Poupard, J. A., Husain, J. and Norris, R. F. (1973). *Bacteriol. Rev.* **37**, 136–165.

Prévot, A. R. (1972). "Les bactéries anaérobies". Crouan et Roques, Lille.

Rohlf, F. J. (1964). *Systematic Zool.* **13**, 102–104.

Schofield, G. M. and Schaal, K. P. (1981). *J. Gen. Microbiol.* **127**, 237–259.

Slack, J. M. (1968). *Int. J. Syst. Bacteriol.* **18**, 253–262.

Slack, J. M. (1974). *In* "Bergey's Manual of Determinative Bacteriology" (R. E. Buchanan and N. E. Gibbon, Eds), 8th edn, pp. 659–667. Williams & Wilkins, Baltimore, Maryland.

Slack, J. M. and Gerencser, M. A. (1970). "The Actinomycetales", (H. Prauser, Ed.), pp. 19–27. Fischer, Jena.

Slack, J. M. and Gerencser, M. A. (1975). "Actinomyces, Filamentous Bacteria: Biology and Pathogenicity". Burgess, Minneapolis, Minnesota.

Slack, J. M., Winger, A. and Moore, D. W., Jr (1961). *J. Bacteriol.* **82**, 54–65.

Sneath, P. H. A. (1957). *J. Gen. Microbiol.* **17**, 201–226.

Sneath, P. H. A. (1968). *J. Gen. Microbiol.* **54**, 1–11.

Sneath, P. H. A. (1970). 'The Actinomycetales" (H. Prauser, Ed.), pp. 371–377. Fischer, Jena.

Sneath, P. H. A. (1971a). "Recent Advances in Microbiology" (A. Pérez-Miravete and D. Paláez, Eds), pp. 581–586. Azociacion Mexicana de Microbiologia, Mexico City.

Sneath, P. H. A. (1971b). *In* "Methods in Microbiology" (J. R. Norris and D. W. Ribbons, Eds), Vol. 7A, 29–98. Academic Press, London and New York.

Sneath, P. H. A. and Collins, V. G. (1975). *Antonie van Leeuwenhoek* **40**, 481–527.

Sneath, P. H. A. and Johnson, R. (1972). *J. Gen. Microbiol.* **72**, 377–392.

Sneath, P. H. A. and Sokal, R. R. (1973). "Numerical Taxonomy". Freeman, San Francisco, California.

Sokal, R. R. and Michener, C. D. (1958). *Kansas Univ. Sci. Bull.* **38**, 1409–1438.

Sokal, R. R. and Rohlf, F. J. (1962). *Taxon* **II**, 33–40.

Sokal, R. R. and Sneath, P. H. A. (1963). "Principles of Numerical Taxonomy". Freeman, San Francisco, California.

Waksman, S. A. (1967). "The Actinomycetes. A Summary of Current Knowledge". The Roland Press Company, New York.

Williams, W. T. and Dale, M. B. (1965). *Adv. Bot. Res.* **2**, 35–68.

Willcox, W. R., Lapage, S. P., Bascomb, S. and Curtis, M. A. (1973). *J. Gen. Microbiol.* **77**, 317–330.

Willcox, W. R., Lapage, S. P. and Holmes, B. (1980). *Antonie van Leeuwenhoek* **46**, 233–299.

Wolff, M. and Israel, J. (1891). Ueber reincultur des *Actinomyces* und seine uebertragbarkeit auf thiere. *Arch Pathol. Anat. Physiol. Klin. Med.* **126**, 11–59.

15

Stimulus Space in Olfaction

K. B. DØVING

The Institute of Zoophysiology, University of Oslo, Blindern, Oslo, Norway

I.	Introduction	.	.	.	.	.	.	.	361
II.	Purpose of classification	.	.	.	.	.	.	.	361
III.	Model Application	.	.	.	.	.	.	.	362
	References .	.	.	.	.	.	.	.	367

ABSTRACT   A three-dimensional model of a stimulus space for odours is presented. The model is based upon the data matrix of Wright and Michels. The odours distribute themselves in three main groups, thus confirming an earlier solution of the same data made with hierarchical clustering analysis. The arrangement of the odours in this three-dimensional model is discussed.

I. Introduction

The number of odorous chemicals that man can discriminate with the olfactory sense seems unlimited. Even some enantiomers have been shown to give the impression of different smells. The seemingly unlimited ability to perceive differences in odour quality has not inhibited attempts of classification. The earlier classifications were often based on personal and subjective experience, like those proposed by Linné (1752), Henning (1916), Zwaardemaker (1925) and Crocker and Henderson (1927). The recent descriptions of odour classifications have been based upon refined methods, such as similarity ratings, confusion data or semantic usage (Amoore, 1962; Harper *et al.*, 1968; Dravnieks, 1974). These authors have also used modern methods for data analysis.

II. Purpose of classification

The aims of a grouping and classification system for odours vary. One aim might be to achieve a reference model within which an unknown odour can

METHODS IN MICROBIOLOGY VOL. 16
ISBN 0-12-521516-9

be fitted. A model might also be used in attempts to correlate the axis of the model with physical parameters. Such attempts have had limited success (Davies, 1970; Døving, 1974; Schiffman, 1974). Models can also provide reasonable suggestions on odour representatives from a cluster which later can be used in matching standard methods for characterizing odour qualities (Schutz, 1964).

III. Model application

Wright and Michels (1964) have presented a data matrix giving the relationship between 45 different odours (plus five repetitions) based upon ratings to nine standard odours. In total 84 subjects were used in their analysis. The 45 different odours and their abbreviations are listed in Table I.

Previous analysis of this data by a hierarchical clustering method showed that the odours were distributed into three clusters; hydrogen sulphide fell outside the three groups (Døving, 1970). A reconstruction of this model is presented in Fig. 1. A non-parametric multi-dimensional scaling method applied to these data revealed two, possibly three, groups in a two-dimensional reconstruction. This solution has been confirmed by Schiffman (1974). The solution in two dimensions has relative high "stress": 8%.

To investigate whether a three-dimensional model of the same data had a more distinct clustering of the odours a physical reconstruction was made.

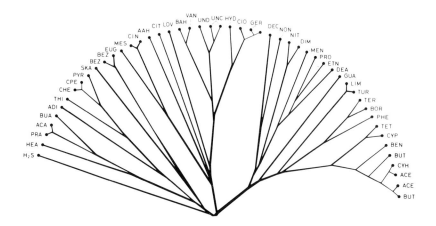

Fig. 1.   The hierarchical clustering scheme obtained by analysing the data of Wright and Michels (1964). Reproduced with permission from the Ciba Foundation. From Døving (1970).

TABLE I

List of odours used by Wright and Michels (1964), a is the angle to the x-y plane, β is the angle to the x- axis (see text.)

Abbreviation	Odour	No.	a (degrees)	β (degrees)
AAH	Acctic acid hexyl ester	6	−49	50
ACA	Acetic acid	36	74	−37
ACE	Acetone	30	−6	26
ADI	Allyl disulphide	41	50	168
BAH	Benzaldehyde	4	−61	6
BEN	Benzene	25	−5	31
BEZ	Benzothiazole	3	41	72
BOR	Borneol acetate	23	−15	26
BUA	Butanoie acid	40	79	−97
BUT	l-Butanol	17	3	24
CHE	Cyclohexene	28	68	141
CIN	Cinnamaldehyde	34	−70	51
CIO	Citronellol	13	−57	−1
CIT	Citral	21	−37	−25
CPE	Cyclopentene	29	68	151
CYH	Cyclohexane	26	−8	25
CYP	Cyclopentane	27	−2	35
DEA	Captric acid (?)	38	−16	16
DEC	l-Decanol	15	−20	4
DIM	Diphenyl methane	10	−38	21
ETH	Ethanol	19	−27	17
EUG	Eugenol	32	−64	99
GER	Geraniol	12	−56	−8
GUA	Guaiacol	2	−12	50
HEA	Capric acid	39	63	4
HYD	Hydroxycitronellal	9	−59	−19
H₂S	Hydrogen sulphide	43	40	−145
LIM	Limonene	22	3	29
LOV	S 8001 (Schimmel) (Love of the valley)	8	−58	−52
MEN	Menthol	31	−27	25
MES	Methyl salcylate	33	−70	44
NIT	Nitrobenzene	1	−41	21
NON	l-Nonanol	16	−37	2
PHE	Phenol	37	−14	37
PRA	Propanoic acid	35	71	−51
PRO	l-Propanol	18	−19	20
PYR	Pyridine	45	74	154
SKA	Skatole	44	60	109
TER	a-Terpineol	11	−21	31
TET	Carbon tetrachloride	20	−2	32
THI	Thiophene	42	61	165
TUR	Turpentine	24	8	31
UNC	γ-Undecalactone (4-pentyl butanolide)	7	−53	6
UND	l-Undecanol	14	−50	−5
VAN	Vanillin	5	−65	9

The three-dimensional solution has a "stress" of 5.1% and it turned out that such a reconstruction in space revealed several new features compared to the two-dimensional model. As can be seen from Fig. 2 the odours are divided in three main groups. The impression of three clusters was confirmed by measuring the distances from a point (BOR) in the middle of one cluster to the other odours. The distances are distributed in three groups as can be seen

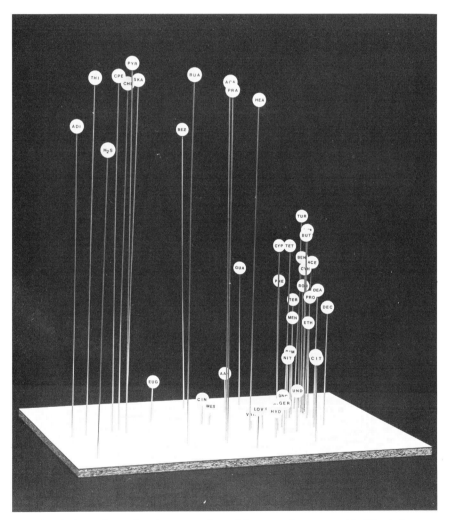

Fig. 2. A three-dimensional construction of the same data as used in Fig. 1, from a solution of a non-parametric clustering analysis. Stress is 5.1%. BAH is behind LOV, CIO behind HYD, NON behind CIT and LIM behind BUT.

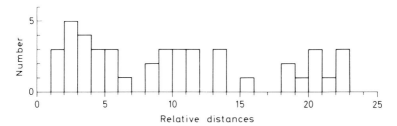

Fig. 3. The relative distances in the three-dimensional model in Fig. 2, between BOR and the 44 other substances.

in Fig. 3. The hydrogen sulphide was situated outside any of the other clusters in this reconstruction, as was the case in the hierarchical model. Thus two different methods have shown that the 45 odours listed in Table I can be grouped into three classes.

There are some features of the model that might be of interest. The three-dimensional model gives a more distinct distribution and thus simplifies the extraction of clusters. The "unpleasant" odours in the upper part of the model might be divided in two subgroups: the acids with short chain-length where acetic acid (ACA) is in the middle; and the "garlic" or "burnt" odours where cyclochexene (CHE) might be chosen as a representative. In the two lower clusters citronellol (CIO) might represent the "flowery" odours and borneol acetate (BOR) the "fruity" odours. Hydrogen sulphide is the only representative of a "foul" odour.

If the arrangement of the odours in space could be described by a simple geometrical shape, it would facilitate our perception of the stimulus space. Two possible distributions emerge upon a visual inspection of the physical model.

It appears that the odours are distributed in a spiral. Without trying to satisfy any mathematical proof for this opinion the idea was tested by measuring the distances to all odours from a chosen axis perpendicular to the plate (*x*-*y* plane). The distances of the odours from this axis was distributed as shown in Fig. 4; indicating that a spiral-like arrangement might be one rational expression of the position of the odours in space.

A better description of the stimulus space for these odours seems to be a sphere. Using a program for the best fitting of a sphere to the points (Døving and Price, 1976, 2. ECRO congress) the model was evaluated in this respect. For the best solution the distances from the centre of the sphere to the points were distributed as shown in Fig. 5. The mean relative distance, later used as radius in the sphere, was 1.18 ± 0.057 (mean $\pm$ one standard deviation). The points can be described in polar coordinates, and these are given in Table I.

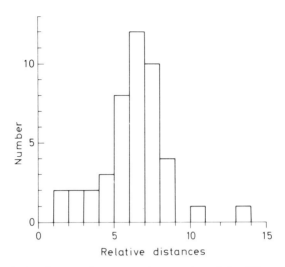

Fig. 4. The relative distances between a chosen axis and the different odours in the model of Fig. 2.

Henning (1916) in his volume "Der Geruch" develops the interesting and elegant concept of an odour prism, a proposal of the stimulus space in olfaction. Before he arrives at the odour prism he has discussed the possibility of an odour sphere, and even illustrates the point. This figure is reproduced in Fig. 6.

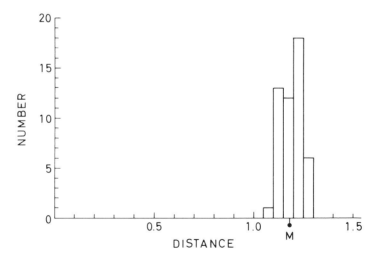

Fig. 5. The relative distances between the centre for the sphere with the best fit and the various odours placed as shown in Fig. 2.

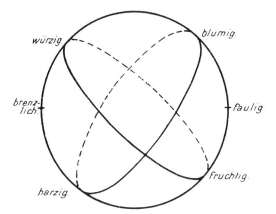

Fig. 6.   Reproduction of the stimulus space suggested by Henning (1916).

It is known that data on colour similarities give a two-dimensional solution where the colours lie on the circumference of a circle (Shepard, 1962). In the so-called colour circle the axes do not have a simple relationship to any physical parameter.

The results of the psychophysical studies made by Wright and Michels (1964) are probably the most extensive and reliable data available for evaluation of a stimulus space in olfaction. The three-dimensional model based upon this data shows properties which might be valuable for our understanding of the stimulus space in olfaction. It might be of interest to evaluate more rigorously other data on odour similarities to substantiate the impression left by the present study.

The elegant model of odour space proposed by Henning in 1916, with the methods available at that time, might prove to suit a stimulus space in olfaction better than hitherto appreciated.

References

Amoore, J. E. (1962). *Proc. Sci. Sec. Toilet Goods Ass. Suppl.* **37**, 1–12.

Crocker, E. C. and Henderson, L. F. (1927). *Am. Perf. Essent. Oil Rev.* **22**, 325–327.

Davies, J. T. (1970). *In* "Taste and Smell in Vertebrates" (G. E. W. Wolstenhome and J. Knight, Eds), pp. 265–281. Churchill, London.

Døving, K. B. (1970). *In* "Taste and Smell in Vertebrates" (G. E. W. Wolstenhome and J. Knight, Eds), pp. 197–225. Churchill, London.

Døving, K. B. (1974). *Ann. N.Y. Acad. Sci.* **237**, 184–192.

Dravnieks, A. (1974). *Ann. N.Y. Acad. Sci.* **237**, 144–163.

Harper, R., Bate-Smith, E. C. and Land, D. G. (1968). "Odour Description and Odour Classification". Churchill, London.

Henning, H. (1916). "Der Geruch". Leipzig.

Linné, C. von (1752). "Odors medicamentorum. Translated to Swedish in: Valda avhandlingar av Carl von Linné" (H. Hedfors, Ed.), no. 15. Stockholm 1954.

Schiffman, S. S. (1974). *Ann. N.Y. Acad. Sci.* **237**, 104–183.

Schutz, H. G. (1964). *Ann. N.Y. Acad. Sci.* **116**, 517–526.

Shepard, R. N. (1962). *Psychometrika* **27**, 219–246.

Wright, R. H. and Michels, K. M. (1964). *Ann. N.Y. Acad. Sci.* **116**, 535–551.

Zwaardemaker, H. (1925). "L'Odorat", p. 178. Paris, Octave Doin.

Index

Abbreviation: p.m.r. proton magnetic resonance

Acinetobacter, 45 (table)
Actinobacterium, 344, 346
Actinobacterium meyerii, 346, 354
Actinomyces, 342, 343, 344, 346
Actinomyces bovis, 346, 354
Actinomyces israelii, 346, 353
Actinomyces naeslundii, 345 (fig), 346
Actinomyces odontolyticus, 346, 353–4
Actinomyces viscosus, 344, 345 (fig), 346
Actinomycetales, 343
Actinomycetes, microaerophilic, 341–60
  classical taxonomy, 342–4
  comparison of numerical and classical taxonomy, 355–6
  comparison of numerical phenetic classification and serological, genetic, phylogenetic classifications, 356
  numerical taxonomy, 344–56
    clustering analysis, 351–2
    cluster parameters, 352–3
    coding data methods, 348–9
    coefficients of resemblance, 348
    natural clusters significance, 355
    selection of test characters, 347
    selection of test strains, 346
    stability, 353–4
    taxonomic structures and ranks, 352
    test errors, 349–50
    theoretical basis, 346–53
    validity, 353–5
    vigour and pattern statistics, 350–1
  taxonomic criteria in identification, 357–8
Aerococcus, 179–210
  biochemical–cultural reactions, 206–7 (table)
  carbohydrate moieties (A–O), 194–5 (table)
  dendrogram, fatty acid/carbohydrate contents, 200–1 (figs)

fatty acid content of bacteria, 186–7 (table)
fatty acid distribution, 197 (table)
identification flow chart, 210 (fig)
isolates characteristics, 209 (table)
saturated/unsaturated fatty acid ratio, 192 (table)
serological scheme, 210
strains studied for fatty acids and carbohydrates, 184–210
  gas chromatography, 196–205
  isolates identification, 205–6
  methods: cell culture, 182–3
    chemical procedure, 183–8
    numerical analysis, 188
  results, 188–95
    carbohydrates, 193
    clustering, 193
    fatty acids, 188–93
  taxonomic status, 180–1
Aerococcus catalasicus, 181
Aerococcus viridans, 181, 182, 184 (table)
  carbohydrate content, 193
  dendrogram, 200 (fig), 201 (fig)
  fatty acid content, 190, 196
Aerococcus viridans NCT8251, 188, 189 (fig), 196 (fig)
Aeromonas, 127–45
  enterotoxin, 130
  growth and isolation, 130–1
  motile species, 129–31
    antigenic structure, 137–8
    cell wall composition, 136, 137 (table)
    DNA base composition/polynucleotide sequence relatedness, 135–6
    ecological significance of serogrouping, 138
    phenotypic characteristics, 134
    serotyping methods, 138–40
  pathogenicity, 130

Aeromonas—cont.
phenotype tests, 132–4
    cysteine-iron-agar, 132
    elastin-agar, 132–3
    nutritional, 133–4
    serotyping, 137–40
taxonomy evolution, 128
virulence, 130
Aeromonas caviae, 128
DNA hybridization groups, 135
phenotypic characteristics, 134, 135
    (table)
Aeromonas hydrophila, 128
DNA hybridization groups, 135
haemolysis on agar, 131
phenotypic characteristics, 134, 135
    (table)
virulence, 130
Aeromonas salmonicida, 128, 131–2
antigenic characterization, 140
fish furunculosis caused by, 131
lysotyping, 142 (table)
phage typing, 140–1
phenotypic characteristics, 134–5
polynucleotide sequence relatedness,
    136
serological characterization, 140
Aeromonas salmonicida subsp. *masou-
    cida*, 136
Aeromonas salmonicida subsp. *salmoni-
    cida*, 136
Aeromonas sobria, 128
DNA hybridization groups, 136
haemolysis on agar, 130
phenotypic characteristics, 134, 135
    (table)
virulence for fish, 130
Anti-*Cr. neoformans* serum, 96
Arachnia, 343, 346
Arachnia propionica, 354
Aspergillus, antigenic structures, 81
Automated identification of bacteria,
    329–39
system flow chart, 331 (fig)

Bacillus sp., 291–328
API identification system, 303–14
    advantages/disadvantages of classical/
        miniaturized methods, 308–9
    identification, 310–14

microcomputer-assisted identification,
    314
morphology, 309
taxonomy, 309–10
test reproducibility, 303–8
bacteriophage typing, 293–9
alkalophilic strains, 298
Bacillus anthracis, 297
B. cereus, 295–7
B. sphaericus, 298
B. stearothermophilus, 298
B. thuringiensis, 297
biochemical characterization, 293
enterotoxins, 302
flow cytometry, 322
identification of typical strains, 294
    (table), 295 (table)
instrumented approaches to identifica-
    tion, 315–23
laser pyrolysis, 322
morphological characterization, 293
percentage positive results matrix,
    304–7 (table)
pyrolysis, 315–16
analysis of data, 319
pyrolysis gas-liquid chromatography
    315, 317
Bacillus sp. identification, 320–1
standardization, 318, 319–20
pyrolysis mass spectrometry, 315,
    319–20
Bacillus sp. identification, 321–2
standardization, 319–20
serotyping, 299–302
H-antigens, 301–2
O-antigens, 302
spore antigens, 299–301
sources of type strains, 296 (table)
Bacillus "amyloliquefaciens", 320, 322
Bacillus anthracis
agglutination reactions, 299
bacteriophage, 297
laser pyrolysis, 322
Bacillus brevis, 300
Bacillus caldolyticus, 298
Bacillus cereus
bacteriophage, 295–7
enterotoxins, 302
food poisoning, 301, 302
strains, 321

H-antigens serotyping, 301
laser pyrolysis, 322
O-antigens, 302
spore antigens, 300–1
Bacillus firmus, 298
Bacillus licheniformis, 298, 300
Bacillus megaterium, 321
Bacillus mycoides, 300, 321
Bacillus polymyxa, 300
Bacillus pumilus, 298
Bacillus sphaericus
  bacteriophage typing, 298
  H-antigen, 301
Bacillus stearothermophilus
  bacteriophage, 298
  O-antigens, 302
  spore agglutination tests, 302
Bacillus subtilis, 298, 300
  H-antigens, 302
Bacillus subtilis subsp. *niger* WM, 322
Bacillus thuringiensis
  bacteriophage, 297
  crystals production in culture, 301
  H-antigens serotyping, 301
Bacteriocin, 214
Bacterionema, 343, 344
Bacterionema matruchotii, 345 (fig)
Beneckea nereida, 134
Beneckea splendida biotype 1, 134
Bifidobacterium, 344, 346
Bordetella pertussis, 45
Bovine mastitis, 6, 20
Branhamella, 227, 228 (table); *see also*
  Neisseria
Branhamella catarrhalis, 227
  agglutination, 235
  differential characteristics, 228 (table)
  isolation, 229
Branhamella caviae, 227
  agglutination, 235
  differential characteristics, 228 (table)
Branhamella cuniculi, 226, 227
  differential characteristics, 228 (table)
Branhamella ovis, 227
  agglutination, 235
  differential characteristics, 228 (table)
Brucella
  cell wall defective forms, 36
  characteristics: antigenic structure,
    33–4

cell structure, 27–30
  genetics, 34–6
  phage receptors, 32–3
  properties, general, 26–7
  taxonomy, 26–7
  variation, 36
classification of biotypes, 29 (table)
classification into species, 28 (table)
distribution of precipitating antigens,
    34 (table)
L forms, 36
non-smooth strain differentiation by
    phage typing, 71 (table)
phage propagating media, 52–3
phage propagating strains, 50–2
safety precautions, 49–50
seed strains, 51
Brucella phages
  Group 1 (Tbilisi, Tb), 25
    adsorption to bacterial cells, 44, 45
      (table), 46
    origins, 37
    plaques, 38–40
  Group 2 (Firenze, Fi), 25
    adsorption to bacterial cells, 44, 45
      (table), 46
    origins, 37
    plaques, 40
  Group 3, 25–6
    adsorption to bacterial cells, 44, 45
      (table), 46
    origins, 37
    plaques, 40
  Group 4, 26
    adsorption to bacterial cells, 44, 45
      (table), 46
    origins, 37
    plaques, 40
  Group 5, 26
    adsorption to bacterial cells, 44–5, 46
    origins, 37–8
    plaques, 42
  isolation, 53
    introduction of lysogenic cultures, 56
    selection of host range mutants, 56–8
    wild strains, 53–5
  lytic patterns, 69 (table)
  origins, 37, 38 (figs)
  preservation, 64
  procedure for phage typing, 65–71

Brucella phages *cont.*
 propagation, 58–60
  agar overlay cultures, 59–60
  agar surface cultures, 60
  liquid culture, 58–9
 properties, 37–49
  adsorption to bacterial cells, 44–6
  growth cycle, 42–4
  phage genetics, 48–9
  plaque morphology, 38–42
  resistance, 46–8
  serological, 49
  structure, 38
 purification, 65
 reference, 50
 relative efficiency of plating on *Brucella*
  species, 70 (table)
 release detection procedures, 55
 standardization, 60–4
  culture medium composition effects on
   host range, 63–4
  efficiency of plating on host strains,
   63
  host range determination, 62–3
  plaque formation and lysis by standard phage concentrations, 62–3
  routine test dilution (RTD) determination, 60–1
Brucella abortus, 24, 25, 26, 28 (table)
 antigen A, 33
 antigen M, 33
 biotypes, 27, 29 (table)
 firmly bound lipids, 31
 phage plaques on, 40
 phage propagating strain, 51 (table)
 reference strain, 62, 63
 rough strain, 62
 routine test dilution (RTD), 61 (fig)
Brucella canis, 25, 26, 28 (table)
 biotypes, 29 (table)
 phage propagating strain, 51 (table)
 reference strain, 62, 63
Brucella melitensis, 24, 25, 26, 28 (table)
 antigen A, 33
 antigen M, 33
 atypical strains, 66
 biotypes, 27, 29 (table)
 dye-resistant strains, 68
 phage propagating strain, 51 (table)
 plaque formation on, 64

 reference strain, 62, 63
 rough strain, 62
Brucella neotomae, 25, 26, 28 (table)
 antigen A, 33
 antigen M, 33
 biotypes, 29 (table)
 phage plaques on, 40
 phage receptors, 32–3
 reference strain, 62, 63
Brucella ovis, 25, 26, 28 (table)
 biotypes, 29 (table)
 phage propagating strain, 51 (table)
 reference strain, 62, 63
Brucella suis, 24, 26, 28 (table)
 antigen A, 33–4
 antigen M, 33–4
 atypical, 28 (table)
 biotypes, 27, 29 (table)
 dye-resistant variants of biotype 1, 68
 phage plaques on, 40
 phage propagating strain, 51
 reference strain, 62, 63
 rough strain, 62

Candida, 76
 antigenic analyses, 82–4
 antigenic structures, 76, 79, 79–81
 frequency: isolates from animals, 101
  (table)
  isolates from chickens, 101 (table)
  isolates from man, 99 (table)
 opportunistic infections, 94
 p.m.r. spectra, 105
 serological grouping application, medically important species, 94–6
 slide agglutination reactions, factor sera
  6 and 11, 90 (fig)
Candida albicans
 antigenic analysis, 82–4, 86 (table)
 antigenic structure, 80 (table)
 frequency: isolates from animals,
  98–100, 101 (table)
  isolates from chickens, 98–100, 101
   (table)
  isolates from man, 98, 99 (table)
Candida albicans type A
 antigenic factor 6 specific for, 112,
  113
 antigenic structure, 80 (table), 87 (table)
 characteristics, 95 (table)

frequency: isolates from animals, 98–100, 101 (table)
isolates from chickens, 98–100, 101 (table)
isolates from man, 98, 99 (table)
identification by slide agglutination 92 (fig)
p.m.r. spectra, 105, 107 (table)
Candida albicans type B
antigenic structure, 80 (table), 87 (table)
characteristics, 95 (table)
frequency: isolates from animals, 98–100, 101 (table)
isolates from man, 98, 99 (table)
identification by slide agglutination 92 (fig)
p.m.r. spectra, 105, 107 (table)
Candida albicans type C, 94
p.m.r. spectra, 105
"Candida check", 91, 107
Candida guillermondii
antigenic analysis, 82–4
antigenic structure, 80 (table), 87 (table)
characteristics, 95 (table)
frequency: isolates from animals, 98–100, 101 (table)
isolates from chickens, 98–100, 101 (table)
isolates from man, 98, 99 (table)
identification by slide agglutination 92 (table)
p.m.r. spectra, 105, 107 (table)
Candida krusei
antigenic analysis, 82–4
antigenic structure, 80 (table), 87 (table)
characteristics, 95 (table)
frequency: isolates from animals, 101 (table)
isolates from man, 99 (table)
identification by slide agglutination 92 (fig)
p.m.r. spectra, 105, 107 (table)
Candida lusitaniae, 94
Candida macedoniensis, 80 (table)
Candida maltosa, 109
Candida mycoderma, 80 (table)
Candida natalensis, 109
Candida parapsilosis
antigenic analysis, 82–4, 86 (table)
antigenic structure, 80 (table), 87 (table)

characteristics, 95 (table)
frequency: isolates from animals, 98–100, 101 (table)
isolates from man, 98, 99 (table)
identification by slide agglutination 92 (fig)
p.m.r. spectra, 105, 107 (table)
Candida pseudotropicalis
antigenic analysis, 82–4, 86 (table)
antigenic factor-8, 112
antigenic structure, 80 (table), 87 (table)
characteristics, 95 (table)
frequency: isolates from animals, 98–100, 101 (table)
isolates from chickens, 98–100, 101 (table)
isolates from man, 98, 99 (table)
identification by slide agglutination 92 (fig)
p.m.r. spectra, 105, 107 (table)
Candida pulcherrima, 80 (table)
Candida rugosa, 95 (table)
Candida sake, 109
Candida sloofi, 101 (table)
Candida stellatoidea
antigenic analysis, 82–4
antigenic structure, 80 (table), 87 (table)
characteristics, 95 (table)
frequency: isolates from man, 99 (table)
identification by slide agglutination using factor sera, 92 (fig)
p.m.r. spectra, 105, 107 (table)
Candida tropicalis
antigenic analysis, 82–4
antigenic structure, 80 (table), 87 (table)
frequency: isolates from animals, 98–100, 101 (table)
isolates from chickens, 98–100, 101 (table)
isolates from man, 98, 99 (table)
identification by slide agglutination 92 (fig)
p.m.r. spectra, 105, 107 (table)
principal characteristics, 95 (table)
Candida viswanathii, 94
principal characteristics, 95 (table)
Candida zeylanoides, 80 (table)
Candidiasis, 76
serological tests, 103 (table)
Ceratocystis, p.m.r. spectra, 105

Classification *v.* identification, 331 (table)
Corynebacterium, 344, 354
Cryptococcosis, serological tests, 103 (table)
Cryptococcus, 94, 96–7
  antigenic structure, 79
  biochemical characteristics, 97
  DNA content, 97
  frequency: isolates from animals, 101 (table)
   isolates from man, 99 (table)
Cryptococcus diffluens, 95 (table)
Cryptococcus laurentii, 95 (table)
Cryptococcus neoformans
  antiserum, 96
  characteristics (type A), 95 (table)
  frequency: isolates from animals, 101 (table)
   isolates from man, 99 (table)
  isolation, 96
  meningitis induced by, 96
  serotypes A, B, C, D, 96

Debaryomyces
  antigenic structures, 79
  isolation, 96
Debaryomyces hansenii, 95 (table)
Debaryomyces subglobusus, 109
Dextrans, 172
Diplococci, Gram-negative aerobic taxonomy, 226
Diplococcus, 226
Diplococcus crassus, 226, 230
Diplococcus mucosus, 226
Diplococcus pharyngis flavus I–III, 226, 227, 230
Diplococcus pharyngis siccus, 226

Enterotoxins, bacillary, 302
Escherichia coli (0:157 serogroups), 34, 45
Eubacterium, 346

Francisella tularensis, 32, 45
Fructose-1,6-diphosphate aldolase, 159
Fungi, yeast-like, classification, 77 (table)

Furunculosis, 131

Gaffkyahomari (P. homari), 181
Glucose-6-phosphate dehydrogenase, 159
Glucose-6-phosphoglucanate, 159

Hanseniaspora, antigenic structures, 79
Hansenula, antigenic structures, 79
Hansenula beijerinkii, 105
Hansenula californica, 105
Hansenula mrakii, 105
Hansenula saturnus, 105
  subsp. *subsufficiens*, 105

Identification *v.* classification, 331 (table)
Identification states, 333
Information numerically arranged, utilization differentiation, 332–3
International Streptomyces Project (ISP), 335, 336 (table)

Jaeger's meningococcus, 229–30

Klebecins, 214
Klebocins, 214
Klebsiella, 213
  markers for sub classifying, 214
  typing by bacteriocins, 214–22
  typing procedure, 216–20
   characterization of bacteriocin producers, 221
   feasibility of typing clinical isolates, 219–20
   reproducibility of results, 218–29
   technique, 216–17
Klebsiella oxytoca, 213, 220
Klebsiella ozaenae, 213, 220
Klebsiella pneumoniae, 213, 220
Klebsiella rhinoscleromatis, 213, 220
Kloeckera, antigenic structures, 79
Kluyveromyces lactis, 112

Lactate dehydrogenases, 159–61
  NAD-dependent, 161

Lactic acid bacilli, glycolytic pathways, 148, 149 (table)
Lactic dehydrogenases, 159–61
Lactobacillus, 346
Lactobacillus bifidus, 344
Lactobacillus brevis
  lactate dehydrogenase, 160 (table)
  peptidoglycan, 164 (table)
Lactobacillus buchnerii
  lactate dehydrogenase, 160 (table)
  peptidoglycan, 164 (table)
Lactobacillus confusus, 154, 155 (table), 156, 157
  commercial use, 172
  dehydrogenase electrophoretic patterns, 162 (table)
  DNA: content, 165, 167 (table)
    hybridization, 168 (table)
  glucose-6-phosphate dehydrogenase, 161
  lactate dehydrogenase, 160 (table), 164
  peptidoglycan, 164 (table)
Lactobacillus fermentum
  lactate dehydrogenase, 160 (table)
  peptidoglycan, 164 (table)
Lactobacillus reuteri
  lactate dehydrogenase, 160 (table)
  peptidoglycan, 164 (table)
Lactobacillus viridescens, 154, 155 (table), 156, 157
  DNA: content, 165, 167 (table)
    hybridization, 168 (table)
  lactate dehydrogenase, 160 (table), 161, 164
  peptidoglycan of cell wall, 164 (table)
Lancefield Group A phages, 5
Lancefield Group C phages, 5
Leuconostoc spp., 147–78
  amino acid requirements, 158–9
  bacteriophage typing, 169
  carbohydrate fermentation, 156
  cheese starters, 150, 151 (table), 173
  commercial importance, 171–4
    dairy industry, 173
    dextran formation, 172
    sugar industry, 171–2
    wine industry, 173–4
  cultivation, 151–2
  dextran production, 156–7
  differentiation from other lactic acid bacteria, 153 (table)

glucose-6-phosphate dehydrogenase, 161, 162 (fig)
hybrid enzymes, 163–4
identification of species, 153–69
  media used, 174–6
immunological studies using dehydrogenases, 163
isolation from natural habitats, 149–51
lactate dehydrogenase, 159–61
lactic acid production, 157–8
metabolism, 159–64
nucleic acid, 165–9
  DNA content, 165–6, 167 (table)
  DNA/DNA hybridization, 167
  RNA/DNA hybridization, 167–9
peptidoglycan, 164–5
serology, 169
taxonomy, 169–70
vitamin requirements, 158–9
Leuconostoc cremoris, 148
  acetate agar inhibition, 150
  carbohydrate fermentation, 156
  cheese starter, 151 (table)
  commercial use, 173
  DNA: content, 165, 167 (table)
    hybridization, 168 (table)
  growth conditions, 152
  peptidoglycan, 164 (table)
Leuconostoc dextranicum, 148
  commercial uses, 171–2
  DNA: content, 165, 167 (table)
    hybridization, 168 (table)
  growth on acetate agar, 150
  peptidoglycan, 164 (table)
Leuconostoc lactis
  carbohydrate fermentation, 156
  cheese starter, 151 (table)
  commercial use, 173
  differentiation, 155 (table)
  DNA: content, 165, 167 (table)
    hybridization, 168 (table)
  growth on acetate agar, 150
  lactate dehydrogenase, 164
  peptidoglycan, 164 (table)
Leuconostoc mesenteroides
  carbohydrate fermentation, 156
  commercial uses, 171–3
  cultivation, 152
  dextran production, 156–7, 172
  DNA: content, 165, 166, 167 (table)
    hybridization, 168 (table)

Leuconostoc mesenteroides — cont.
  growth on acetate agar, 150
  lactate dehydrogenase, 159–61
  L-malate fermentation, 158
  peptidoglycan, 164 (table)
Leuconostoc mesenteroides subsp. *cremoris*, 155 (table)
Leuconostoc mesenteroides subsp. *dextranicum*, 155 (table)
Leuconostoc mesenteroides subsp. *mesenteroides*, 155 (table)
Leuconostoc oenos, 151–2
  , commercial use, 173–4
  dehydrogenase electrophoretic patterns, 162 (fig)
  differentiation, 154–6, 155 (table)
  DNA: content, 165, 167 (table)
    hybridization, 168 (table)
  glucose-6-phosphate dehydrogenase, 161
  growth conditions, 152
  lactate dehydrogenase, 161, 163–4
  lactic acid production, 157
  malate conversion to lactate, 158
  peptidoglycan, 164 (table), 165
Leuconostoc paramesenteroides
  carbohydrate fermentation, 156
  DNA: content, 165, 167 (table)
    hybridization, 168 (table)
  growth on acetate agar, 150
  peptidoglycan, 164 (table)
  riboflavin requirement, 159

Micrococcus, 226
Micrococcus catarrhalis, 226, 227
Micrococcus cinereus, 226, 230

NAG vibrios, 285
Necromonas salmonicida, 129
Neisseriae, non-gonococcal non-meningococcal, 225–45
  agglutination, 230–5
  antigen preparation, 236
  antisera preparation, 241–2
  complement fixation reaction, 240–1
  differential characteristics, 228 (table)
  immunodiffusion, 236–7
  immunoelectrophoresis, 239–40
  isolation, 227–9
  non-growth on agar, 229
  serology, 229–30
  tube precipitation, 235–6
Neisseria animalis, 226, 228 (table)
Neisseria canis, 228 (table)
Neisseria catarrhalis, 227; *see also Branhamella catarrhalis*
Neisseria caviae, 226, 227; *see also Branhamella caviae*
Neisseria cineriae, 228 (table), 229
Neisseria cuniculi, 226, 227; *see also Branhamella cuniculi*
Neisseria cuniculi subsp. *gigantea*, 236 (footnote)
Neisseria denitrificans, 228 (table)
Neisseria elongata, 226, 227, 230
  agglutination, 234–5
  isolation, 229
Neisseria elongata subsp. *elongata*, 228 (table), 234
Neisseria elongata subsp. *glycolitica*, 228 (table), 234–5
Neisseria flava, 233
Neisseria flavescens, 226, 228 (table), 229
  agglutination, 231
Neisseria lactamica, 226, 228 (table)
  agglutination, 234
Neisseria mucosa, 226, 230
  agglutination, 232–3
  isolation, 229
  pigmented strains, 232–3
Neisseria mucosa subsp. *heidelbergensis*, 228 (table), 232
Neisseria mucosa subsp. *mucosa*, 228 (table), 232
Neisseria ovis, 227; *see also Branhamella ovis*
Neisseria perflava, 227, 228 (table)
  agglutination, 233
Neisseria sicca, 227, 228 (table)
  agglutination, 233
Neisseria subflava, 227, 230
  agglutination, 233–4
Neisseria subflava subsp. *flava*, 228 (table)
Neisseria subflava subsp. *subflava*, 228 (table)

Odontomyces, 344

Odontomyces viscosus, 344
Odour classification, 361–8
Odour prism, 366
Odours, model of stimulus space for, 361–8

Pasteurella multocida, 45, 247–58
  capsular substances identification, 251–6
  capsular type A culture identification, 258
  capsular type A strains identification with hyaluronidase, 255–6
  capsular type B identification by counterimmunoelectrophoresis, 255
  capsular type D cultures identification with acriflavine, 256
  capsular type E identification by counterimmunoelectrophoresis, 255
  colony variation, 250–1
  culture handling, 248–50
  designated serotypes, 258
  indirect haemagglutination test (IHA), 251–5
  O-antigens identification test, 258
  serotyping, 247–58
   steps followed in, 249 (fig)
  somatic serogroup antigens identification, 256–7
   antigen, 257
   antisera, 257
   gel diffusion tests, 257
  species characteristics, 248
  type B and E strains, coagglutination test, 258
Pediococcus, 179–210
  biochemical–cultural reactions, 206–7 (table)
  carbohydrate moieties (A–O), 194–5 (table)
  dendrogram, fatty acid/carbohydrate content, 200–1 (figs)
  fatty acid content of strains, 186–7 (table)
  fatty acid distribution, 197 (table)
  identification flow chart, 210 (fig)
  isolates characteristics, 209 (table)
  saturated/unsaturated fatty acid ratio, 192 (table)
  serological scheme, 210 (fig)

  strains studied for fatty acids and carbohydrates, 184–210
   gas chromatography, 196–205
   isolates identification, 205–6
   methods: cell culture, 182–3
    chemical procedure, 183–8
    numerical analysis, 188
   results, 188–95
    carbohydrates, 193
    clustering, 193
    fatty acids, 188–93
   taxonomic status, 180–2
Pediococcus acidilactici, 182, 202, 204
  beer as source, 203
  dendrogram, 200 (fig)
  fatty acid/carbohydrate profile, 203
  Group III strain, 190, 198, 206
  strains examined, 185 (table)
Pediococcus cerevisiae, 181, 182
  fatty acid composition, 199
  Group III strain, 190, 191, 202
  isolation from beer, 203
Pediococcus damnosus, 181, 182
  beer destroyed by, 180
  dendrogram, 200 (fig), 201 (fig)
  gas chromatography findings, 199, 202–5 *passim*
  Group II strain, 189, 198, 205
  melizitose acidification, 208
  strains: homogeneity, 193
   examined, 184–5 (table)
  taxonomic states, 181, 182
Pediococcus damnosus DSM20291
  carbohydrate profile, 198 (fig)
  fatty acid profile, 189, 190 (fig)
Pediococcus dextrinicus, 182, 185 (table)
  dendrogram, 200–1 (fig)
  Group II strain, 189, 198, 205
   deviation from, 191
  type strain, 205
Pediococcus halophilus, 181, 182, 185 (table)
  dendrogram, 200 (fig), 201 (fig)
  fatty acid composition, 199, 204, 205
  Group II strain, 198, 205
  melizitose acidification, 208
  pentoses acidified by, 208
  type strain, 193
Pediococcus homari (G. homari), 181, 208
Pediococcus inopinatus, 182, 184 (table)

Pediococcus inopinatus—cont.
  dendrogram, 200 (fig), 201 (fig)
  Group II strain, 189, 198, 203, 205
  type strains, 193
Pediococcus parvulus, 181, 182, 185 (table), 193
  dendrogram, 200 (fig), 201 (fig)
  fatty acid content, 193
  Group II strain, 189, 205
Pediococcus pentosaceous, 182, 185 (table), 202
  in beer, 203
  dendrogram, 200 (fig), 201 (fig)
  fatty acid composition, 191, 199, 202
  Group III strain, 190, 198, 208
  pentoses acidified by, 208
Pediococcus soyae, 181
Pediococcus sp., 185 (table), 202
  dendrogram, 200 (fig)
Pediococcus urinae-equi, 182, 184 (table), 202, 204
  carbohydrate content, 193
  dendrogram, 200 (fig), 201 (fig)
  fatty acid content, 188, 191
Phage-typing system
  discrimination, 2
  pathogenic streptococci, 3
  reproducibility, 2
6-Phosphogluconate, 159
Pichia
  antigenic structures, 79
  p.m.r. spectra of mannans, 105
Plesiomonas shigelloides, 259–69
  antigenic schema, 265, 266 (table)
  antiserum absorption, 262
  biochemical characteristics, 261 (table)
  cultural characteristics, 260
  ecology, 265–8
  epidemiology, 265–8
  H-antigen determination, 263–5
  H-antigenic relationships, 264 (table)
  H-antisera preparation, 262
  isolation from animals, 267
  O-antigen determination, 263
  O-antigenic relationships, 264 (table)
  O-antiserum preparation, 260–2
  serology, 260–5
Pneumocins, 214
Polysaccharides of medically important yeasts, p.m.r. spectra, 106 (fig)

Prophage typing, 4
Propionibacterium, 346
Proteus OX:2/OXL19/OX:K, 45
Proteus vulgaris, 45
Proton magnetic resonance spectroscopy, 104–5
  polysaccharides of medically important yeasts, 106 (fig)
  polysaccharides of *Saccharomyces* spp., 110–11 (figs)
  signals from *Candida* spp and *T. glabrata* mannans, 107 (table)
Pseudomonas aeruginosa, 45, 215
  phage-typing system, 4, 338
Pseudomonas maltophila, 32, 45 (table), 45

Red-leg disease, 130
Rhodotorula, 94
  antigenic structures, 79
  biochemical characteristics, 97
  DNA content, 97
  frequency: isolates from animals, 101 (table)
    isolates from man, 99 (table)
  isolation, 96
Rhodotorula glutinis, 95 (table)
Rhodotorula minuta, 101 (table)
Rhodotorula mucilaginosa, 99 (table)
Rhodotorula rubra, 95 (table), 99 (table)
Ribulose-5-phosphate, 159
Rothia, 343, 344, 354
Rothia dentocariosa, 345 (fig)

Saccharomyces
  antigenic analyses, 84–8
    antigenic structure analysis, 84–8
    known antigens, 84
    new antigens, 84–8
  antigenic factor 10 common to, 112
  antigenic structures, 76, 79
  frequency: isolates from animals, 101 (table)
    isolates from man, 99 (table)
  p.m.r. spectra, 105
Saccharomyces bayanus, 109
Saccharomyces carlsbergensis
  antigenic analysis, 84–8
  antigenic structure, 87 (table)

Saccharomyces cerevisiae, 109
  antigenic factor 18 specific for serotype I, 112
  antigenic structure, 80 (table), 87 (table)
  characteristics, 95 (table)
  frequency: isolates from animals, 101 (table)
    isolates from man, 99 (table)
  isolation, 96
Saccharomyces cerevisiae-like species, 109
Saccharomyces chevalieri, 109
  antigenic analysis, 84–8
  antigenic structure, 87 (table)
Saccharomyces dairensis, 105, 109
Saccharomyces delbrueckii, 109
Saccharomyces diastaticus, 109
Saccharomyces ellipsoideus, 80 (table)
Saccharomyces exiguus, 105, 109
Saccharomyces fermentati, 109
Saccharomyces florenturius, 109
Saccharomyces fragilis, 80 (Table)
Saccharomyces heterogenicus, 109
Saccharomyces inconspicuus, 109
Saccharomyces italicus
  antigenic analysis, 84–8
  antigenic structure, 87 (table)
Saccharomyces logos, 109
Saccharomyces mellis
  antigenic analysis, 84–8
  antigenic structure, 87 (table)
Saccharomyces oviformis, 109
  antigenic analysis, 84–8
  antigenic structure, 87 (table)
Saccharomyces rosei, 109
Saccharomyces rouxii
  antigenic analysis, 84–8
  antigenic structure, 87 (table)
Saccharomyces saitoanus, 109
Saccharomyces steineri
  antigenic analysis, 84–8
  antigenic structure, 87 (table)
Saccharomyces unisporus, 109
Saccharomyces uvarum, 109
Saccharomyces vafer, 109
Salmonella serotypes, Kauffman-White Group N, 32, 34, 45
Salmonella dublin, 45
Salmonella typhimurium, 45
Salmonella urbana, 45 (table)

Shigella sonnei, 215
Spore antigens, 299–300
Sporobolomyces, 79
Sporosarcina ureae, O-antigens, 302
Staphylococcal phage-typing system, 20
Staphylococcus aureus, 45
Stimulus space in olfaction, 361–8
Streptococcus (i)
  bacteriophages, 3–6
  cheese starter, 151 (table)
  pathogenic, 2
Streptococcus (i), Group A
  lysotyping scheme, 5
  M types, 4
  phage susceptibility, 3–4
Streptococcus (i), Group B
  additional phage-typing method, need for, 7–8
  animal origin, 7
  evaluation of combined sero-phage typing system, 14–18
    discrimination, 16
    reproducibility, 15–16
    results interpretation and recording, 17–18
    storage and stability, 17
  female genital tract infection, 6–7, 18–19
  human pathogens, 5–6, 18–19
  medium and conditions of growth, 8
  neonatal infection, 6–7, 19–20
  phage isolation, 8
  phage propagation, 9
  phage testing, 9–11
    lytic spectrum, 11, 12 (table), 13 (table)
    routine test dilution, 11
  phage-typing, 12–14
    phage application, 14
    phage examination, 14
    reagent preparation, 12–14
  sero-phage type association, 18
  serotyping systems, 3
  virulent sero-phage types, 18
Streptococcus (i), Group C
  lysotyping scheme, 5
Streptococcus (i), Group D
  phage activity, 5
Streptococcus agalactiae, 6
  evaluation of combined sero-phage typing system, 14–18

Streptococeus agalactiae—*cont.*
 discrimination, 16
 reproducibility, 15–16
 results interpretation and reporting,
  17–18
 storage and stability, 17
 medium and conditions of growth, 8
 phage isolation, 8
 phage propagation, 9
 phage testing, 9–11
  lytic spectrum, 11, 12 (table), 13 (table)
  routine test dilution, 11
 phage-typing, 12–14
  phage application, 14
  phage examination, 14
  reagent preparation, 12–14
 sero-phage type association, 18
 virulent sero-phage types, 18
Streptococcus bovis, 2
Streptococcus equi, 5
Streptococcus equisimilis, 2
Streptococcus faecalis, 2
 phage-typing system, 5
Streptococcus faecium, 2, 5
Streptococcus lactis, 151 (table)
Streptococcus lactis subsp. *diacetylactis*,
  151 (table), 151
Streptococcus MG, 45
Streptococcus pyogenes, 2
Streptomyces, identification, 334–5

Torulospora delbrueckii, 104, 105, 109
Torulopsis
 antigenic structures, 79
 frequency: isolates from animals, 101
  (table)
  isolates from chickens, 101 (table)
  isolates from man, 99 (table)
 p.m.r. spectra, 105
Torulopsis candida, 109
Torulopsis colliculosa, 109
Torulopsis famata
 characteristics, 95 (table)
 frequency: isolates from man, 99 (table)
Torulopsis glabrata
 antigenic structures, 80 (table), 87
  (table)
 characteristics, 95 (table)

frequency: isolates from animals, 100,
 101 (table)
 isolates from chickens, 100, 101 (table)
 isolates from man, 98, 99 (table)
 identification by slide agglutination
  using factor sera, 92 (table)
 p.m.r. spectra, 105, 107 (table)
Torulopsis sphaerica, 112
Torulopsis stellata subsp. *cambresieri*,
  109
Trichosporon
 frequency: isolates from animals, 101
  (table)
 isolates from man, 99 (table)
Trichosporon cutaneum
 frequency: isolation from animals, 101
  (table)
 isolation from chickens, 101 (table)
 isolation from man, 99 (table)

Vibrio albensis, 274
Vibrio anguillarum, 134, 275, 287 (table)
Vibrio cholerae, 271
 biochemical characteristics, 272–4
 cultural characteristics, 272
 epidemiology, 284–6
 H-antigen, 275–7
 H-agglutination test, 276-7
 H-antisera, 275–7
 relationship, 276 (table)
 slide agglutination test, 277–8
 tube agglutination test, 277
 O-agglutination tests, 279–80
 micro-titre agglutination, 280
 slide agglutination, 279
 tube agglutination, 279–80
 O-antigen, 278–80
 intrarelationship, 282 (table)
 test strains, 281 (table)
 variation in O1 *V. cholerae*, 281–4
 O-antiserum, 278–9
 serology, 274–84
 serovars comparison, 284, 284 (table)
 toxigenicity of strains, 287 (table)
Vibrio cholerae biovar *classical*, 271
Vibrio cholerae biovar *eltor*, 271
Vibrio comma, 32, 45
Vibrio mimicus, 272

biochemical characteristics, 272–4
cultural characteristics, 272
epidemiology, 284–7
serology, 284
toxigenicity of strains, 287 (table)

Yeasts
antigenic analyses methods, 81–94
antigenic analyses of *Saccharomyces*, 84–8
antigenic structure(s), 79–81
  medically important yeasts, 80 (table)
antigenic structure(s), coenzyme Q and (G + C) mole % DNA, 115–18 (appendix)
antigenic structure significance in taxonomy, 107–12
  change of genus, 112
  rapid identification, 107
  separation of species, 109
  unification of species, 109
brewing: detection of contamination, 100–2
  fluorescent antibody technique, 102–3
  slide agglutination, 100–2
classification, 76, 77 (table), 78
frequencies, serological method identified, 97–100
  isolates from animals, 98–100, 101 (table)
  isolates from chickens, 98–100, 101 (table)
  isolates from man, 97–8, 99 (table)
identification, 78
immunochemical basis of serological specificity, 112–13

immunochemical determinants of antigenic factors, presumptive structures, 113 (fig)
interrelationships among grouping methods
  chemical properties, 104
  DNA base sequence relationships, 104
  physiochemical properties, 104–5
medically important, 94–6
  identification flow chart, 92 (fig)
serogroup determination, 81–8
  absorption tests, 81–2
  antigenic analyses of *Candida*, 82–4
  antigen preparation, 81
  antisera preparation, 81
serological characterization, 75–126
serological diagnosis of systemic diseases due to, 103
serological grouping, 78–81
  application, 94–112
serological groups, 97 (table)
serological identification, 88–9
  antisera preparation, 91–3
  antisera specificities, 93
  commercial sources of factor sera, 90–1
  factor sera preparation, 88, 89 (table)
  isolation, 88
  refined antibody fractions preparation, 93–4
  slide agglutination, 88–9
species names in lists of culture collections, 108 (table)
Yersinia enterocolitica, 32, 34, 45
Yersinia pseudotuberculosis, 45

Contents of published volumes

Volume 1

E. C. Elliott and D. L. Georgala. Sources, handling and storage of media and equipment

R. Brookes. Properties of materials suitable for the cultivation and handling of microorganisms

G. Sykes. Methods and equipment for sterilization of laboratory apparatus and media

R. Elsworth. Treatment of process air for deep culture

J. J. McDade, G. B. Phillips, H. D. Sivinski and W. J. Whitfield. Principles and applications of laminar-flow devices

H. M. Darlow. Safety in the microbiological laboratory

J. G. Mulvany. Membrane filter techniques in microbiology

C. T. Calam. The culture of micro-organisms in liquid medium

Charles E. Helmstetter. Methods for studying the microbial division cycle

Louis B. Quesnal. Methods of microculture

R. C. Codner. Solid and solidified growth media in microbiology

K. I. Johnstone. The isolation and cultivation of single organisms

N. Blakebrough. Design of laboratory fermenters

K. Sargeant. The deep culture of bacteriophage

M. F. Mallette. Evaluation of growth by physical and chemical means

C. T. Calam. The evaluation of mycelial growth

H. E. Kubitschek. Counting and sizing micro-organisms with the Coulter counter

J. R. Postgate. Viable counts and viability

A. H. Stouthamer. Determination and significance of molar growth yields

Volume 2

D. G. MacLennan. Principles of automatic measurement and control of fermentation growth parameters

J. W. Patching and A. H. Rose. The effects and control of temperature

A. L. S. Munro. Measurement and control of pH values

H.-E. Jacob. Redox potential

D. E. Brown. Aeration in the submerged culture of micro-organisms

D. Freedman. The shaker in bioengineering

J. Bryant. Anti-foam agents

N. G. Carr. Production and measurement of photosynthetically usable light evolution in stirred deep cultures

G. A. Platon. Flow measurement and control

Richard Y. Morita. Application of hydrostatic pressure to microbial cultures

D. W. Tempest. The continuous cultivation of micro-organisms: 1. Theory of the chemostat

C. G. T. Evans, D. Herbert and D. W. Tempest. The continuous cultivation of micro-organisms: 2. Construction of a chemostat

J. Ričica. Multi-stage systems

R. J. Munson. Turbidostats

R. O. Thomson and W. H. Foster. Harvesting and clarification of cultures—storage of harvest

Volume 3A

S. P. Lapage, Jean E. Shelton and *T. G. Mitchell*. Media for the maintenance and preservation of bacteria

S. P. Lapage, Jean E. Shelton, T. G. Mitchell and *A. R. Mackenzie*. Culture collections and the preservation of bacteria

E. Y. Bridson and *A. Brecker*. Design and formulation of microbial culture media

D. W. Ribbons. Quantitative relationships between growth media constituents and cellular yields and composition

H. Veldkamp. Enrichment cultures of prokaryotic organisms

David A. Hopwood. The isolation of mutants

C. T. Calam. Improvement of micro-organisms by mutation, hybridization and selection

Volume 3B

Vera G. Collins. Isolation, cultivation and maintenance of autotrophs

N. G. Carr. Growth of phototrophic bacteria and blue-green algae

A. T. Willis. Techniques for the study of anaerobic, spore-forming bacteria

R. E. Hungate. A roll tube method for cultivation of strict anaerobes

P. N. Hobson. Rumen bacteria

Ella M. Barnes. Methods for the gram-negative non-sporing anaerobes

T. D. Brock and *A. H. Rose*. Psychrophiles and thermophiles

N. E. Gibbons. Isolation, growth and requirements of halophilic bacteria

John E. Peterson. Isolation, cultivation and maintenance of the myxobacteria

R. J. Fallon and *P. Whittlestone*. Isolation, cultivation and maintenance of mycoplasmas

M. R. Droop. Algae

Eve Billing. Isolation, growth and preservation of bacteriophages

Volume 4

C. Booth. Introduction to general methods

C. Booth. Fungal culture media

D. M. Dring. Techniques for microscopic preparation

Agnes H. S. Onions. Preservation of fungi

F. W. Beech and *R. R. Davenport*. Isolation, purification and maintenance of yeasts

G. M. Waterhouse. Phycomycetes

E. Punithalingham. Basidiomycetes: Heterobasidiomycetidae

Roy Watling. Basidiomycetes: Homobasidiomycetidae

M. J. Carlile. Myxomycetes and other slime moulds

D. H. S. Richardson. Lichens

S. T. Williams and *T. Cross*. Actinomycetes

E. B. Gareth Jones. Aquatic fungi

R. R. Davies. Air sampling for fungi, pollens and bacteria

George L. Barron. Soil fungi

Phyllis M. Stockdale. Fungi pathogenic for man and animals: 1. Diseases of the keratinized tissues

Helen R. Buckley. Fungi pathogenic for man and animals: 2. The subcutaneous and deep-seated mycoses

J. L. Jinks and *J. Croft*. Methods used for genetical studies in mycology

R. L. Lucas. Autoradiographic techniques in mycology

T. F. Preece. Fluorescent techniques in mycology

G. N. Greenhalgh and *L. V. Evans.* Electron microscopy

Roy Watling. Chemical tests in agaricology

T. F. Preece. Immunological techniques in mycology

Charles M. Leach. A practical guide to the effects of visible and ultraviolet light on fungi

Julio R. Villanueva and *Isabel Garcia Acha.* Production and use of fungi protoplasts

Volume 5A

L. B. Quesnel. Microscopy and micrometry

J. R. Norris and *Helen Swain.* Staining bacteria

A. M. Paton and *Susan M. Jones.* Techniques involving optical brightening agents

T. Iino and *M. Enomoto.* Motility

R. W. Smith and *H. Koffler.* Production and isolation of flagella

C. L. Oakley. Antigen-antibody reactions in microbiology

P. D. Walker, Irene Batty and *R. O. Thomson.* The localization of bacterial antigens by the use of fluorescent and ferritin labelled antibody techniques

Irene Batty. Toxin-antitoxin assay

W. H. Kingham. Techniques for handling animals

J. De Ley. The determination of the molecular weight of DNA per bacterial nucloid

J. De Ley. Hybridization of DNA

J. E. M. Midgley. Hybridization of microbial RNA and DNA

Elizabeth Work. Cell walls

Volume 5B

D. E. Hughes, J. W. T. Wimpenny and *D. Lloyd.* The disintegration of microorganisms

J. Sykes. Centrifugal techniques for the isolation and characterization of sub-cellular components from bacteria

D. Herbert, P. J. Phipps and *R. E. Strange.* Chemical analysis of microbial cells

I. W. Sutherland and *J. F. Wilkinson.* Chemical extraction methods of microbial cells

Per-Åke-Albertsson. Biphasic separation of microbial particles

Mitsuhiro Nozaki and *Osamu Hayaishi.* Separation and purification of proteins

J. R. Sargent. Zone electrophoresis for the separation of microbial cell components

K. Hannig. Free-flow electrophoresis

W. Manson. Preparative zonal electrophoresis

K. E. Cooksey. Disc electrophoresis

O. Vesterberg. Isolectric focusing and separation of proteins

F. J. Moss, Pamela A. D. Rickard and *G. H. Roper.* Reflectance spectrophotometry

W. D. Skidmore and *E. L. Duggan.* Base composition of nucleic acids

Volume 6A

A. J. Holding and *J. G. Collee.* Routine biochemical tests

K. Kersters and *J. de Ley.* Enzymic tests with resting cells and cell-free extracts

E. A. Dawes, D. J. McGill and *M. Midgley.* Analysis of fermentation products

S. Dagley and *P. J. Chapman.* Evaluation of methods to determine metabolic pathways

Patricia H. Clarke. Methods for studying enzyme regulation

G. W. Gould. Methods for studying bacterial spores

W. Heinen. Inhibitors of electron transport and oxidative phosphorylation

Elizabeth Work. Some applications and uses of metabolite analogues in microbiology

W. A. Wood. Assay of enzymes representative of metabolic pathways

H. C. Reeves, R. Rabin, W. S. Wegener and *S. J. Ajl*. Assays of enzymes of the tricarboxylic acid and glyoxylate cycles

D. T. Gibson. Assay of enzymes of aromatic metabolism

Michael C. Scrutton. Assay of enzymes of CO_2 metabolism

Volume 6B

J. L. Peel. The use of electron acceptors, donors and carriers

R. B. Beechley and *D. W. Ribbons*. Oxygen electrode measurements

D. G. Nicholls and *P. B. Garland*. Electrode measurements of carbon dioxide

G. W. Crosbie. Ionization methods of counting radio-isotopes

J. H. Hash. Liquid scintillation counting in microbiology

J. R. Quayle. The use of isotopes in tracing metabolic pathways

C. H. Wang. Radiorespirometric methods

N. R. Eaton. Pulse labelling of micro-organisms

M. J. Allen. Cellular electrophysiology

W. W. Forrest. Microcalorimetry

J. Marten. Automatic and continuous assessment of fermentation parameters

A. Ferrari and *J. Marten*. Automated microbiological assay

J. R. Postgate. The acetylene reduction test for nitrogen fixation

Volume 7A

G. C. Ware. Computer use in microbiology

P. H. A. Sneath. Computer taxonomy

H. F. Dammers. Data handling and information retrieval by computer

M. Roberts and *C. B. C. Boyce*. Principles of biological assay

D. Kay. Methods for studying the infectious properties and multiplication of bacteriophage

D. Kay. Methods for the determination of the chemical and physical structure of bacteriophages

Anna Mayr-Harting, A. J. Hedges and *R. C. W. Berkeley*. Methods for studying bacteriocins

W. R. Maxted. Specific procedures and requirements for the isolation, growth and maintenance of the L-phase of some microbial groups

Volume 7B

M. T. Parker. Phage-typing of *Staphylococcus aureus*

D. A. Hopwood. Genetic analysis in micro-organisms

J. Meyrath and *Gerda Suchanek*. Inoculation techniques—effects due to quality and quantity of inoculum

D. F. Spooner and *G. Sykes*. Laboratory assessment of antibacterial activity

L. B. Quesnel. Photomicrography and macrophotography

Volume 8

L. A. Bulla Jr, G. St Julian, C. W. Hesseltine and *F. L. Baker*. Scanning electron microscopy

H. H. Topiwala. Mathematical models in microbiology

E. Canale-Parola. Isolation, growth and maintenance of anaerobic free-living spirochetes

O. Felsenfeld. Borrelia

A. D. Russell, A. Morris and *M. C. Allwood*. Methods for assessing damage to bacteria induced by chemical and physical agents

P. J. Wyatt. Differential light scattering techniques for microbiology

Volume 9

R. R. Watson. Substrate specificities of aminopeptidases: a specific method for microbial differentiation

C.-G. Heden, T. Illeni and *I. Kuhn*. Mechanized identification of micro-organisms

D. B. Drucker. Gas-liquid chromatographic chemotaxonomy

K. G. Lickfield. Transmission electron microscopy of bacteria

P. Kay. Electron microscopy of small particles, macromolecular structures and nucleic acids

M. P. Starr and *H. Stolp*. *Bdellovibrio* methodology

Volume 10

T. Meitert and *Eugenia Meitert*. Usefulness, applications and limitations of epidemiological typing methods to elucidate nosocomial infections and the spread of communicable diseases

G. A. J. Ayliffe. The application of typing methods to nosocomial infections

J. R. W. Govan. Pyocin typing of *Pseudomonas aeruginosa*

B. Lányi and *T. Bergan*. Serological characterization of *Pseudomonas aeruginosa*

T. Bergan. Phage typing of *Pseudomonas aeruginosa*

N. B. McCullough. Identification of the species and biotyes within the genus *Brucella*

Karl-Axel Karlsson. Identification of *Francisella tularensis*

T. Omland. Serotyping of *Haemophilus influenzae*

E. L. Biberstein. Biotyping and serotyping of *Pasteurella haemolytica*

Shigeo Namioka. *Pasteurella multocida*—Biochemical characteristics and serotypes

Neylan A. Vedros. Serology of the meningococcus

Dan Danielsson and *Johan Maeland*. Serotyping and antigenic studies of *Neisseria gonorrhoeae*

B. Wesley Catlin. Characteristics and auxotyping of *Neisseria gonorrhoeae*

Volume 11

Frits Ørskov and *Ida Ørskov*. Serotyping of *Enterobacteriaceae*, with special emphasis on K-antigen determination

R. R. Gillies. Bacteriocin typing of *Enterobacteriaceae*

H. Milch. Phage typing of *Escherichia coli*

P. A. M. Guinée and *W. J. van Leeuwen*. Phage typing of *Salmonella*

S. Ślopek. Phage typing of *Klebsiella*

W. H. Traub. Bacteriocin typing of clinical isolates of *Serratia marcescens*

T. Bergan. Phage typing of *Proteus*

H. Dikken and *E. Kmety*. Serological typing methods of leptospires

Volume 12

S. D. Henriksen. Serotyping of bacteria

E. Thal. The identification of *Yersinia pseudotuberculosis*

T. Bergan. Bacteriophage typing of *Yersinia enterocolitica*

S. Winblad. *Yersinia enterocolitica* (synonyms "*Paxteurella X*", *Bacterium enterocoliticum* for serotype O–8)

S. Mukerjee. Principles and practice of typing *Vibrio cholerae*

H. Brandis. Vibriocin typing

P. Oeding. Genus *Staphylococcus*

J. Rotta. Group and type (groups A and B) identification of haemolytic streptococci

H. Brandis. Bacteriocins of streptococci and bacteriocin typing

Erna Lund and *J. Henricksen.* Laboratory diagnosis, serology and epidemiology of *Streptococcus pneumoniae*

Volume 13

D. E. Mahony. Bacteriocin, bacteriophage, and other epidemiological typing methods for the genus *Clostridium*

H. P. R. Seeliger and *K. Höhne.* Serotyping of *Listeria monocytogenes* and related species

I. Stoev. Methods of typing *Erysipelothrix insidiosa*

A. Saragea, P. Maximescu and *E. Meitert. Corynebacterium diphtheriae.* Microbiological methods used in clinical and epidemiological investigations

T. Bergan. Bacteriophage typing of *Shigella*

M. A. Gerencser. The application of fluorescent antibody techniques to the identification of *Actinomyces* and *Arachnia*

W. B. Schaefer. Serological identification of atypical mycobacteria

W. B. Redmond, J. H. Bates and *H. W. Engel.* Methods of bacteriophage typing of mycobacteria

E. A. Freund, H. Ernø and *R. M. Lemecke.* Identification of mycoplasmas

U. Ullmann. Methods in *Campylobacter*

Volume 14

T. Bergan. Classification of Enterobacteriaceae

F. Ørskov and *I. Ørskov.* Serotyping of *Escherichia coli*

W. H. Ewing and *A. A. Lindberg.* Serology of the Shigella

I. Ørskov and *F. Ørskov.* Serotyping of *Klebsiella*

R. Sakazaki. Serology of *Enterobacter* and *Hafnia*

P. Larsson. Serology of *Proteus mirabilis* and *Proteus vulgaris*

Volume 15

A. A. Lindberg and *L. Le Minor.* Serology of *Salmonella*

B. Lanyi. Biochemical and serological characterization of *Citrobacter*

T. L. Pitt and *Y. J. Erdman.* Serological typing of *Serratia marcescens*

R. Sakazaki. Serological typing of *Edwardsiella tarda*

M. B. Slade and *A. I. Tiffen.* Biochemical and serological characterization of *Erwinia*

G. Kapperud and *T. Bergan.* Biochemical and serological characterization of *Yersinia enterocolitica*

T. Bergan and *K. S. Ørheim.* Gas-liquid chromatography for the assay of fatty acid composition in Gram-negative bacilli as an aid to classification